Climate Change and Hydrological Processes, 2nd Edition

Climate Change and Hydrological Processes, 2nd Edition

Guest Editors

Alina Bărbulescu
Romulus Costache
Cristian Ștefan Dumitriu

MDPI

Basel • Beijing • Wuhan • Barcelona • Belgrade • Novi Sad • Cluj • Manchester

Guest Editors

Alina Bărbulescu
Department of Civil Engineering
Transilvania University of Brașov
Brașov
Romania

Romulus Costache
Flash-Flood Forecast Department
National Institute of Hydrology and Water Management
Bucharest
Romania

Cristian Ștefan Dumitriu
Faculty of Mechanical Engineering and Robotics in Constructions
Technical University of Civil Engineering of Bucharest
Bucharest
Romania

Editorial Office
MDPI AG
Grosspeteranlage 5
4052 Basel, Switzerland

This is a reprint of the Special Issue, published open access by the journal *Water* (ISSN 2073-4441), freely accessible at: https://www.mdpi.com/journal/water/special_issues/0H31T1E7FV.

For citation purposes, cite each article independently as indicated on the article page online and as indicated below:

Lastname, A.A.; Lastname, B.B. Article Title. *Journal Name* **Year**, *Volume Number*, Page Range.

ISBN 978-3-7258-5739-5 (Hbk)
ISBN 978-3-7258-5740-1 (PDF)
https://doi.org/10.3390/books978-3-7258-5740-1

Contents

water

MDPI

Editorial

Climate Change and Hydrological Processes, 2nd Edition

Alina Bărbulescu

Department of Civil Engineering, Transilvania University of Brașov, 5 Turnului Str, 500152 Brașov, Romania; alina.barbulescu@unitbv.ro

1. Introduction

Climate change is reshaping the global water cycle in profound and often unpredictable ways. Rising temperatures, shifting precipitation regimes, and intensifying extreme events are altering the distribution, availability, and quality of water resources across regions [1,2]. These changes have direct implications for ecosystems, agriculture, energy production, urban water supply, and human well-being. Understanding the hydrological consequences of a warming climate is therefore a central challenge in both scientific research and policy development. Hydrological processes—such as evapotranspiration, infiltration, groundwater recharge, and river discharge—are highly sensitive to climatic drivers. Small shifts in temperature and precipitation patterns can trigger nonlinear responses in catchment behavior, water storage, and flood or drought risks [3,4]. Moreover, feedback among the land surface, atmosphere, and human interventions complicates the prediction of future water dynamics. Bridging the gap between climate projections and hydrological impacts requires advances in modeling, long-term monitoring, and the integration of interdisciplinary perspectives [5–8].

Research over the past two decades has demonstrated that climate change is exerting significant and multifaceted impacts on global and regional water resources. Observational evidence shows alterations in precipitation regimes, with many areas experiencing increases in the intensity and frequency of heavy rainfall events, even in regions where the mean precipitation has remained stable or declined [9]. These changes amplify flood risks and challenge existing water management infrastructure designed under assumptions of stationarity. Conversely, prolonged dry spells and shifts in seasonal precipitation have contributed to more severe and persistent droughts in diverse settings, reducing surface water availability and placing stress on groundwater resources [10,11].

Declining snowpack, earlier snowmelt, and accelerated glacier retreat are altering runoff seasonality in mountain catchments worldwide [12]. While some basins are experiencing a temporary increase in meltwater supply, many are projected to face substantial reductions in summer flows as glaciers diminish, with cascading consequences for ecosystems, agriculture, and hydropower generation. Groundwater systems, often relied upon to buffer surface water variability, are increasingly recognized as vulnerable: recharge is affected by altered precipitation and evapotranspiration patterns, while over-abstraction in water-stressed regions amplifies depletion trends [13,14].

Methodologically, advances have been achieved through detection and attribution studies, which identify anthropogenic signals in streamflow, snowpack, and drought indices, although disentangling climate-driven effects from direct human activities such as land use change, reservoir regulation, and irrigation remains complex [15,16]. Coupled climate–hydrological models, remote sensing datasets such as GRACE for terrestrial water storage, and socio-hydrological approaches have significantly enhanced our understanding of climate–water interactions on multiple scales [17,18]. Nevertheless, major uncertainties persist, particularly regarding the projection of compound extremes, feedback between climate and human water use, and the representation of groundwater processes in global and regional models [19].

Overall, the research record highlights both the urgency of addressing climate-induced water challenges and the limitations of our current knowledge. Future work is needed to strengthen the integration of physical and social sciences, improve the representation of human–environment feedback, advance water quality modeling under climate extremes, and develop actionable, uncertainty-informed guidance for policy and management [20]. As hydrological nonstationarity becomes the defining condition of the twenty-first century, the study of climate change and water resources stands at the center of global sustainability challenges [21].

2. Main Contributions to This Special Issue

The papers showcased in this second edition of "Climate Change and Hydrological Processes" cast a wide net as they explore these issues, combining case studies, methodological advances, and regional comparisons. Together, they underscore both the progress that we have made in diagnosing hydrological change and the challenges that still lie ahead.

Using historical cartography, remote sensing, statistical trend tests, and GIS-based morphometric indices, Radu and Comănescu (contribution 1) document a long-term simplification of the Ialomița riverbed (Romania): a shift from braided/multi-threaded channel patterns toward a sinuous single-thread channel, accompanied by narrowing, incision, a reduced area of active and bankfull channels, reduced migration of the thalweg, and so on. Global warming (increasing temperature, rising precipitation) is evident in climatic data; human interventions (dams, embankments, gravel extraction, land use expansion) are intensifying these transitions. The authors project that unless river restoration or management is implemented, degradation will continue in the short-term.

In their study, Ajin et al. (contribution 2) develop flood susceptibility maps for the Buzău River catchment in Romania. They use four modern ensemble boosting methods (AdaBoost, CatBoost, LightGBM, and XGBoost), combined with multi-tier feature selection and explainable AI (XAI) tools (e.g., SHAP values) to pinpoint which factors matter most. The models perform very well (AUC ~0.97 for CatBoost), and key predictors turn out to be the slope, distance from rivers, topographic wetness index, and land use/land cover. The novelty is in combining high predictive performance with interpretability, which helps in applying such models for planning and risk mitigation.

In their article, Dobrică et al. (contribution 3) examine a lake in Romania (part of the Razim–Sinoe complex) that has undergone substantial hydrological modifications, including canalization, polder construction, changing connections to the Black Sea, and irrigation infrastructure. After human intervention and then the cessation of irrigation, the lake dried in 2020. The authors trace chemical evolution: decreasing salinity, a decline in sapropelic mud production, and changing inputs of fresh vs. saline water. It presents a cautionary tale of how hydrological connectivity and human changes (both direct infrastructure and indirect management) can alter chemical and ecological processes to the same extent that volumetric changes can.

The paper by Popescu and Bărbulescu (contribution 4) focuses on the Vărbilău River catchment (Romania), which is prone to flash floods and terrain instability. Using GIS-based travel-time modeling, they assess how accessibility to emergency intervention units is affected under normal and hazard-constrained conditions (e.g., blocked roads, steep terrain). The study quantifies changes in response times, identifies vulnerable zones in terms of accessibility, and points to how infrastructure planning should consider dynamic hazard conditions. In effect, the paper extends the flash-flood context from susceptibility mapping to operational readiness and resilience.

Ma et al. (contribution 5) investigate patterns in rapid alternations between drought and flood (DFAAs) from 1970 to 2019 in the Heilongjiang River Basin (China) and how vegetation growth responds with delays. Key findings: drought→flood events increased over time; flood→drought events decreased. Vegetation responds with lags (3–4 months in spring–summer, ~3 months in summer–autumn). These lags are important for predicting ecosystem impacts, achieving food security, and designing early warning systems.

Ande et al. (contribution 6) compare multiple precipitation datasets (satellite, gauge, merged) for the Godavari basin (India), apply bias correction, and build predictive models (Random Forest, M5P, etc.) to forecast streamflow under CMIP6 scenarios. They find that merged products such as MSWEP often outperform others and that machine learning models offer competitive performance if carefully tuned and validated. Their methodological comparative approach offers practical guidance for hydrological modeling under climate change.

In their article, Taheri et al. (contribution 7) review AI-based methods (neural networks, tree-based, kernel/support vector methods, and hybrid models) for the estimation of evapotranspiration (ET). They discuss the limitations of classical methods (e.g., Penman–Monteith) in data-poor contexts and the promise of data-driven models. They warn, however, of the dangers of overfitting, variable selection inconsistency, interpretability, and the need to embed physical constraints. They call for the standardization of inputs, better integration of remote sensing, and hybrid approaches combining AI with physical models.

Wavelet analysis, Granger causality, and singular value decomposition (SVD) are used in (contribution 8) to determine statistical time-lagged relationships between winter Arctic sea ice anomalies and subsequent spring precipitation anomalies over China. The authors identify key sea ice regions (Barents, Kara, East Siberian, Chukchi Seas) whose anomalies influence atmospheric circulation, jet stream positioning, and precipitation patterns in China. They find that including sea ice variability enhances the explanatory power of precipitation models beyond classical climate indices such as ENSO, PDO, and AO. This study adds an intriguing large-scale teleconnection dimension to the hydrology–climate nexus.

3. Concluding Remarks

Through the quality and diversity of its eight articles, this Special Issue, "Climate Change and Hydrological Processes, 2nd Edition" paints a remarkably balanced and forward-looking portrait of hydrological science under climate stress. The combination of retrospective diagnostics, methodological innovation, and forward projections enables both a deeper understanding of this topic and paths toward actionable forecasting.

The key takeaway is that hydrological vulnerability is dynamic. Risk zones shift, channel forms evolve, extreme behaviors change, human interventions modulate and sometimes redirect climate signals, and the very tools that we use to map susceptibility must evolve, too.

This Special Issue reminds us that future progress will be achieved not only by using stronger predictive models but by integrating dynamical change, uncertainty, intervention modeling, and socioecological relevance. As we move forward, dialog between modelers, field scientists, decision makers, and stakeholders will become ever more critical.

Funding: This research received no external funding.

Conflicts of Interest: The author declares no conflicts of interest.

List of Contributions:

1. Radu, A.; Comănescu, L. Historical Evolution and Future Trends of Riverbed Dynamics Under Anthropogenic Impact and Climatic Change: A Case Study of the Ialomița River (Romania). *Water* **2025**, *17*, 2151. https://doi.org/10.3390/w17142151.
2. Ajin, R.S.; Costache, R.; Bărbulescu, A.; Fanti, R.; Segoni, S. Flood Susceptibility Assessment Using Multi-Tier Feature Selection and Ensemble Boosting Machine Learning Models. *Water* **2025**, *17*, 2041. https://doi.org/10.3390/w17142041.
3. Dobrica, G.; Maftei, C.E.; Carazeanu Popovici, I.; Lupascu, N. Evolution of Nuntași-Tuzla Lake Chemistry in the Context of Human Intervention. *Water* **2025**, *17*, 1482. https://doi.org/10.3390/w17101482.
4. Popescu, C.; Bărbulescu, A. GIS-Based Accessibility Analysis for Emergency Response in Hazard-Prone Mountain Catchments: A Case Study of Vărbilău, Romania. *Water* **2025**, *17*, 2803. https://doi.org/10.3390/w17192803.
5. Ma, H.; Jing, J.; Dai, C.; Xu, Y.; Qi, P.; Song, H. Spatiotemporal Dynamics of Drought–Flood Abrupt Alternations and Their Delayed Effects on Vegetation Growth in Heilongjiang River Basin. *Water* **2025**, *17*, 1419. https://doi.org/10.3390/w17101419.
6. Ande, R.; Pandugula, C.; Mehta, D.; Vankayalapati, R.; Birbal, P.; Verma, S.; Azamathulla, H.M.; Nanavati, N. Understanding Climate Change Impacts on Streamflow by Using Machine Learning: Case Study of Godavari Basin. *Water* **2025**, *17*, 1171. https://doi.org/10.3390/w17081171.
7. Taheri, M.; Bigdeli, M.; Imanian, H.; Mohammadian, A. An Overview of Evapotranspiration Estimation Models Utilizing Artificial Intelligence. *Water* **2025**, *17*, 1384. https://doi.org/10.3390/w17091384.
8. Wang, H.; Wang, W.; Guo, F. Time-Lag Effects of Winter Arctic Sea Ice on Subsequent Spring Precipitation Variability over China and Its Possible Mechanisms. *Water* **2025**, *17*, 1443. https://doi.org/10.3390/w17101443.

References

1. Bărbulescu, A.; Maftei, C.E. Evaluating the Probable Maximum Precipitation. Case study from the Dobrogea region, Romania. *Rom. Rep. Phys.* **2023**, *75*, 704. [CrossRef]
2. Albulescu, A.C.; Minea, I.; Boicu, D.; Larion, D. Comparative multi-criteria assessment of hydrological vulnerability—Case study: Drainage basins in the Northeast Region of Romania. *Water* **2022**, *14*, 1302. [CrossRef]
3. Birsan, M.-V.; Nita, I.-A.; Amihăiesei, V.-A. Influence of large-scale atmospheric circulation on Romanian snowpack duration. *Rom. Rep. Phys.* **2024**, *76*, 708.
4. Cristian, A.; Zuzeac, M.; Ciocan, G.; Iorga, G.; Antonescu, B. A thunderstorm climatology of Romania (1941—2022). *Rom. Rep. Phys.* **2024**, *76*, 710.
5. Wang, L.; Cui, S.; Li, Y.; Huang, H.; Manandhar, B.; Nitivattananon, V.; Fang, H.; Huang, W. A review of the flood management: From flood control to flood resilience. *Heliyon* **2022**, *8*, e11763. [CrossRef] [PubMed]
6. Viviroli, D.; Archer, D.R.; Buytaert, W.; Fowler, H.J.; Greenwood, G.B.; Hamlet, A.F.; Huang, Y.; Koboltschnig, G.; Litaor, M.I.; López-Moreno, J.I.; et al. Climate change and mountain water resources: Overview and recommendations for research, management and policy. *Hydrol. Earth Syst. Sci.* **2011**, *15*, 471–504. [CrossRef]
7. Bertoldi, G.; Camporese, M.; Sulis, M. Advances in Catchment Science through Integrated Hydrological Modelling and Monitoring. *Water* **2021**, *13*, 2013. [CrossRef]
8. John, A.; Nathan, R.; Horne, A.; Stewardson, M.; Webb, J.A. How to incorporate climate change into modelling environmental water outcomes: A review. *J. Water Clim. Change* **2020**, *11*, 327–340. [CrossRef]
9. IPCC. *Climate Change 2022: Impacts, Adaptation, and Vulnerability. Contribution of Working Group II to the Sixth Assessment Report of the Intergovernmental Panel on Climate Change*; Cambridge University Press: Cambridge, UK, 2022.
10. WMO. *State of Climate Services 2021: Water*; World Meteorological Organization: Geneva, Switzerland, 2021.
11. Gudmundsson, L.; Seneviratne, S.I.; Zhang, X. Anthropogenic climate change detected in European drought observations. *Environ. Res. Lett.* **2016**, *11*, 044005. [CrossRef]
12. Hock, R.; Rasul, G.; Adler, C.; Cáceres, B.; Gruber, S.; Hirabayashi, Y.; Jackson, M.; Kääb, A.; Kang, S.; Kutuzov, S.; et al. High Mountain Areas. In *IPCC Special Report on the Ocean and Cryosphere in a Changing Climate*; Pörtner, H.-O., Roberts, D.C., Masson-Delmotte, V., Zhai, P., Tignor, M., Poloczanska, E., Mintenbeck, K., Alegría, A., Nicolai, M., Okem, A., et al., Eds.; Intergovernmental Panel on Climate Change: Geneva, Switzerland, 2019; pp. 131–202.
13. Famiglietti, J.S. The global groundwater crisis. *Nat. Clim. Change* **2014**, *4*, 945–948. [CrossRef]

14. Cuthbert, M.O.; Taylor, R.G.; Favreau, G.; Todd, M.C.; Shamsudduha, M.; Villholth, K.G.; MacDonald, A.M.; Scanlon, B.R.; Kotchoni, D.O.; Vouillamoz, J.M.; et al. Observed controls on resilience of groundwater to climate variability in sub-Saharan Africa. *Nature* **2019**, *572*, 230–234. [CrossRef] [PubMed]
15. Gudmundsson, L.; Leonard, M.; Do, H.X.; Westra, S.; Seneviratne, S.I. Observed trends in global indicators of mean and extreme streamflow. *Geophys. Res. Lett.* **2019**, *46*, 756–766. [CrossRef]
16. Dumitriu, C.Ș.; Bărbulescu, A.; Maftei, C.E. IrrigTool—A New Tool for Determining the Irrigation Rate Based on Evapotranspiration Estimated by the Thornthwaite Equation. *Water* **2022**, *14*, 2399. [CrossRef]
17. Tapley, B.D.; Watkins, M.M.; Flechtner, F.; Reigber, C.; Bettadpur, S.; Rodell, M.; Sasgen, I.; Famiglietti, J.S.; Landerer, F.W.; Chambers, D.P.; et al. Contributions of GRACE to understanding climate change. *Nat. Clim. Change* **2019**, *9*, 358–369. [CrossRef] [PubMed]
18. Popescu-Bodorin, N.; Bărbulescu, A. A ten times smaller version of CPC Global Daily Precipitation Dataset for parallel distributed processing in Matlab and R. *Rom. Rep. Phys.* **2024**, *76*, 703.
19. van Dijk, A.I.J.M.; Brakenridge, G.R.; Kettner, A.J.; Beck, H.E.; De Groeve, T.; Schellekens, J. River gauging at global scale using optical and passive microwave remote sensing. *Water Resour. Res.* **2016**, *52*, 6404–6418. [CrossRef]
20. Pahl-Wostl, C. An evolutionary perspective on water governance: From understanding to transformation. *Water Resour. Manag.* **2017**, *31*, 2917–2932. [CrossRef]
21. UNESCO-WWAP. *The United Nations World Water Development Report 2020: Water and Climate Change*; UNESCO: Paris, France, 2020.

water

MDPI

Article

Historical Evolution and Future Trends of Riverbed Dynamics Under Anthropogenic Impact and Climatic Change: A Case Study of the Ialomița River (Romania)

Andrei Radu [1,2,*] and Laura Comănescu [1]

[1] Faculty of Geography, University of Bucharest, 1 Nicolae Bălcescu Boulevard, 010041 Bucharest, Romania; laura.comanescu@geo.unibuc.ro

[2] National Institute of Hydrology and Water Management, 97E București-Ploiești Road, 013686 Bucharest, Romania

* Correspondence: andrei.radu@s.unibuc.ro

Abstract: Riverbed dynamics are natural processes that are strongly driven by human and climatic factors. In the last two centuries, the anthropogenic influence and impact of climate change on European rivers has resulted in significant degradation of riverbeds. This research paper aims to determine the historical evolution (1856–2021) and future trends of the Ialomița riverbed (Romania) under the influence of anthropogenic impact and climate change. The case study is a reach of 66 km between the confluences with the Ialomicioara and Pâscov rivers. The localisation in a contact zone between the Curvature Subcarpathians and the Târgoviște Plain, the active recent tectonic uplift of the area, and the intense anthropogenic intervention gives to this river reach favourable conditions for pronounced riverbed dynamics over time. To achieve the aim of the study, we developed a complex methodology which involves the use of Geographical Information System (GIS) techniques, hierarchical cluster analysis (HCA), the Mann–Kendall test (MK), and R programming. The results indicate that the evolution of the Ialomița River aligns with the general trends observed across Europe and within Romania, characterised by a reduction in riverbed geomorphological complexity and a general transition from a braided, multi-thread into a sinuous, single-thread fluvial style. The main processes consist of channel narrowing and incision alternating with intense meandering. However, specific temporal and spatial evolution patterns were identified, mainly influenced by the increasingly anthropogenic local influences and confirmed climate changes in the study area since the second half of the 20th century. Future evolutionary trends suggest that, in the absence of river restoration interventions, the Ialomița riverbed is expected to continue degrading on a short-term horizon, following both climatic and anthropogenic signals. The findings of this study may contribute to a better understanding of recent river behaviours and serve as a valuable tool for the management of the Ialomița River.

Keywords: riverbed dynamics; anthropogenic impact; climate change; evolutionary trend; Ialomița River; Romania

1. Introduction

Riverbed dynamics are natural processes that are highly influenced by climatic and anthropogenic factors [1,2]. In this context, climate change and direct human interventions within the fluvial system could cause major imbalances in the natural dynamic geomorphological behaviour of the river.

Currently, climate change is a fact, driven by anthropogenic influence through greenhouse emissions, already influencing weather and climate extremes all over the world [3]. As an effect of anthropogenic climate change, the streamflow indicates alteration on a local and regional scale across various parts of the world. However, on a global scale, no consistent trends are observed [4]. Regarding Europe, the analysis of the river discharges between 1950 and 2010 shows various trends within the streamflow of decreasing in the southern part, mixed in the central part, and increasing in the northern part [5].

The study of river channel dynamics over the last two centuries (100–150 years) is considered a fundamental tool from a river management perspective which can provide essential information to understand past and present dynamics and, more importantly, to identify possible future evolutionary trends [6,7].

Throughout the last two decades, many studies have used cartographic materials such as topographic maps and remote sensing products to reconstruct river behaviour over time, assessing qualitative and quantitative morphological changes in riverbeds worldwide [8–17].

In Europe, studies have investigated the geomorphic behaviour of the rivers over the last two centuries in response to natural and anthropogenic factors among the majority of European countries: Italy [2,7,18–29], Ukraine [30], Spain [31–34], Poland [19,35–38], Scotland [39], Slovenia [40], Slovakia [41], France [6,42–44], Germany [45,46], Russia [47], Albania [48], Croatia [49], Hungary [50–54], Czech Republic [55], and Austria [56]. Among the European rivers, there is observed a degradation trend of riverbeds, mainly highlighted by narrowing and incision processes, as a result of both human intervention and climate change.

Authors of most studies of this type consider that river channel evolution in the last up to 250 years is a consequence of the climate changes that occurred after the Little Ice Age (LIA), specifically a global warming, associated with anthropogenic impact [57].

In Romania, the fluvial processes align with the evolutionary trend observed for European rivers [57], but with a delay in the channel response to long-term anthropogenic influence [58]. In terms of vertical dynamics, represented by the changes in elevation, the riverbed dynamics were analysed at the geologic scale, focusing on the longitudinal evolution of the rivers [59], or on a contemporary scale [57,60–64], over the past century. Regarding riverbed planform dynamics, the studies made on Romanian rivers consist of diachronic analysis based on historical cartographic data and remote sensing over approximately 150 years [57,63,65–79].

Regardless of the geographical location, the majority of studies take into account increasingly accentuated anthropogenic intervention along with the effects of the climate regime and climate change on riverbed dynamics. Channel narrowing and incision primarily highlight a categorical trend of riverbed complexity reduction.

We select for analysis the Ialomița River reach between the confluences with the Ialomicioara and Pâscov rivers (66 km) because we consider it to have favourable conditions for an accentuated riverbed dynamic, such as the location at the contact between the lower Subcarpathians and the high plain units, active recent tectonic uplift of the area, a sedimentary lithology within the watershed, and a high degree of anthropisation. Furthermore, the Prahova River, a tributary of the Ialomița which evolved in similar geographic conditions, presents intense riverbed dynamics over time and a general trend of degradation, a fact highlighted by several studies [66,69,75,76]. Based on this reasoning, we consider that the studied river reach may present the same evolution pattern and trends.

The selection of the 166-year period of analysis was motivated and conditioned by the early cartographic sources required, the first qualitative map of the study area being realised starting with the second half of the 19th century.

The aim of the study is to determine the historical evolution (1856–2021) and future trends of the Ialomița riverbed under the influence of anthropogenic impact and climate change. The specific objectives are to (a) collect and integrate the historical cartographical sources into the GIS environment; (b) define a uniform, standardised, and semi-automated method for river planform measurements; (c) assess the historical geomorphic planform riverbed evolution between 1856 and 2021; (d) perform hierarchical cluster analysis (HCA) to identify spatial patterns in riverbed dynamics; (e) identify trends in time series of hydrological and climatic parameters using the Mann–Kendall (MK) test; and (f) assess the anthropogenic influence on the riverbed dynamics.

2. Study Area

2.1. General Setting

The study area covers the upper part of the Ialomița river basin, upstream of the confluence with the Pâscov river, situated in the central-eastern part of Romania (Figure 1a,b).

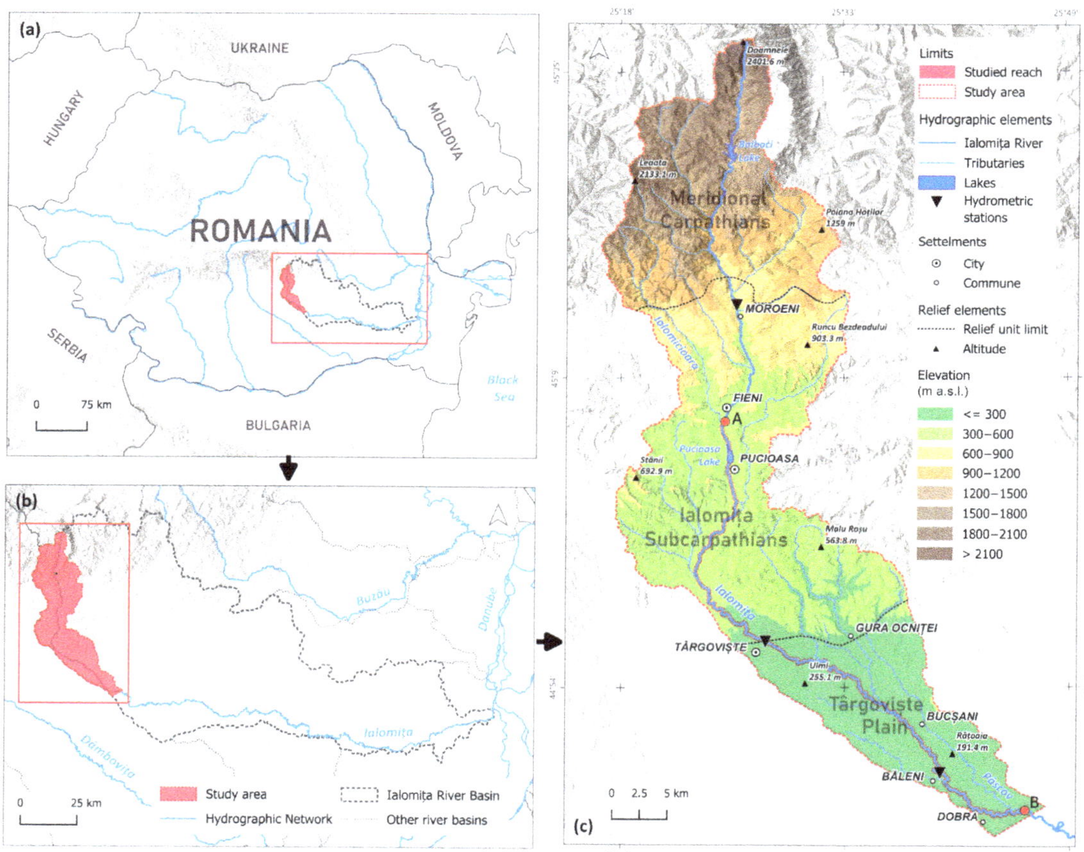

Figure 1. Location of the study area within (**a**) Romania, (**b**) Ialomița catchment, and (**c**) the hydrographic basin.

It has a surface of 1045 km^2 representing around 10% of the entire catchment area of 10,350 km^2 [80]. Within the study area, the investigated river reach crosses the southern part of the catchment, between the confluences with the rivers Ialomicioara at the north (25°25′ E, 45°07′ N) and Pâscov at the south (25°45′ E 44°47′ N) (Figure 1c).

The Ialomița River is one of the southern Romanian Carpathian rivers, gathering its springs in the Bucegi Mountains from a maximum altitude of 2406 m a.s.l. under the name of the Obârșiei Valley. Downstream, the main course of the Ialomița river is formed by the convergence with two other valleys, Șugările and Doamnele. The length of the river is 417 km, with a watershed area of 10,350 km^2 [80], an altitude difference of 2272 m, a sinuosity coefficient of 1.98, and an average slope of 5.35‰. Regarding the analysed river sector, it has a length of 66.3 km (in 2021), a sinuosity coefficient of 1.48, and a level difference of only 272 m, a value due to the sector development in approximately equal proportions in the Subcarpathian and the plain areas. The average slope is 4.12‰.

According to palaeographic evolution of the area, the current course of the Ialomița River took shape in the last 2.5 million years under the action of Wallachian orogenic movements, along with the other rivers in the Subcarpathian bend area located south of the Trotuș Fault. Until the Upper Pliocene, the major morphological changes in the relief, interruption, elevation, or subsidence would not have allowed the finalisation of the Ialomița Valley [57].

The geotectonic structure presents high complexity through the localisation at the contact between the large morpho-structural units of the territory of Romania, the Carpathian Orogen, and the Moesic Platform. The area is divided into different parts: the Infrabucovinic Nappe and the Ceahlău Nappe in the mountains; the Curbicortical Flysch Nappe, the Macla Nappe, the Tarcău Nappe, and the Subcarpathian Nappe in the hills; and the Carpathian Avantfosse in the lower hills and especially in the plains [81,82]. The positioning of the upper river sector within the Curvature Subcarpathians implies more intense riverbed dynamics due to the specific tectonic uplift of the area and the sedimentary lithology, which lead to high sediment loading of the water [76].

The study area covers all the major relief steps, starting with the Bucegi and Leaota Mountains in the north, the hilly Ialomița Subcarpathians area in the centre, and the Târgoviște and Cricov Plains in the south. The altitude range is between 157 m a.s.l., located near the confluence zone with the Pâscov River, and 2406 m a.s.l. at the Mecetul Turcesc Peak. Regarding the studied river reach, it crosses only the Subcarpathian and Plain units between the altitudes of 157 and 432 m a.s.l.

The climate is temperate–continental of transitional type [83], mainly determined by the Mediterranean influences [84]. Among the climatic parameters, the mean precipitation and temperature (1950–2022) over the study area are 745.3 mm and 9.01 °C, respectively. However, these vary significantly depending on altitude because of the relief amplitude of over 2200 m.

From a hydrological point of view, the flow regime of the Ialomița river observed at the hydrometric stations within the study area (Figure 1c) is characterised by mean multiannual discharges of 4.25 m^3/s at Moroeni (1961–2022), 7.61 m^3/s at Târgoviște (1976–2022) and 8.37 m^3/s at Băleni (1961–2022). These discharges fluctuate between a minimum and maximum of 0.88 m^3/s (2011) and 9.26 m^3/s (2005) at Moroeni, 3.11 m^3/s (1992) and 16.59 m^3/s (2005) at Târgoviște, and 1.92 m^3/s and 18.27 m^3/s at Băleni.

Regarding seasonal variability, the Ialomița River is highly influenced by climatic factors and is characterised by a pluvio-nival flow regime with distinct seasonal characteristics. The mean monthly values indicate the highest discharges in Spring, determined by snowmelt combined with precipitation, and in Summer, mainly influenced by convective rainfall. Peak mean monthly discharges are observed in May and June of 8.08 m^3/s and 6.18 m^3/s at Moroeni, 15.51 m^3/s and 12.03 m^3/s at Târgoviște, and 15.52 m^3/s and 12.85 m^3/s at Băleni. The discharges gradually decrease in Autumn and Winter to minimum values of 1.90 m^3/s in January at Moroeni, 4.52 m^3/s in October at Târgoviște, and 5.01 m^3/s in January at Băleni.

The mean annual suspended-sediment discharge is 7.08 kg/s (1976–2022) at Târgoviște and 15.17 kg/s (1961–2022) at Băleni. The seasonal variability is correlated with the flow regime, with the highest mean values in Spring–Summer in May and June of 12.57 kg/s and 15.56 kg/s at Târgoviște and 33.95 kg/s and 28.14 kg/s at Băleni. The sediment supply decreases towards Autumn and Winter, with minimum values of 1.94 kg/s in October at Târgoviște and 5.50 kg/s in January at Băleni.

2.2. River Management and Anthropogenic Intervention

The first river management activities in the study area started between 1928 and 1930 with the construction of the Scropoasa–Dobrești hydrotechnical complex in the upper sector of the watershed. Major hydrotechnical projects continued until 1988, resulting in a total of eight dam reservoirs, of which four were placed on the main course of the Ialomița River [85]. Among these, Lake Pucioasa is the only one located on the river reach (Figure 1c). A brief technical characterisation of the dams and associated reservoirs is presented in Table 1 [80,86].

Table 1. Technical characterisation of the dams and associated reservoirs.

Dam/Lake Name	Construction Period	Put into Service	Dam Height (m)	Total Volume of Water (mil. m^3)	Reservoir Use [1]
Bolboci	1976–1985	1988	55	19.4	W, H, F, O
Scropoasa	1928–1930	1930	26	0.55	H
Dobrești	1928–1930	1930	10	0.04	H
Pucioasa	-	1975	30.5	10.6	W, H, F, O

Note(s): [1] W—water supply, H—hydro energy, F—flood defence, O—other uses.

In addition to dams, the main channelisation structures on the Ialomița River include levees, embankments, and gabions to control lateral erosion along with grade control structures to manage vertical erosion.

Most of these structures were built at the same time as the construction of the Pucioasa Reservoir in 1975, which raised the local erosion level on average by 15 m. This caused intense sedimentation in the sector upstream of the dam, which is in dynamic equilibrium, and downward erosion downstream to the contact with the plain area (at Târgoviște), where were built successive simple and cascade weirs [87].

Gravel mining in Romania was at its peak during the communist period, especially in the period 1970–1989, when large-scale construction projects were in progress [62]. In recent times, although mining activity is permitted and regulated under specific conditions, some companies and individuals extract gravel illegally, as Armaș [76] mentions for the Prahova River.

The development of settlements adjacent to the river influences the dynamics of the riverbed. The greatest pressure is exerted by the urban centres of Fieni, Pucioasa, and Târgoviște, located in the Subcarpathian area (Figure 1c). The expansion of anthropogenic space along the river leads to the narrowing of the riverbed's freedom space. Additionally, water withdrawals for economic activities and for supplying the population have contributed to a reduction in river flow.

A particular case of economic activity's influence on riverbed dynamics is the thermal power station from Doicești, functional between the 1950s and the 2010s. During this period of time, ash deposits resulting from the burning of fuels for electricity production were formed in the floodplain zone, restricting the lateral dynamics of the riverbed.

3. Materials and Methods

We approached a complex methodology, the entire workflow following the steps depicted in the flowchart of the study (Figure 2).

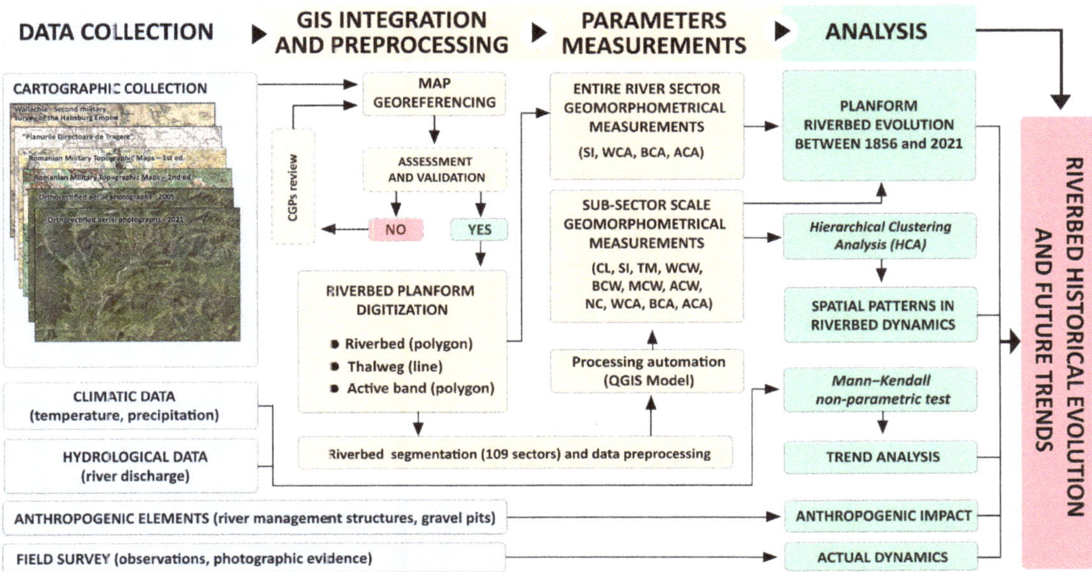

Figure 2. Flowchart of the study.

Data collection was the first step, gathering cartographic, climatic, hydrologic, anthropogenic, and field survey observation data. The integration and processing in the GIS environment were carried out using QGIS software version 3.22, some work steps being semi-automated in Model Designer. We processed and analysed the data to create graphic materials using Microsoft Excel and R version 4.2.2. Statistical analysis was also performed using R. At the end of the workflow chart, all the analysis results converge in order to assess the Ialomița riverbed's historical evolution (1856–2021) and future trends.

3.1. Data Collection

3.1.1. Cartographic Data Description

For the study, six cartographic materials were collected, four of them representing the most qualitative historical topographic maps available for the territory of Romania and the other two orthorectified aerial photographs. These materials were published between 1856 and 2021, covering a time period of 166 years. An overview of cartographic data sources, including the names of the maps, the publishers and publishing periods, scale, projection, and land survey periods can be observed in Table S1.

Wallachia—The second military survey of the Habsburg Empire can be considered the first qualitative topographic map of the former Wallachia, the southern region of present-day Romania. The map gives valuable and detailed information of the study area in the middle of the 19th century, taking into account the relative map sheet high scale of 1:57,600. Its quality is considered satisfactory, as the topographic survey was conducted using a triangulation network consisting of 468 points. The error between the original and the current measurements is estimated to be less than 20 m [88,89].

The "Plan Director de Tragere", translated as Firing Master Plan, represents the first topographic map of the entire territory of Romania, in Lambert projection, drawn up as a result of the military needs arising from the outbreak of the First World War [90]. A total of 2118 maps were made [91] at a scale of 1:20,000, reprinted between 1954 and 1959 without significant changes such as updating the names of settlements or redrawing with actualised conventional symbols [90].

Starting in the second half of the 20th century, The Romanian Military Topographic Directorate created a new set of topographic maps realised in the transverse cylindrical projection Gauss–Krüger at scale 1:25,000. There are 1818 map sheets with dimensions of 5′ latitude and 7′30″ longitude, published in 2 editions [92]. The first edition used land surveys between 1954 and 1955 and was printed between 1958 and 1961 [93]. After more than 20 years, the second edition of the topographic maps was produced, the data collection being made from aerial images and topographical surveys carried out from 1974 to 1978. The map sheets were edited between 1974 and 1986 and printed between 1975 and 1987 [92].

Map sheets of the topographical maps described above used in the study and their sources are presented in Table 2.

Table 2. Topographic maps sheets used in the study.

Map Name	No. of Map Sheets	Map Sheets Nomenclature	Source
Wallachia—Second military survey of the Habsburg Empire	3	Section 25.26, East Column V.VI; Section 27.28, East Column V.VI; Section 27.28, East Column VII.VIII	Austrian State Archives through the Arcanum platform: https://maps.arcanum.com/en/map/secondsurvey-wallachia (accessed on 16 August 2023). [94]
"Plan Director de Tragere"	9	3850-Voinești, 3849-Dragomirești, 3951-Fieni, 3950-Glodeni, 3949-Târgoviște, 3948-Văcărești, 4049-Ghirdoveni, 4048-Bucșani, and 4047-Cornățelul	Geospatial community, available online at https://www.geo-spatial.org/vechi/maps/download-planuri-tragere.php (accessed on 15 August 2024). [95]
Romanian Military Topographic Maps—1st ed.		L-35-099-D-b, L-35-099-D-d, L-35-11-B-b, L-35-111-B-d, L-35-112-A-a, L-35-112-A-c, L-35-112-A-d, L-35-112-C-b, L-35-112-D-a	The Defence Geospatial Intelligence Agency "Division General Constantin Barozzi"
Romanian Military Topographic Maps—2nd ed.			University of Bucharest, Faculty of Geography

The most recent cartographic data used in the study are the orthophoto maps from 2005 and 2021, at 50 cm spatial resolution and ±1.5 m precision [96]. Unlike the other sources, the orthophoto plans are natively generated in digital format and are projected in the Romanian national coordinate reference system, Stereo 70 (EPSG:3844). These were provided by the Faculty of Geography, University of Bucharest.

3.1.2. Climatic and Hydrological Data

Regarding the climatic data, we set out to analyse climate parameters of precipitation and temperature at daily time steps, yearly averaged over the study area between 1950 and 2022.

To cover the entire time interval, we selected and combined 3 open-access climatic databases: European Climate Assessment and Dataset (ECAD) [97], Romanian Daily Gridded Climatic Dataset (ROCADA) [98], and European Meteorological Observations

(EMO-1) [99]. Although the ECAD data cover the whole period, the other datasets were selected for the interval 1961–2022, because they are more reliable, several gauging stations being used for the interpolation. Table 3 provides details about these datasets.

Table 3. Climatic data used in the study.

Climatic Data Source	Spatial Resolution	Dataset Time Interval	Selected Time Interval	Stations in Romania	
				Prec.	Temp.
ECAD	0.1°	1950–2024	1950–1960	29	26
ROCADA	0.1°	1961–2013	1961–2013	188	150
EMO-1	0.01667°	1990–2022	2014–2022	Include data from ECAD and other European databases	

From a hydrological point of view, we collected information from the hydrometric station Băleni, located in the southern part of the study area (Figure 1c), about 10 km upstream from the end of the studied river sector. The data span the period from 1961 to 2022 and consist of river discharge (Q) and suspended load discharge (R), aggregated at a yearly scale, and the most significant floods. The National Institute of Hydrology and Water Management provided these data.

3.1.3. Anthropogenic Elements and Field Survey

Regarding the human factor in the riverbed dynamics, we documented and inventoried the anthropogenic interventions within the Ialomița River in the study area, which include dams, embankments, low-head dams/sills, and gravel pits. This was performed mainly based on the interpretation of cartographic materials and aerial orthorectified imagery used in the study (Table S1) and of satellite imagery obtained from Google Earth Pro.

To validate the results of the study, a field campaign was carried out in September 2024 along the studied river reach. We focused mainly on the locations of bridges, which are the most accessible, and some areas where we noticed radical changes in the riverbed.

3.2. GIS Processing and Integration

3.2.1. Georeferencing

Except for the orthorectified aerial photographs, georeferencing is necessary for the other cartographic materials, bringing them all into a unitary coordinate system. Elements such as the age and the quality of the topographic survey and the map drawing and the projection system of the historical sources require the use of different georeferencing methods depending on the characteristics of each map. It is essential to choose the optimal method for this process, minimising errors and obtaining results as close as possible to the reality in the field.

Thus, we processed Wallachia—Second military survey of the Habsburg Empire and "Plan Director de Tragere" using the raster-to-raster georeferencing type [100]. Taking into account the age and the general poor quality and accuracy of the maps, we selected the Thin Plate Spline (TPS) transformation algorithm, the most appropriate for this type of map [101,102]. Considering that TPS optimises local accuracy despite global accuracy [102], we focused on placing as many control points as possible near the river in order to maintain the accuracy in the riverbed area. We used churches and main road intersections, common elements found in the 2021 orthophoto plan, as control ground points (CGPs).

For both editions of the Military Topographic Maps, which have a generally good quality and a quite accurate coordinate system, a different approach was utilised. The georeferencing was performed using coordinates at the intersections of the map grid in the

Gauss–Krüger projection, zone 5 (EPSG:28405), later reprojected in Stereo 70. We choose the projective transformation algorithm, considered particularly useful for scanned maps [102].

All the maps were brought in the Romanian national coordinate reference system—Stereo 70 (EPSG:3844) using the QGIS Georeferencing Plugin. Table 4 shows that the total number of CGPs used varies for each map.

Table 4. Georeferencing parameters for the topographic maps.

Map Name	Georeferencing Algorithm	Total GCPs	RMSE (Mean)
Wallachia—Second military survey of the Habsburg Empire	Thin Plate Spline (TPS)	98	0.0020
"Plan Director de Tragere"		115	0.00041
Romanian Military Topographic Maps—1st ed.	Projective	51	0.49
Romanian Military Topographic Maps—2nd ed.	Already georeferenced		

A manual georeferencing validation was performed by overlaying the georeferenced maps on recent orthophotos. Thus, we visually assessed the accuracy and correctness of the georeferencing by comparing stable planimetric elements over time, such as roads and railway intersections, bridges, buildings (especially churches), lakes, and land parcel boundaries. Altimetric features such as river terraces were also compared with a Digital Elevation Model (DEM). If the georeferencing was not satisfactory, we reviewed the placement of the CGPs in areas with significant errors.

3.2.2. Riverbed Planform Mapping

Riverbed configuration planforms were extracted from all six cartographic sources (Table S1) using QGIS 3.22 version. In order to calculate the parameters from the next work stage, the following initial elements were digitised:

1. The riverbed channel planform (polygon) or wetted channel [6,103], which represents the space covered by the river's water, including secondary channels.
2. The thalweg of the river (line) or central axis of the main channel [6], digitised as the approximative centreline of the main channel. In the case of orthophotos, this can vary. The imagery clearly distinguishes between shallow and deep water, indicating that the thalweg corresponds to the deeper portion of the channel.
3. The active channel (polygon) [7,21], riverbed [22], fluvial area [6], or active belt/band [104] representing the area including the riverbed channel and the sedimentation zones near the channel, including bare and less vegetated bars. It may correspond with the bankfull stage of the channel, sometimes named the low floodplain.

The digitisation process was manually performed by a single operator at a constant scale in order to maintain a high level of accuracy [7,105].

3.2.3. Riverbed Segmentation

Riverbed segmentation is necessary to carry out a unitary and systematic analysis. To do this, we aggregated the channel planform from all six periods, resulting in a historical expansion area of the riverbed between 1856 and 2021. Based on this, there was delineated a polygon which follows the generalised extent and direction of the riverbed in time (Figure 3). Using the v.voronoi.skeleton QGIS tool, a central axis was automatically computed and then smoothed to avoid sudden changes in river flow direction. Thus, it resulted in a centreline of the historical extent of the riverbed over a length of 54 km.

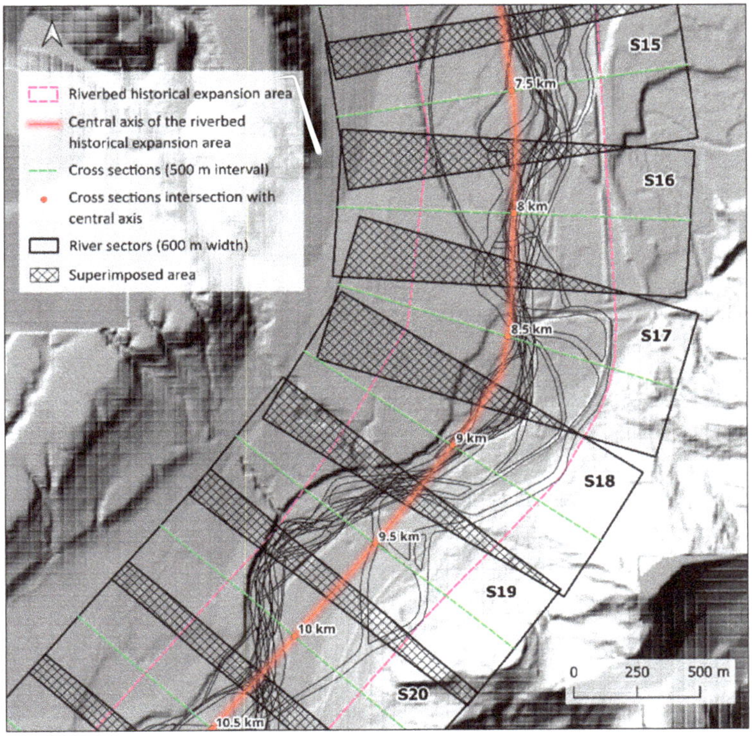

Figure 3. Sector delineation procedure.

The riverbed segmentation was performed by generating perpendicular lines at 500 m along the historical centreline, resulting in a total of 109 transects, including the beginning and end of the studied reach. These were used for the calculation of the widths and number of channels. For the calculation of areas and sinuosity, polygon sectors were generated having in the centre the transects, with a width of the same 500 m and resulting in the same number of transects (109).

To avoid excluding from the analysis parts of the riverbed where the river changes direction significantly, the polygons were expanded by 50 m on both sides, resulting in a new width of 600 m instead of the initial value of 500, to encompass the entire historical expansion area within the sectors. This approach leads to the overlapping of polygons (Figure 3), but they remain perpendicular to the historical centreline, preserve the same areas [6], and provide full coverage.

Both segmentation elements, the transects (T) and the sector polygons (S), were used as distinct methods (Table 5) to calculate geomorphometric parameters and indices in the next step.

Table 5. Geomorphometric parameters and indices assessed in the study.

Parameter/Index	Abbreviation	Method [1,2]	Measure Unit
Channel length	CL	S	m
Sinuosity Index	SI	S	dimensionless
Thalweg migration	TM	T	m
Wetted channel width	WCW	T	m
Bankfull channel width	BCW	T	m
Main channel width	MCW	T	m
Active channel width	ACW	T	m
Number of channels	NC	T	dimensionless
Wetted channel area	WCA	S	km^2
Bankfull channel area	BCA	S	km^2
Active channel area	ACA	S	km^2

Note(s): [1] S—sector, [2] T—transect.

3.3. *Geomorphometric Measurements*

3.3.1. Parameters and Indices Measurements

Based on the planforms digitised in the previous stage and the segmentation methodology (Figure 3), 9 parameters and 2 indices were measured and calculated. A summary of these is presented in Table 5.

Channel length (CL) was already presented in the previous section, representing the sinuous length of the main channel within a river sector. This parameter is used for the calculation of the sinuosity index (SI) [106], defined as the ratio between CL and the valley length (VL)—the straight length between the endpoints of the CL (Figure 4a). The SI was calculated according to Equation (1):

$$SI = \frac{CL}{VL} \quad (1)$$

In terms of widths, the measurements were performed using the transects which intersect the river channel planform and the active channel area. Thus, four parameters resulted: the main channel width (MCW) (Figure 4b); the wetted channel width (WCW), including the MCW and the width of the secondary channels (Figure 4c); the bankfull river channel width (BCW), which is formed from the WCW and the sand bars or isles between the channels (Figure 4d). These measurements need to be perpendicular to the central axis of the intersected elements. Consequently, the initial transects were manually adjusted.

The areas were calculated using the method explained earlier (Figure 3), by looking at where the sector polygons meet the riverbed shape. Three area parameters resulted: wetted channel area (WCA) (Figure 4c), derived from the initial manually digitised river planform; bankfull river channel area (BCA), resulting from dissolving the WCA into a single polygon, without gaps (Figure 4d); and active channel area (ACA) (Figure 4e), previously digitised.

In addition to these, the number of channels (NC) was as measured using the transects, counting the points of intersection between the cross-section line and the wetted channel of the river (Figure 4f).

The thalweg migration (TM) was assessed by measuring the distance between the intersection point of the transect with the thalweg line for successive time periods.

These parameters were carried out for all the cartographic sources analysed in the study (Table 1).

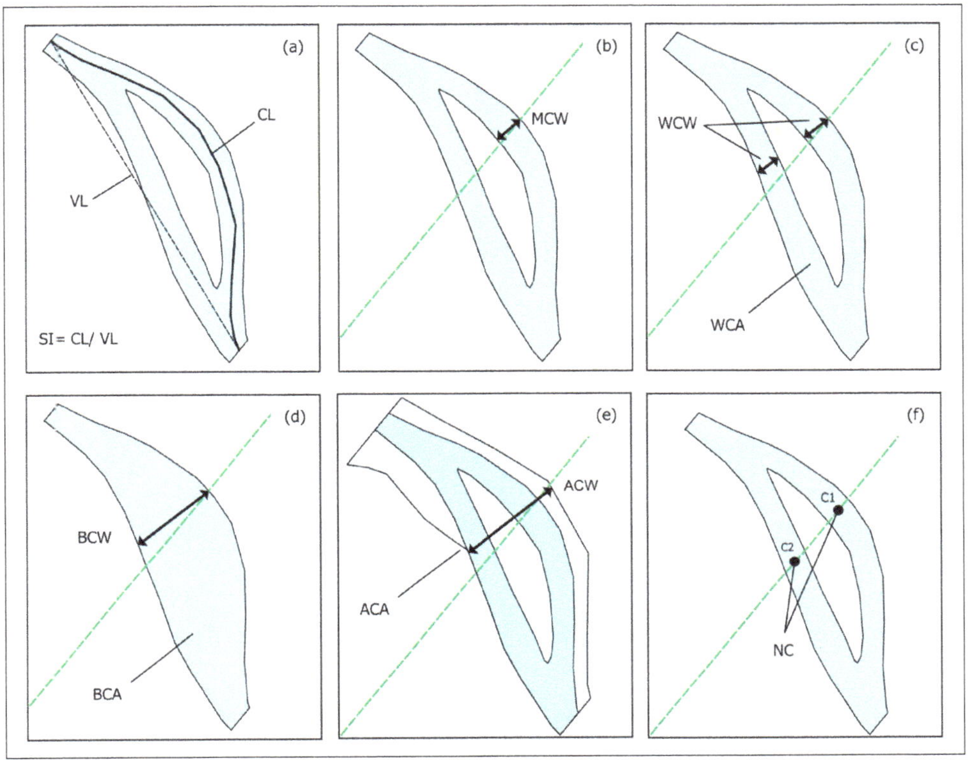

Figure 4. Geomorphometric parameters and indices measurement: (**a**) sinuosity index (SI), defined as the ratio between channel length (CL) and valley length (VL); (**b**) main channel width (MCW); (**c**) wetted channel width (WCW) and area (WCA); (**d**) bankfull channel width (BCW) and area (BCA); (**e**) active channel width (ACW) and area (ACA); (**f**) number of channels (NC). The green dashed line represents the transect. The riverbeds in the images represent the area contained in a sector.

3.3.2. Processing Automation

Considering that the cross-sections do not always cross the channel perpendicularly, the width calculation cannot be fully automated; transects should be manually adjusted for every time period and sector in order to maintain the accuracy of the measurements. Furthermore, all model inputs must be prepared and reviewed in advance.

Subsequently, the parameter calculation was automated in the GIS environment using QGIS Model Designer (Figure 5). The result consists of a spreadsheet at both sector and transect scales for each parameter or index calculated across the study time span (1856–2021).

The model was applied one period at a time.

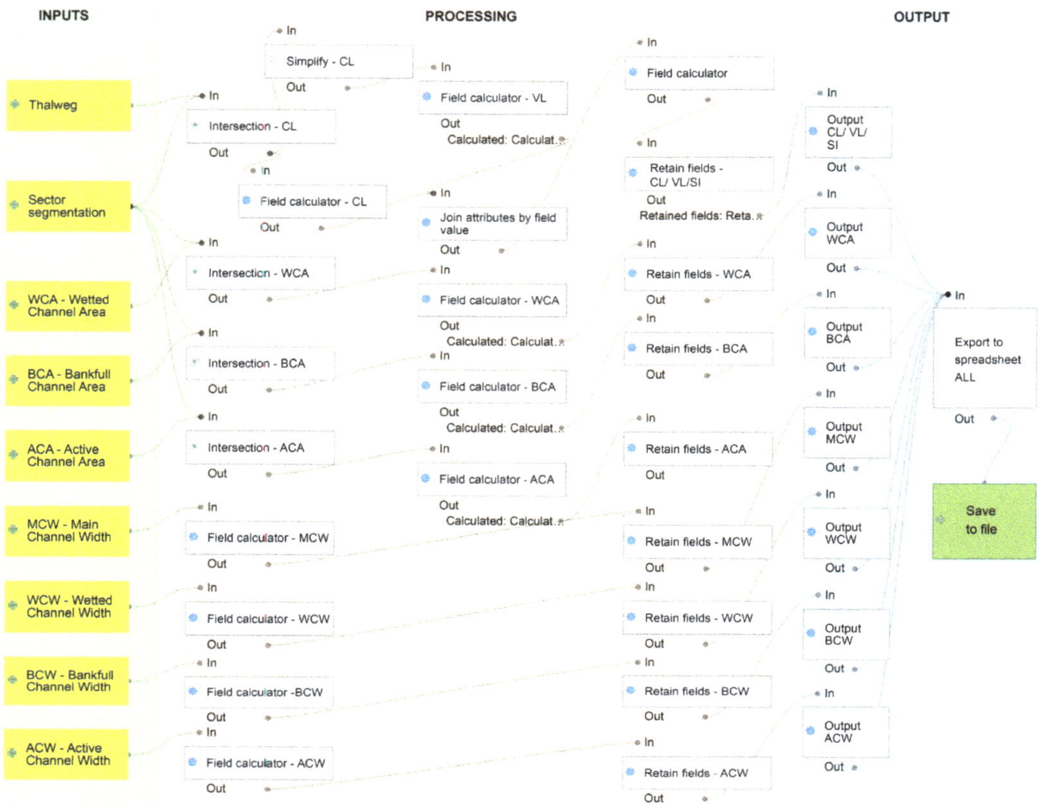

Figure 5. Automated model for parameters and indices calculation.

3.4. Data Analysis and Postprocessing

All the riverbed geomorphological parameters and indices calculated in the previous section of the study, along with the climatic, hydrological, and human interventions, were statistically processed and graphically represented in order to provide a simple and better visualisation and interpretation. These actions were made using R 4.2.2 through RStudio GUI, QGIS 3.22, and Microsoft Excel.

For the climatic data, we extracted daily mean areal precipitation over the studied watershed between 1950 and 2022 (73 years) using the "terra" R package, version 1.6.53 [107]. Then, they were aggregated into yearly mean values.

We used the Mann–Kendall (MK) non-parametric test [108,109] to detect monotonic trends in the yearly time series of precipitation and temperature over the study area and river flow gauged at the hydrometric station. The test was performed using the "Kendall" R package, version 2.2.1 [110]. The key equations of the MK test are provided as follows, starting with the MK statistic S, which is calculated as

$$S = \sum_{i=1}^{n-1} \sum_{j=i+1}^{n} sgn(x_j - x_i), \quad (2)$$

where x_j and x_i are the values in years j and k, with $j > k$ and n the total number of years.

The *sgn()* is the sign function, defined as

$$sgn(x_j - x_i) = \begin{cases} 1, & \text{if } x_j - x_k > 0 \\ 0, & \text{if } x_j - x_k = 0 \\ -1, & \text{if } x_j - x_k < 0 \end{cases} \quad (3)$$

The distribution of the MK statistic S is approximated by a normal distribution for large n, with mean = 0 and standard deviation given by Equation (4).

$$Var(S) = \sqrt{\frac{n(n-1)(2n+5) - \sum_{i=1}^{t} t_i(t_i - 1)(2t_i + 5)}{18}} \quad (4)$$

where n is the number of observations, t is the number of tied groups in the time series, t_i the number of ties of extent i. To test the statistical significance of the trend, the standardised Z test was calculated according to Equation (5). For this analysis, we set the significance level at $\alpha = 0.10$ for a two-tailed hypothesis test.

$$Z = \begin{cases} \frac{S-1}{Var(S)} \; if \; S > 0 \\ 0 \;\; if \; S = 0 \\ \frac{S+1}{Var(S)} \; if \; S < 0 \end{cases} \quad (5)$$

The strength and the direction of the trend is described by the Kendall's Tau correlation coefficient (τ). Assuming there are no ties into the data series, this is calculated according to Equation (6):

$$\tau = \frac{\mathrm{C} - \mathrm{D}}{\mathrm{n}(\mathrm{n}-1)/2} \quad (6)$$

where C is the number of concordant pairs, D is the number of discordant pairs, and n is the number of observations.

The R package "changepoint", version 2.3, [111] was utilised to identify changes and breakpoints in data series of precipitation, temperature, and discharges.

HCA was used to group the river sectors into clusters which have common characteristics and the same spatial pattern of evolution. The analysis was performed using the base statistics package in R [112], the Euclidean method for the distance matrix, and Ward's D2 method for the clustering algorithm. A similar approach HCA was also performed for the Garonne River, France [6].

4. Results

4.1. Morphometric Analysis of River Reach Between 1856–2021

4.1.1. Entire River Reach

The SI, ACA, BCA, and WCA (Figure 6a,b) were calculated for the entire river reach, and several changes occurred during 1856—2021. We observed a general downward trend in all the parameters, with some upward oscillations in 1902, 1957, and 1978.

The sinuosity index describes the river reach as moderately sinuous. Its values reduce from 1.24 in 1856 to 1.21 in 2021, the general trend being of decrease (Figure 6a). However, 1902 deviates significantly, reaching a value of 1.29.

The areas occupied by different riverbed parts show the same downward trend, as can be observed in Figure 6b. BCA undergoes the most significant changes, becoming almost five times smaller, from 10.6 km^2 in 1856 to 2.3 km^2 in 2021. The period 1902–1978 indicates a stationary trend, with areas between 5 and 7 km^2. In terms of ACA, the area drops continuously from 15 6 km^2 in 1856 to 4.3 km^2 in 2021, with a growth in 1957. WCA shows a weak decrease compared with the other parameters, from 2.94 km^2 in 1856 to 2.12 km^2 in

2021. It can also be observed that the proportion between the parameters becomes smaller (Figure 6b). In 1856, BCA and WCA are 68% and 17% from the area of ACA. This changed in 2021, where BCA is 52% and WCA 49%.

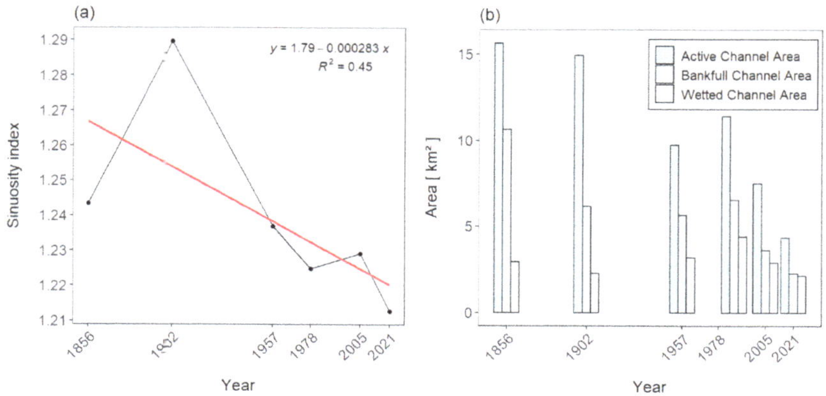

Figure 6. Evolution of parameters calculated for the entire studied river reach: (**a**) sinuosity index; (**b**) comparative view of wetted channel, fluvial, and active band areas.

Yearly rates of change indicate the amplitude of the morphological changes in river parameters depending on each time interval. The sinuosity index (SI) decreases at a rate of 0.02% per year, with the most active periods occurring between 1856 and 1902, 1903 and 1957, and 2006 and 2021 (Table 6).

Table 6. Yearly rate of change (%) in calculated parameters.

Index/Parameters	1856–1902	1903–1957	1958–1978	1988–2005	2006–2021	1856–2021
SI	+0.08	−0.07	−0.03	+0.02	−0.08	−0.02
WCA	−0.49	+0.74	+1.26	−1.91	−1.66	−0.23
BCA	−0.90	−0.15	+0.52	−2.49	−2.31	−2.23
ACA	−0.09	−0.63	+0.57	−1.92	−2.62	−1.57

The area measurements indicate, in general, negative values between 1856 and 2021 (Table 6). The bankfull channel area (BCA) has the highest rate of change of −2.23% per year, followed by the active channel area (ACA), with −1.57% per year. The changes in wetted channel area (WCA) are significantly lower (−0.23 per year). The period 1958–1978 is the only one in which all the parameters show a positive rate of change.

4.1.2. Sector Scale Analysis

The analysis of the riverbed at the sector scale between 1856 and 2021 shows a general downward trend in most of the parameters and indices values, which indicates a reduction in the geomorphological complexity of the riverbed (Figure 7a–i).

Regarding the sinuosity index (SI), the median values indicate a slight tendency of the riverbed to become less sinuous. The mean values confirm this, oscillating between 1.136 in 1856, 1.59 in 1902, and reaching 1.143 in 2021. The outliers represent, in most of the cases, meandering sectors of the river (SI > 1.5) (Figure 7a).

The absolute and mean number of riverbed channels is represented in Figure 7b. The mean values decrease slowly from 1.72 in 1865 to 1.60 in 1957, followed by a sudden drop to 1.38 in 1978 and 1.17 in 2021. This indicates the transition to a single-thread channel.

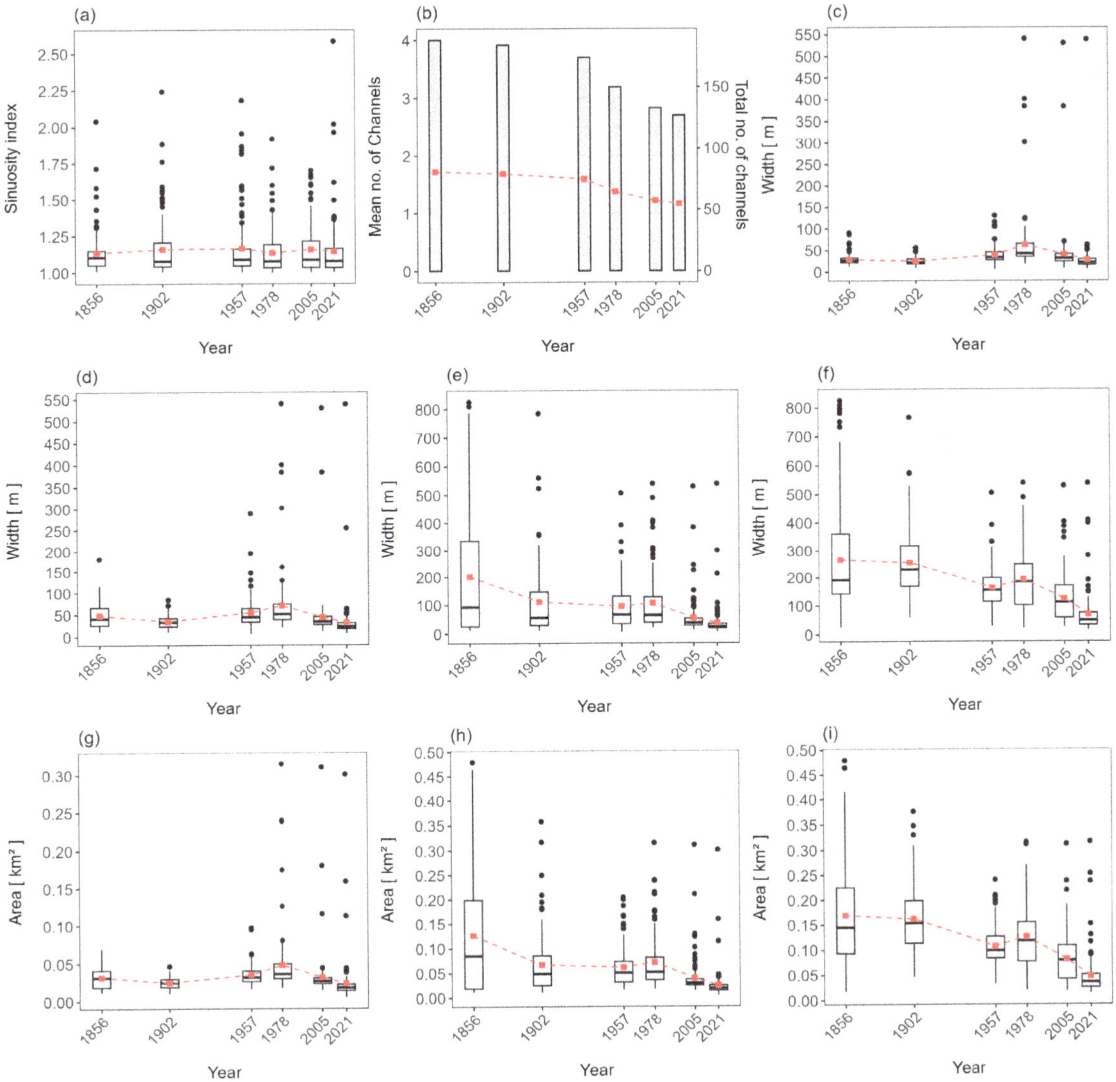

Figure 7. Evolution of parameters and indices calculated for the studied river reach: (**a**) Sinuosity index—SI, (**b**) Number of river channels—NC, (**c**) Main channel width—MCW, (**d**) Wetted channel width—WCW, (**e**) Bankfull channel width—BCW, (**f**) Active channel width—ACW, (**g**) Wetted channel area—WCA, (**h**) Bankfull channel area—BCA, (**i**) Active channel area—ACA. The box shows the first quartile (Q1) and the third quartile (Q3), the black horizontal line is the median value, the red square shows the mean value, and the red dotted line shows the mean values evolution.

We analysed the widths in terms of the main channel (MCW), wetted channel, bankfull channel (BCW), and active channel (ACW) (Figure 7c–f).

MCW is, in general, low between 1856 and 1902 (<30 m), rises between 1902 and 1978 (mean over 50 m), and then drops to similar values as initially. Also, the spread of the middle 50% indicates a more homogenous main channel in 1956, 1902, and 2021 than in the other periods, especially 1978 (Figure 7c).

WCW follows the MCW values and distribution, except in 1856, where the width and the spread of the interquartile range are significantly higher, showing more variability (Figure 7d). This indicates a multithread riverbed, with secondary channels contributing to increasing the value. In both situations, the outliers over 300 m which have appeared since 1978 represent the width of Pucioasa Lake, located in the northern part of the reach.

The bankfull channel width follows an obvious downward trend, highlighted by the changes in mean and median values as well as the data distribution (Figure 7e). Between 1856 and 2021, the mean BCW decrease was from 202 to 37 m, with more than 90 m being lost just in the period 1856–1902. The median width reduces from 93 to 24 m. Regarding the distribution of the data, the BCW spread of the widths becomes smaller in time: the middle 50% decreases from ~300 m in 1856 to 18 m in 2021, indicating a general intense shrinking and narrowing of the riverbed. A considerable number of outliers have appeared since 1978.

In terms of ACW, the mean and median values indicate a negative trend, divided into two stages of decrease: 1856–1957 from 267 and 191 m to 163 and 155 m; 1978–2005 from 193 and 184 m to 70 and 40 m. Also, the range of variability decreased, the middle 50%, reaching from ~220 m to 40 m, became more than 5 times smaller in 2021 than in 1856. The same years show significant outlier values (Figure 7f).

The area parameters data distribution is quite similar to their width correspondents (WCW–ACA, BCW–BCA, ACW–ACA), indicating a strong correlation between these parameters. At the same time, this is a confirmation that the measurements are correct and representative, considering that the method used was different: transects for the widths and polygon sectors for the areas (Figure 4).

WCA, BCA, and ACA between 1856 and 2021 maintain the same downward trend as the other parameters/indices. Regarding wetted channel area (WCA), the mean values decrease from 0.032 km^2 in 1856 to 0.023 km^2 in 2021. During this period, a deviation occurs from 1957 to 1978, where the area reaches 0.048 km^2. Looking at the median values (less influenced by extreme values), this deviation from the trend became smaller (Figure 7g). The bankfull channel area (BCA) became, in general, circa five times smaller from 1856 (0.13 km^2—mean and 0.086 km^2—median) to 2021 (0.025 km^2—mean and 0.019 km^2—median). There are observed periods of decreasing, between 1856 and 1902 and 1978 and 2021, and relative stability between 1902 and 1978. Also, the variation in the areas decreased; the interquartile range became considerably smaller (Figure 7h). The active channel area (ACA) drops from 0.169 km^2 (mean) and 0.145 km^2 (median) in 1856 to 0.048 km^2 (mean) and 0.036 km^2 (median) in 2021. The year 1978 deviates from the trend (Figure 7i).

4.1.3. Riverbed Migration

We assessed the riverbed migration by measuring the displacement of the thalweg for each sector between successive periods, which became smaller along with the advance in time, as shown in Figure 8. For the entire period of the study (1856–2021), the riverbed has moved, on average, by 190 m and the median by 142 m. The interquartile range extends between approx. 60 and 260 m, showing a moderate spread of 200 m. Some outliers exceed 600 m, indicating drastic changes in the riverbed that are most likely caused by anthropogenic intervention (Figure 8a).

Analysing each time period, it is observed that the median migration distance continuously decreases from 167 m in 1856 to 142 m in 2021. The average values keep the same trend except for the period 1988–2005, where the mean is dragged up by high values and outliers. The spread of the migration distance interquartile range gradually decreases from period to period, showing a clear lower variability in time. The middle 50% reduced more than five times, from 284 m between 1856 and 1902 to 55 m between 2006 and 2021. Additionally, Figure 8b shows a positive skew in the distribution across all periods.

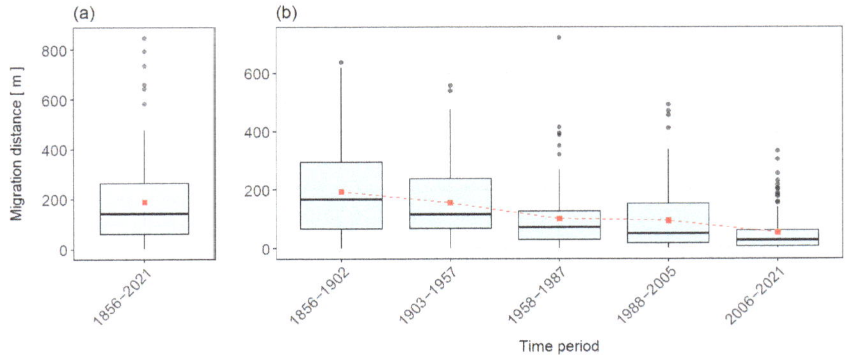

Figure 8. Evolution of riverbed migration distance (**a**) for the entire time span of the study and (**b**) between successive time periods.

We also calculated the rate of migration for each time span and the full period. The average rate of riverbed migration between 1856 and 2021 is approx. 1.14 m/year. Regarding the changes between each time step, the values are as follows: 1856–1902: 4.15 m/year, 1903–1957: 2.91 m/year, 1958–1978: 3.4 m/year, 1988–2005: 5.27 m/year, and 2006–2021: 3.44 m/year.

4.2. HCA

The clustering analysis of the riverbed was performed in order to identify spatial patterns in riverbed dynamics by grouping the river sectors into clusters with common characteristics or which share the same evolutionary trend. We used for the analysis the indices and parameters presented in Table 5, except for CL, MCW, and TM, which negatively affect the clustering. Therefore, 8 parameters and indices for 109 river sectors were included in the HCA. Following several tests, we considered 5 clusters to be representative and to indicate specific spatial patterns in riverbed dynamics. The hierarchical cluster dendrogram (Figure 9) displays the analysis's result, clusters being highlighted.

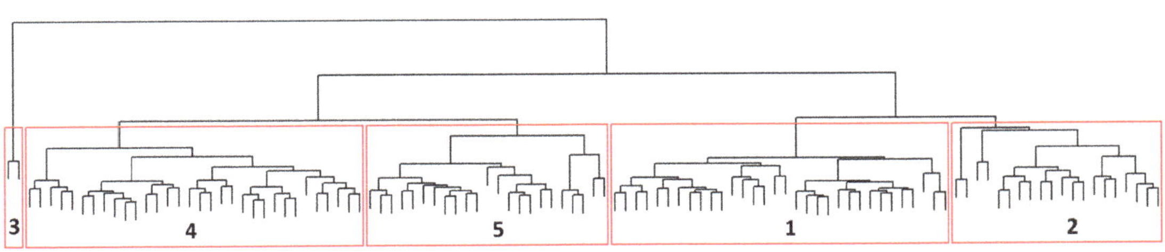

Figure 9. Dendrogram of the HCA applied on river sector data. Red rectangles represent the clusters. The number is the name of the cluster.

The agglomerative coefficient of the HCA is 0.89, which indicates a strong clustering structure. The cluster robustness was assessed trough the silhouette score [113], indicating a low averaged width of 0.11, which suggests a weak separation between the identified clusters. In the context of the fluvial system and the study, this reflects the geomorphic behaviour and complexity of riverbed dynamics. River sectors' gradual morphological transitions and uneven temporal evolutionary trajectories across six different periods lead to an overlapping of the clusters, affecting robustness. However, we consider that the low silhouette score does not necessarily mean a poor clustering structure, but rather, it is a reflection of the river geomorphological behaviour. In order to characterise the clusters, Figure S1 shows the z-scores for the analysed geomorphometric parameters at the sector level. Figure 10 also displays the spatial distribution of the clusters.

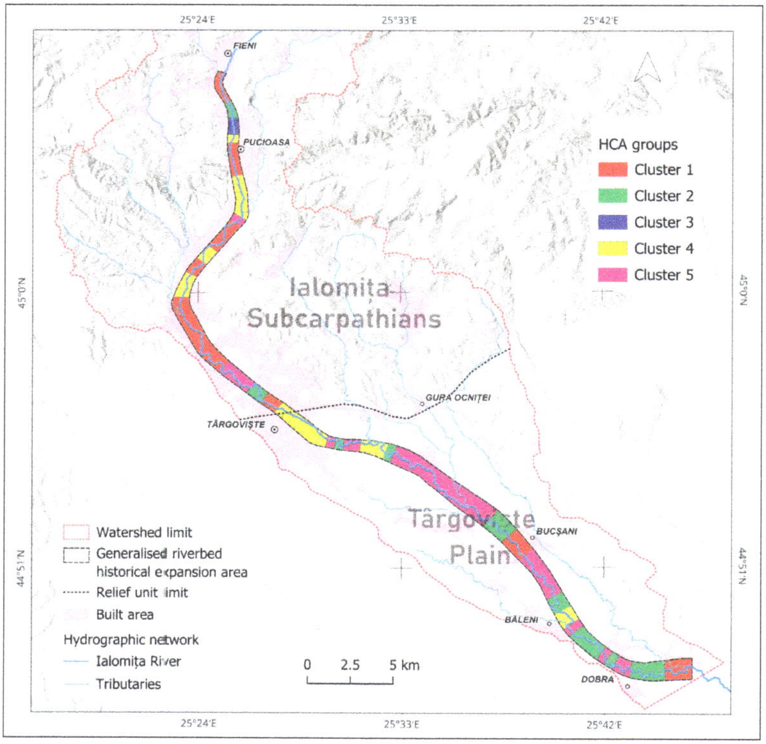

Figure 10. Spatial distribution on riverbed sector clusters.

Cluster 1 includes 32 sectors located predominantly in the northern part of the studied reach (Figure 10). These are characterised by consistent high values for most of the parameters, which persisted throughout the analysed period until 1957 or 2005 for some sectors. The z values are generally up to 2.5, sometimes reaching 5 (Figure S1). In 2005 and 2021, the values of the wetted channel, bankfull channel, and active channel parameters are the same or similar.

Cluster 2 contains 20 highly dynamic sectors over the analysed period, except 1856. During 1902–2021, these sectors exhibit mostly sustained meandering activity. However, in 1957 and 1978, some sectors developed anastomosed fluvial characteristics. Z-scores correspondingly capture these elements, showing generally high values starting with 1902 (Figure S1). Mostly, these sectors are located in the plain area (Figure 10).

Cluster 3 is formed from two sectors, which suffered drastic changes starting in 1978. The riverbed widths and areas increased significantly and remained larger up until 2021. The z-score shows up to eight times standard deviations above the mean in terms of widths and areas (Figure S1). The SI and NC remain almost the same. These correspond with Lake Pucioasa, located near the city with the same name (Figure 10).

Cluster 4 contains 23 sectors, generally highly influenced by anthropogenic activity and mainly located in the middle and upper part of the studied reach (Figure 10). A transition is observed from a multi-thread channel in 1856–1902, or just 1902 in some cases, to a single-thread, low sinuous or straight channel. According to Figure S1, the majority of the parameters have z-scores up to 5 from 1856/1902 to 1978. In subsequent periods leading to 2021, the parameter values exhibit a declining trend, with their z-scores converging under the mean.

The expansion of built space in the riparian zones and hydrotechnical developments has caused an intense narrowing of the riverbed by reducing the river's freedom space [114] gradually over time.

Cluster 5 includes 32 sectors, located predominantly in the southern part of the studied reach (Figure 10), which are generally characterised by a low sinuous or meandered channel, considering the evolution over the entire analysed period (1856–2021). Between 1902/1957 and 2005, certain river sectors exhibited pronounced geomorphological dynamics, as evidenced by high z-scores (generally up to 2.5). Notably, the sinuosity index (SI) reached significantly higher values (z up to 7.5), while the number of channels (NC) also showed marked deviations (Figure S1). These spatial patterns highlight zones of intense fluvial activity, particularly where meandering processes dominate or where wandering prevails in multi-thread channels.

4.3. Climatic Regime and Trend

Figure 11 shows the mean annual precipitation over the study area. The precipitation values range from minimums of 404.3 mm (1950) and 421.8 mm (2000) to maximums of 1146.3 mm (2014) and 1188 mm (2005). The average mean multi-annual precipitation is 745.3 mm. A slightly upward trend is observed, confirmed by the MK test (τ = 0.285, p-value = 0.0004). Based on two changepoints identified in 1965 and 2002, three periods are distinguished which confirm the upward trend of precipitation:

- From 1950 to 1965—with a range between 404 mm (1950) and 743 (1964) and with a mean of 617 mm.
- From 1966 to 2004—with a minimum of 422 mm in 2000 and a maximum of 973 mm in 1972. The mean value is 743 mm, which is more than 100 mm higher than the other period.
- From 2005 to 2022—with a mean value of 864 mm, annual values ranging between 590 mm in 2011 and 1188 mm in 2005. We observe the same increase in the mean value (over 100 mm) as in the previous period.

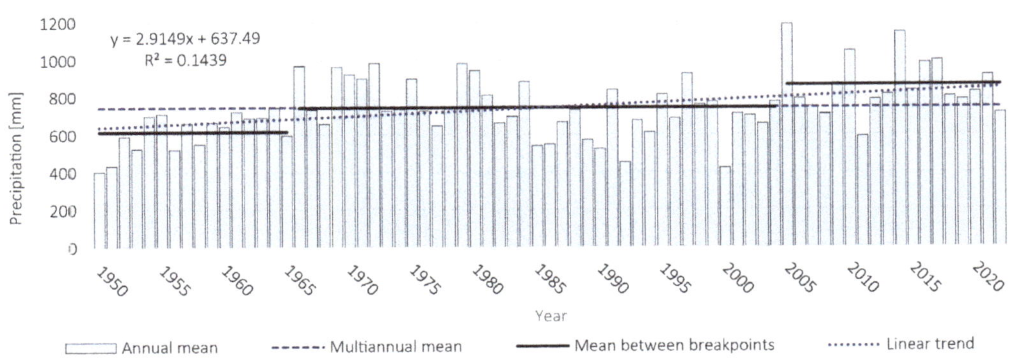

Figure 11. Yearly precipitation evolution within the study area between 1950 and 2022.

The mean annual temperature shows a clear continuous upward trend in the study area between 1950 and 2022 (Figure 12), which confirms there is climate warming in the area. The multiannual mean temperature is 9.01 °C, with annual average values between the minimum of 6.06 °C (1956) and maximum of 10.69 °C (2019). This positive trend is confirmed by the MK test, which indicates a strong rank correlation coefficient of 0.573, statistically significant at p-value = 2.22×10^{-16} A breakpoint was identified in the temperatures in 1998, from which point the annual average is permanently above the multi-annual average. This finding suggests a permanent change in the climatic regime in the study area.

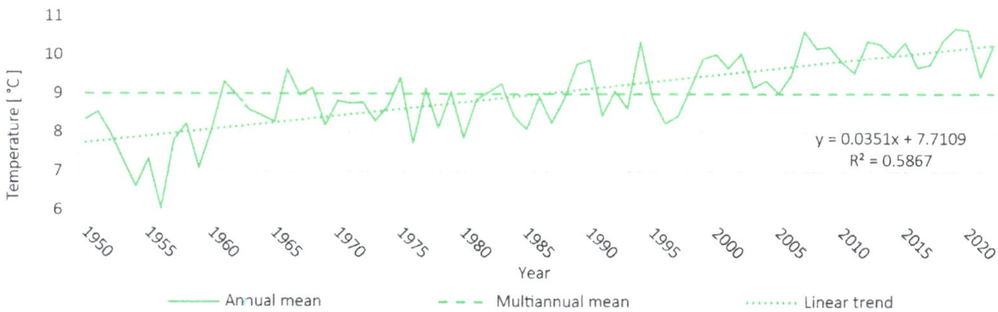

Figure 12. Yearly temperature evolution within the study area between 1950 and 2022.

4.4. Hydrological Regime and Trend

The general evolutionary trend of the mean annual discharge shows a slight decrease (Figure 13a).

The whole period can be split into three parts, divided by the breakpoints detected in 1984 and 2003:

- The first period (1961–1984) has the highest discharge, with a mean of 10.5 m^3/s.
- Starting with 1985, the second period begins and lasts until 2003, characterised by a drastic change in the flow. The mean discharge drops by half, reaching 5.03 m^3/s.
- The third period (2004–2022) shows an equally sudden return of the discharges near the values in the first period. The mean discharge is 8.97 m^3/s.

According to the MK test applied for 1961–2022, a small negative trend is observed, but it is not significant ($\tau = -0.112$, p-value = 0.20).

The maximum annual discharge (Figure 13b) stretches between values of 28 m^3/s in 2000 and 745 m^3/s in 2001. Also, a few peaks in the maximum flow are highlighted, respectively, 1972, 1975, 1982, 1991, 1997, 2001, and 2005, with $Q > 400\ m^3/s$, representing the most significant floods between 1961 and 2022. A weak negative trend is observed, confirmed with the MK test, but is not significant ($\tau = -0.103$, p-value = 0.242).

The minimum discharge (Figure 13c) annual values range between 0.07 m^3/s in 1964 and 3.03 m^3/s in 2005, showing a general slightly upward trend ($\tau = -0.15$, p-value = 0.087).

Regarding the suspended load discharge (R) (Figure 14), the mean multiannual value is 15.17 kg/s, with a minimum of 0.84 kg/s in 1992 and a maximum of 63.54 kg/s in 2005. The general evolutionary trend shows a slight decrease, as in the case of average flows, confirmed by the MK test, not statistically significant ($\tau = -0.105$, p-value = 0.229). Furthermore, the entire period can be divided by a change point detected in 1975:

- Period 1961–1975, with a mean of 22.09 kg/s;

- Period 1976–2022, where the mean value of R is 12.97 kg/s, significantly lower than before the change point, especially between 1985 and 1995, the values being under 10 kg/s.

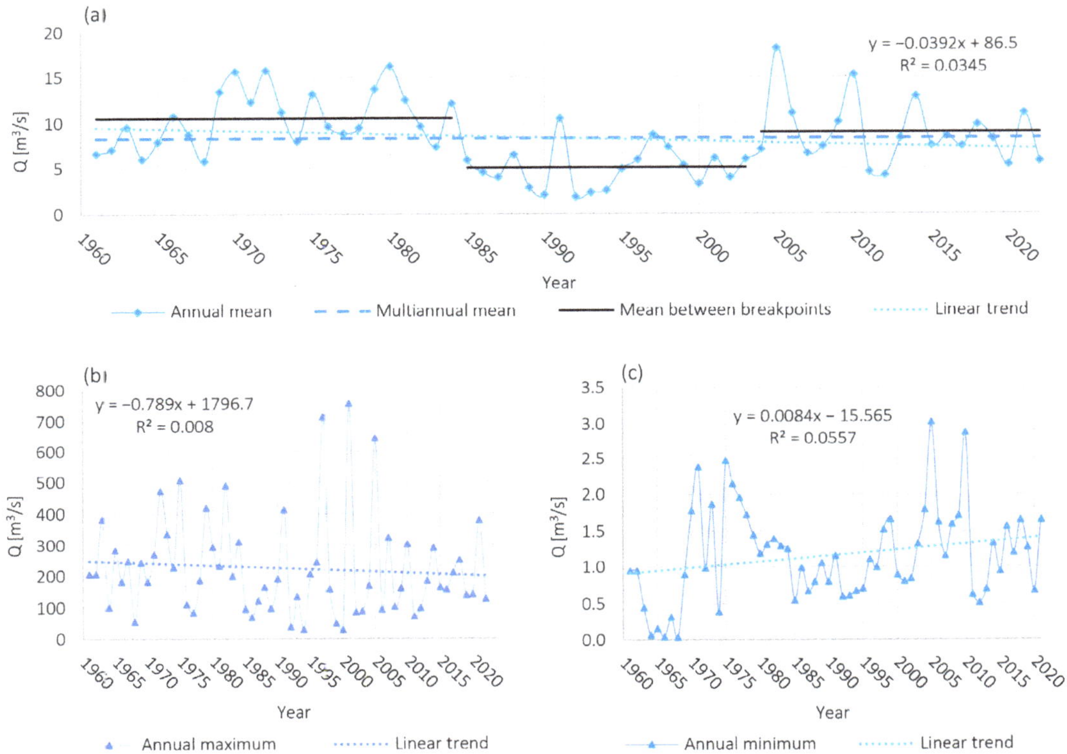

Figure 13. Mean (**a**), maximum (**b**), and (**c**) minimum annual discharge variability of the Ialomița River at the Băleni hydrometric station between 1961 and 2022.

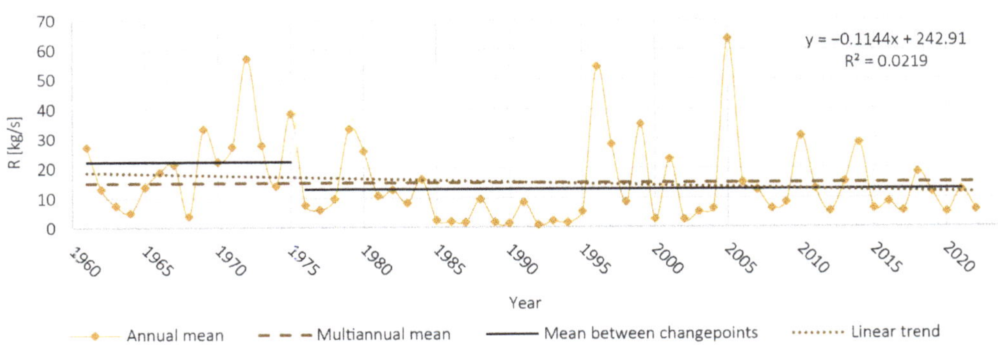

Figure 14. Annual suspended sediment discharge variability of the Ialomița River at the Băleni hydrometric station between 1961 and 2022.

4.5. Anthropogenic Activity Assessment

Analysing the cartographic sources between 1856 and 2021, the river bank protection structures are scarcely present until 2005 (around 2 km), followed by a sudden increase in

2021, reaching 15.7 km, mainly concrete embankments realised to protect the settlements near the river. Regarding bed erosion, we were unable to identify on the historical cartographic sources any hydrotechnical structures prior to 2005, at which point 11 locations were found, reaching 13 by 2021.

Analysing the topographic maps, we were unable to identify any gravel mining activity in the studied Ialomița reach riverbed and floodplain on the map from 1978.

Starting with 2002, by analysing satellite images through Google Earth Pro, we identified 11 gravel pits (GPs) located in the plain area, downstream to Târgoviște city (Figure 15). Among these, six of them were still active in 2021, with a mining activity of up to 19 years for GP3, GP4, GP5, GP11 and up to 13 years for GP8 and GP10. Although GP10 and GP11 are not in the limits of the study area, we included them in the analysis because they can influence the riverbed due to their proximity.

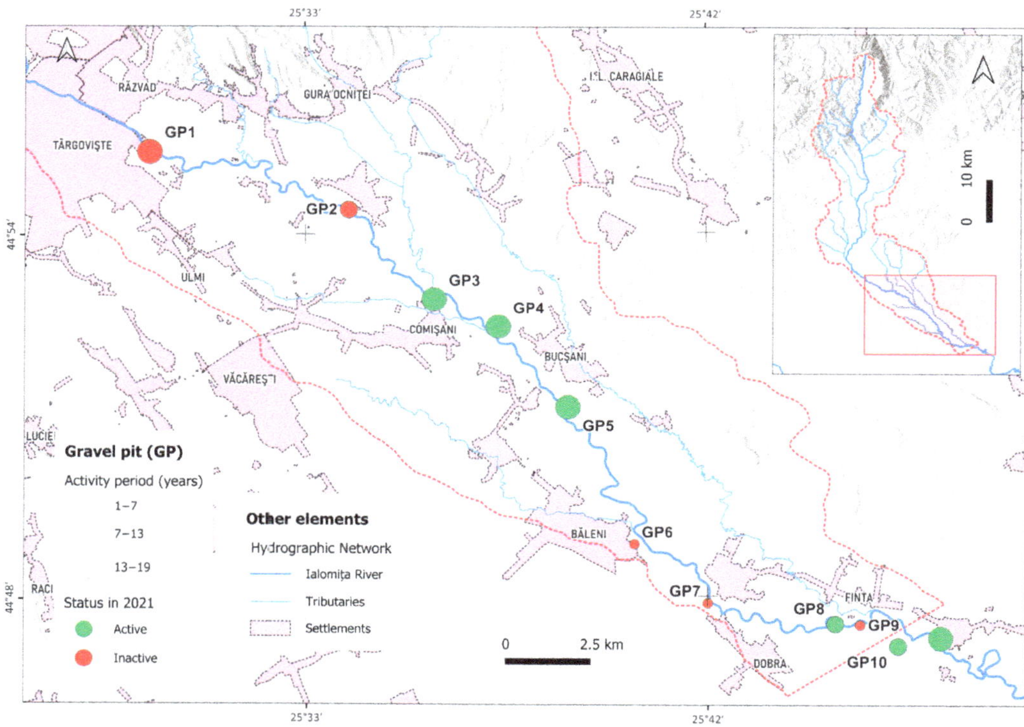

Figure 15. Spatial distribution and temporal evolution of gravel pits on studied river reach.

4.6. Present Riverbed Dynamics

The field survey performed in 2024 shows that the Ialomița River is still becoming narrower and deeper in the area we studied, as shown in Figure 16.

The amplitude of the river deepening process depends on the relief unit, which is greater in the Subcarpathian zone. In some places, very intense incision reaches up to approximately 10 m (Figure 16b,c) and extends significantly in length, taking on the appearance of a canyon (Figure 16d). In this zone, the river channel loses its connectivity with the floodplain, leading to an irreversible geomorphic transition into a new terrace. In the plain area, the downcutting is weaker (Figure 16f) but can still create problems for anthropogenic developments in the riverbed.

Figure 16. Observed riverbed incision and its impact on bridge structures along the studied Ialomița River reach. The images are ordered from north to south as follows: (**a**) bridge pier erosion near Fieni; (**b–d**) very intense riverbed incision in the Subcarpathian zone south to Pucioasa; (**e**) incision affecting 2 of 3 bridge piers on DC136B road near Brănești; (**f**) incision in the exploitation area of a former gravel pit (GP2 in Figure 15); (**g**) intense incision at bridge pier on DJ711B road at Băleni; (**h**) incision of over 1 m (length of the white ruler) at the base of the pillar. The riverbed has silted up due to the ongoing dredging of Lake Pucioasa.

Road infrastructure elements such as bridges are the most affected by the incision into the riverbed, a fact observed during the field survey (Figure 16a,e,g). We observe significant local scour at the base of the bridge's piers, which remain suspended (Figure 16a). In the case of other types of bridges, previously deep foundation piles are now visible above the riverbed (Figure 16e,g) by over 1 metre for the bridge at Băleni (Figure 16h).

On the other hand, we observed intense lateral erosion in dynamic equilibrium zones. In this case, the narrowing and slight incision of the riverbed led to the creation of a single-thread channel that acquired intense meandering processes. As a consequence, concrete embankments were constructed, whose length increased from under 2 km in 2005 to over 15 km in 2021, to protect adjacent riverside settlements.

5. Discussion

The study results show a general decreasing trend in almost all the parameters and indices for the studied river reach, measured as a whole (Figure 6) as well as at the sector level (Figure 7). Also, the thalweg migration distance between successive years follows the same tendency (Figure 8). However, riverbed measurements in 1978 highlight a deviation from the main tendency of decreasing, as can be observed in Table 6 and Figure 7c–i. Consequently, based on the findings of this study, the interval spanning 1856 to 2021 can be divided into three distinct evolutionary periods, as follows.

The first period lasts from the mid-19th century to the beginning of the second half of the 20th century (including 1856, 1902, and 1957) and is characterised by fluvial dynamics occurring under quasi-natural, slightly disturbed conditions, especially for the period 1856–1902. Although the Scropoasa and Dobrești lakes have been formed since 1930, their limited storage capacity (Table 1) and high distance from the studied river reach make them ineffective for producing notable changes in the riverbed.

A large number of channels (Figure 7b), combined with the small values of the main channel width (Figure 7c) and the wetted channel width and area (Figure 7c,g), and with the high values of the bankfull and active channel widths and areas (Figure 7e,f,h,i), indicate that the river had a generally braided fluvial style, where the aggradation process is dominant. It is also notable that the migration of the riverbed between successive periods is accentuated (Figure 8b), which indicates a large free space of the river. Considering that the anthropogenic influence on the river is at a minimal level in this period, we cannot attribute the decreasing trend to this cause. Rather, climate change is the main driving factor in riverbed complexity reduction.

The second period is represented by the second half of the 20th century, a time when the most important anthropogenic developments of the Ialomița River took place, completed in 1988. The analysis of the riverbed in 1978 shows a clear deviation from the direction of evolution, as can be observed in Table 6 and Figures 6 and 7. This is attributed to the direct influence of the construction of Pucioasa Dam (Table 1), finalised in 1975, and of the related downstream riverbed regularisation structures. The dam retains a large portion of sediment, as evidenced by the volume of water that the lake can retain decreasing from 10.6 mil. m^3 in 1975 (Table 1) to 2.22 mil. m^3 in 2016 [80], the lake being silted up in a proportion of 79%. The mean annual silting rate is 1.88%/year, representing a volume of approximately 200,000 m^3/year.

The Pucioasa reservoir caused a decrease in suspended sediment supply (Figure 14), where a breakpoint was identified in the same year as the dam's commissioning. This led to significant changes in riverbed dynamics, forcing the river to transition to a single-thread channel in the Subcarpathian area and a meandering–anastomosed style in the plain area. All of these changes are highlighted by the decrease in the number of channels (Figure 7b) and the increase in the width and area in 1978 (Figure 7c–i).

The finalisation of the Bolboci Dam in 1984, combined with low quantities of precipitation, led to a hydrological dry period, with values generally below the mean, identified by breakpoints from 1984 to 2004 into the mean annual discharges gauged at Baleni

(Figure 13a). The suspended sediment discharge is also low, below the mean from 1985 to 1995 (Figure 14). This period favoured a narrowing process of the riverbed and the transition to a single-thread channel until the end of the century.

The third period represents the beginning of the 21st century, including 2005 and 2021, where the riverbed returns to the general descending trend of evolution. Compared with other periods, the measured parameters have significantly low values and, at the same time, a homogenous and tight distribution (Figure 7). Also, the migration distance of the thalweg is low on average, more than five times compared with 1856–1902. The variability is also lower (Figure 8). This indicates a lack of complexity in the riverbed morphology and a categorical transition to a sinuous, single-thread planform pattern.

The narrowing of the channel is associated with the incision, straightening, and deepening of the riverbed (Figure 16), accentuated by the intense gravel mining activity in the plain area starting in the 2000s. By 2021, we were inventorying 11 gravel extraction sites, half of them being active at the end of the period under review. These also have among the longest operating periods (Figure 15).

The floods could also be a factor which causes the deepening of the river channel, favoured by the instability created by the exploitation of aggregates from the riverbed. In 2001 and 2005, the largest historical floods recorded on the Ialomița River occurred, with peak discharges of 758 m^3/s and 646 m^3/s, respectively (Figure 13b).

HCA indicates some spatial patterns in the evolution of the Ialomița riverbed. Overall, these clusters underscore the interplay between natural fluvial dynamics and human-induced constraints, highlighting zones of both stability and intense morphological activity. Based on the analysis results (Figure 9, Figure 10, and Figure S1), the studied sector can be divided into two distinct zones, separated by the main relief units it traverses, Ialomița Subcarpathians and Târgoviște Plain, as follows:

- The northern sector corresponds to the Subcarpathian and contact with Târgoviște Plain areas, where clusters 1, 3, and 4 are predominant. Here, the riverbed evolution is directly influenced by anthropogenic activity which occurred in the second half of the 20th century: the construction of the Pucioasa Dam and downstream riverbed regularisation structures; the development of the urban centres of Fieni, Pucioasa, and Târgoviște; and the emergence and development of industrial activities such as the Doicești thermal power plant. Also, anthropogenic activities have caused an incision in the bedrock, thus lithologically conditioning the evolution of the riverbed in some areas.
- The southern sector overlapping the plain area generally consists of clusters 2 and 5. This area is characterised by a predominantly natural evolution of the river, the anthropogenic influence being exerted rather indirectly until the 2000s. Unlike the other zone, only rural settlements exist in the vicinity of the river. The riverbed generally has a greater space of freedom, being grafted into friable sedimentary rocks such as gravel and sand. After 2000, the appearance of the gravel pits increased the degree of anthropogenic impact with effects on the riverbed dynamics. Natural events, such as floods in 2001 and 2005, also affected the evolution of the riverbed.

Regarding the climatic component in the study area, an increasing trend in annual temperatures ($\tau = 0.573$, p-value = 2.22×10^{-16}) and precipitation ($\tau = 0.285$, p-value = 0.0004) is observed. Comparatively analysing Figures 11 and 13a, the annual evolution of precipitation correlates with the evolution of discharge, but their trends are opposite. High temperatures could lead to intense evapotranspiration, reducing the proportion of precipitation that becomes runoff and then streamflow by concentrating it into the river channel. Also, there is observed an increasing rain shower frequency in Romania [115,116], which can reduce contribution to the mean annual discharge, favouring surface runoff and limiting groundwater recharge. However, the fact that the Ialomița River's flow regime is anthropogenically influenced is also a good reason why the discharge trend does not respond to the precipitation trend. Here we include the effect of dams and water withdrawals for economic activities and population supply.

Figure 17 presents linear models of the analysed parameters in the study (Table 5) which support the idea of the ongoing and continuing degradation process and transition of the Ialomița riverbed observed in the present (Figure 16) and its short-term future evolution. Strong decreasing trends are observed for the NC, BCW, ACW, BCA, and ACA (Figure 17b,e,f,h,i). A medium negative trend is observed for the median SI. MCW, WCW, and WCA (Figure 17c,d,g) remain relatively stable, being determined by long-term baseflow conditions.

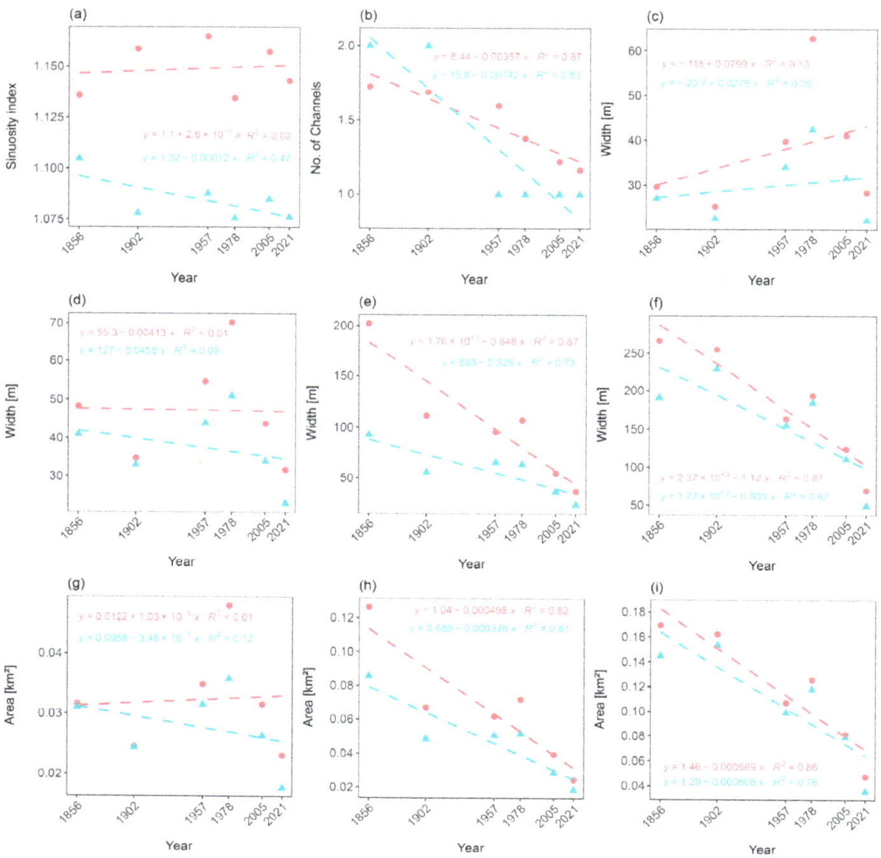

Figure 17. Linear models of parameters and indices versus time calculated for the studied river reach: (**a**) Sinuosity index—SI, (**b**) Number of river channels—NC, (**c**) Main channel width—MCW, (**d**) Wetted channel width—WCW, (**e**) Bankfull channel width—BCW, (**f**) Active channel width—ACW, (**g**) Wetted channel area—WCA, (**h**) Bankfull channel area—BCA, (**i**) Active channel area—ACA. The orange points are mean values, the blue triangles median values, and the dashed lines linear trendlines.

According to the analysis of Rădoane et al. [57], in the past 150 years, there is observed a general tendency of channel narrowing and lateral stability in many European rivers, which facilitates the process of riverbed incision. Also, the rivers in Romania follow this trend, but with a delay of major human interventions.

A representative trend in the fluvial style of the Romanian rivers (Someș [70], Someșul Mic [74], Putna [79], Siret [78], Buzău [57], Moldova [72,73], Prahova [66,69,75,76]) is represented by a transition from complex, meandering, anastomosed, or braiding channel patterns to a simple, straight, sinuous (in certain cases with alternate bars) one. This is reflected in a general reduction in the complexity of the riverbed, also observed in the analysis carried out on the Ialomița River.

There were identified some limitations of the study, related in particular to the cartographic sources and the georeferencing process. Old topographic maps could contain significant mapping errors because of constraints in surveying technology and geodetic control. For the first cartographic source (Wallachia—Second military survey of the Habsburg Empire), the error between the original and the current measurements is estimated to be up to 20 m [88,89]. Also, a map at a scale two times smaller than the others (Table S1) inevitably leads to a greater degree of generalisation of the riverbed characteristics. In addition, the lack of common stable elements preserved over time used as control points makes the georeferencing process difficult and less accurate. For the other historical topographic maps, georeferencing leads to map distortions as well, but they are smaller through the improvements in quality over time.

Human errors in the riverbed features digitisation process are also one of the limitations of the study. These can be objectives related exclusively to the quality elements of the maps, such as map sheet physical degradation (paper degradation, fading ink and discolouration, human markings) and distortions from scanning handling and scanning artefacts. On the other hand, some errors are derived from subjective interpretation of the river morphological features, especially for orthorectified aerial images.

Other limitations are related to the uneven temporal coverage of the cartographic (1856–2021), hydrological (1961–2022), climatic (1950–2022), and anthropogenic (gravel pits inventory after 2000) data sources available for this study. This gap in temporal coverage led us to interpret our results within the broader context of climate change and anthropogenic influences based on similar studies conducted in Europe and Romania.

6. Conclusions

This paper presents the first results on the study of historical evolution and future trends of riverbed dynamics of the Ialomița River, Romania, mainly based on the analysis of successive sets of cartographic sources between 1865 and 2021. We approach a complex methodology, which allows the measurement of parameters and indices in a standardised and partially automated manner. The morphometric analysis reveals a general downward trend in riverbed sinuosity, number of channels, widths, surfaces, and thalweg migration distance. All of these indicate major changes in the Ialomița riverbed behaviour and fluvial style, its channel morphology undergoing a simplification and a transition from a complex, braided, multi-thread channel to a simple sinuous, single-thread one. The main processes consist of narrowing, incising, and straightening, alternating with intense meandering.

Changes in Ialomița River behaviour can be primarily attributed to climate change that occurred after the Little Ice Age (LIA). Our analysis over the last ~70 years confirms that climate change is underway in the study area, as reflected in upward trends in temperature and precipitation. Since the second half of the 20th century, anthropogenic influence directly on the riverbed has intensified significantly, assuming a dominant role in shaping the riverbed's evolution.

Using HCA, we managed to identify spatial patterns in riverbed dynamics. Their evolutionary trajectory is highly dependent on the relief crossed by the river and anthropogenic influence over time.

The studied reach of the Ialomița River shares a similar historical evolution trend in riverbed dynamics with other European and Romanian rivers, which have reduced their geomorphological complexity by transition from a braided, multi-thread into a sinuous, single-thread riverbed pattern.

Field-based observations confirm the ongoing trend of riverbed degradation within the studied reach, mainly represented by the actions of narrowing and incision processes. Some bridges are already affected by downcutting at the base of the support pillars, endangering road traffic safety. Conversely, in areas affected by lateral erosion, concrete embankments have been constructed to protect the riparian settlements.

Regarding the future evolution trends, we consider the studied river reach will continue to degrade and reduce its morphological complexity on a short-term horizon, following the climate change signal and increased anthropogenic intervention. Ongoing in-channel aggregate extraction activities may intensify the riverbed incision process across the plain region. Further channel deepening in the Subcarpathian region will lead to a permanent geomorphic disconnection from the floodplain, transforming it into a new terrace. On-site observations have already identified this process. In certain zones, channel narrowing and confinement into a single-thread course may lead to intense local meandering, which could significantly impact adjacent human settlements through intense bank erosion, particularly during flood events.

We acknowledge the limitations of the study, which stem from the integration of historical cartographic sources into GIS, particularly through the heterogeneity and quality of the maps, the georeferencing process, and the interpretation of riverbed morphological features. In addition, the uneven temporal coverage of the analysed datasets represents a further constraint. Therefore, some of our statements and conclusions may be subject to a certain degree of uncertainty and generalisation, limiting our ability to fully understand the factors underlying the dynamics of the Ialomița riverbed.

The results of this study may be a useful tool in the management of the Ialomița River in the analysed area, and future anthropogenic interventions should take these into account. In the context of climate change and intensifying human activity, our findings may also contribute to a better understanding of recent behaviour of the riverbeds. Assuming that this study provides only a general overview of the river's historical evolution and future trends, further research is needed to advance the knowledge of Ialomița riverbed dynamics.

Supplementary Materials: The following supporting information can be downloaded at: https://www.mdpi.com/article/10.3390/w17142151/s1, Table S1: Cartographic data overview; Figure S1: The z-score of the HCA.

Author Contributions: Conceptualization, A.R. and L.C.; methodology, A.R. and L.C.; software, A.R.; validation, A.R. and L.C.; formal analysis, A.R. and L.C.; investigation, A.R. and L.C.; data curation, A.R.; writing—original draft preparation, A.R. and L.C.; writing—review and editing, A.R. and L.C.; visualization, A.R.; supervision, L.C. All authors have read and agreed to the published version of the manuscript.

Funding: This paper has received funding for publication through the University of Bucharest.

Data Availability Statement: Restrictions apply to the availability of these data. The Wallachia—The second military survey of the Habsburg Empire map was obtained from Arcanum: https://maps.arcanum.com/en/map/secondsurvey-wallachia (accessed on 16 August 2023); The Romanian Military Topographic Map—1st edition was obtained from The Defence Geospatial Intelligence Agency "Division General Constantin Barozzi"; The Romanian Military Topographic Map—2nd edition and the Orthophoto maps were obtained from the Faculty of Geography, University of Bucharest. The hydrological data were obtained from The National Institute of Hydrology and Water Management. These datasets are available from the authors with the permission of each respective third-party data provider. All other raw data supporting the conclusions of this article will be made available by the authors on request.

Conflicts of Interest: The authors declare no conflicts of interest.

References

1. Langat, P.K.; Kumar, L.; Koech, R. Monitoring River Channel Dynamics Using Remote Sensing and GIS Techniques. *Geomorphology* **2019**, *325*, 92–102. [CrossRef]
2. de Musso, N.M.; Capolongo, D.; Caldara, M.; Surian, N.; Pennetta, L. Channel Changes and Controlling Factors over the Past 150 Years in the Basento River (Southern Italy). *Water* **2020**, *12*, 307. [CrossRef]
3. Calvin, K.; Dasgupta, D.; Krinner, G.; Mukherji, A.; Thorne, P.W.; Trisos, C.; Romero, J.; Aldunce, P.; Barrett, K.; Blanco, G.; et al. *IPCC, 2023: Climate Change 2023: Synthesis Report. Contribution of Working Groups I, II and III to the Sixth Assessment Report of the Intergovernmental Panel on Climate Change*; Lee, H., Romero, J., Eds.; IPCC: Geneva, Switzerland, 2023. [CrossRef]
4. Eyring, V.; Gillett, N.P.; Achuta Rao, K.M.; Barimalala, R.; Barreiro Parrillo, M.; Bellouin, N.; Cassou, C.; Durack, P.J.; Kosaka, Y.; McGregor, S.; et al. Human Influence on the Climate System. In *Climate Change 2021: The Physical Science Basis. Contribution of Working Group I to the Sixth Assessment Report of the Intergovernmental Panel on Climate Change*; Masson-Delmotte, V., Zhai, P., Pirani, A., Connors, S.L., Péan, C., Berger, S., Caud, N., Chen, Y., Goldfarb, L., Gomis, M.I., et al., Eds.; Cambridge University Press: Cambridge, UK; New York, NY, USA, 2021; pp. 423–552. [CrossRef]
5. Caretta, M.A.; Mukherji, A.; Arfanuzzaman, M.; Betts, R.A.; Gelfan, A.; Hirabayashi, Y.; Lissner, T.K.; Liu, J.; Gunn, E.L.; Morgan, R.; et al. Water. In *Climate Change 2022: Impacts, Adaptation and Vulnerability. Contribution of Working Group II to the Sixth Assessment Report of the Intergovernmental Panel on Climate Change*; Pörtner, H.-O., Roberts, D.C., Tignor, M., Poloczanska, E.S., Mintenbeck, K., Alegría, A., Craig, M., Langsdorf, S., Löschke, S., Möller, V., et al., Eds.; Cambridge University Press: Cambridge, UK; New York, NY, USA, 2022; pp. 551–712. [CrossRef]
6. David, M.; Labenne, A.; Carozza, J.-M.; Valette, P. Evolutionary Trajectory of Channel Planforms in the Middle Garonne River (Toulouse, SW France) over a 130-Year Period: Contribution of Mixed Multiple Factor Analysis (MFAmix). *Geomorphology* **2016**, *258*, 21–39. [CrossRef]
7. Mandarino, A.; Pepe, G.; Cevasco, A.; Brandolini, P. Quantitative Assessment of Riverbed Planform Adjustments, Channelization, and Associated Land Use/Land Cover Changes: The Ingauna Alluvial-Coastal Plain Case (Liguria, Italy). *Remote Sens.* **2021**, *13*, 3775. [CrossRef]
8. Agnihotri, A.K.; Ohri, A.; Mishra, S. Channel Planform Dynamics of Lower Ramganga River, Ganga Basin, GIS and Remote Sensing Analyses. *Geocarto Int.* **2020**, *35*, 934–953. [CrossRef]
9. Harmar, O.P.; Clifford, N.J. Planform Dynamics of the Lower Mississippi River. *Earth Surf. Process. Landf.* **2006**, *31*, 825–843. [CrossRef]
10. Basnayaka, V.; Samarasinghe, J.T.; Gunathilake, M.B.; Muttil, N.; Rathnayake, U. Planform Changes in the Lower Mahaweli River, Sri Lanka Using Landsat Satellite Data. *Land* **2022**, *11*, 1716. [CrossRef]
11. Batalla, R.J.; Iroumé, A.; Hernández, M.; Llena, M.; Mazzorana, B.; Vericat, D. Recent Geomorphological Evolution of a Natural River Channel in a Mediterranean Chilean Basin. *Geomorphology* **2018**, *303*, 322–337. [CrossRef]
12. Joyce, H.M.; Warburton, J.; Hardy, R.J. A Catchment Scale Assessment of Patterns and Controls of Historic 2D River Planform Adjustment. *Geomorphology* **2020**, *354*, 107046. [CrossRef]
13. Langat, P.K.; Kumar, L.; Koech, R.; Ghosh, M.K. Characterisation of channel morphological pattern changes and flood corridor dynamics of the tropical Tana River fluvial systems, Kenya. *J. Afr. Earth Sci.* **2020**, *163*, 103748. [CrossRef]
14. Rasheed, N.J.; Al-Khafaji, M.S.; Alwan, I.A. Investigation of Rivers Planform Change in a Semi-Arid Region of High Vulnerability to Climate Change: A Case Study of Tigris River and Its Tributaries in Iraq. *Reg. Stud. Mar. Sci.* **2023**, *68*, 103233. [CrossRef]
15. Schwenk, J.; Khandelwal, A.; Fratkin, M.; Kumar, V.; Foufoula-Georgiou, E. High Spatiotemporal Resolution of River Planform Dynamics from Landsat: The RivMAP Toolbox and Results from the Ucayali River. *Earth Space Sci.* **2017**, *4*, 46–75. [CrossRef]
16. Vercruysse, K.; Grabowski, R.C. Human Impact on River Planform within the Context of Multi-Timescale River Channel Dynamics in a Himalayan River System. *Geomorphology* **2021**, *381*, 107659. [CrossRef]
17. Zhou, M.; Xia, J.; Lu, J.; Deng, S.; Lin, F. Morphological Adjustments in a Meandering Reach of the Middle Yangtze River Caused by Severe Human Activities. *Geomorphology* **2017**, *285*, 325–332. [CrossRef]
18. Magliulo, P.; Bozzi, F.; Pignone, M. Assessing the Planform Changes of the Tammaro River (Southern Italy) from 1870 to 1955 Using a GIS-Aided Historical Map Analysis. *Environ. Earth Sci.* **2016**, *75*, 355. [CrossRef]

19. Rinaldi, M.; Wyżga, B.; Surian, N. Sediment Mining in Alluvial Channels: Physical Effects and Management Perspectives. *River Res. Appl.* **2005**, *21*, 805–828. [CrossRef]
20. Ziliani, L.; Surian, N. Evolutionary Trajectory of Channel Morphology and Controlling Factors in a Large Gravel-Bed River. *Geomorphology* **2012**, *173–174*, 104–117. [CrossRef]
21. Surian, N.; Rinaldi, M.; Pellegrini, L.; Audisio, C.; Maraga, F.; Teruggi, L.; Turitto, O.; Ziliani, L. Channel adjustments in northern and central Italy over the last 200 years. In *Management and Restoration of Fluvial Systems with Broad Historical Changes and Human Impacts*; Geological Society of America Special, Papers; James, L.A., Rathburn, S.L., Whittecar, G.R., Eds.; Geological Society of America: Boulder, CO, USA, 2009; Volume 451, pp. 83–95. [CrossRef]
22. Cencetti, C.; De Rosa, P.; Fredduzzi, A. Geoinformatics in Morphological Study of River Paglia, Tiber River Basin, Central Italy. *Environ. Earth Sci.* **2017**, *76*, 128. [CrossRef]
23. Scorpio, V.; Aucelli, P.P.C.; Giano, S.I.; Pisano, L.; Robustelli, G.; Rosskopf, C.M.; Schiattarella, M. River Channel Adjustments in Southern Italy over the Past 150years and Implications for Channel Recovery. *Geomorphology* **2015**, *251*, 77–90. [CrossRef]
24. Scorpio, V.; Andreoli, A.; Dinkelaker, N.; Marchese, E.; Coviello, V.; Gems, B.; Vignoli, G.; Comiti, F. Multi-decadal Quantification of Interactions between Coarse Sediment Fluxes and Channel Management in South Tyrol, Eastern European Alps. *Earth Surf. Process. Landf.* **2024**, *49*, 1869–1889. [CrossRef]
25. Rinaldi, M. Bed-Level Adjustments in the Po River Catchment (Northern Italy). *Ital. J. Eng. Geol. Environ.* **2021**, *2*, 41–50. [CrossRef]
26. Mandarino, A. Morphological Adjustments of the Lower Orba River (NW Italy) since the Mid-Nineteenth Century. *Geomorphology* **2022**, *410*, 108280. [CrossRef]
27. Brenna, A.; Surian, N.; Mao, L. Alteration of Gravel-Bed River Morphodynamics in Response to Multiple Anthropogenic Disturbances: Insights from the Sediment-Starved Parma River (Northern Italy). *Geomorphology* **2021**, *389*, 107845. [CrossRef]
28. Nones, M.; Guerrero, M.; Schippa, L.; Cavalieri, I. Remote Sensing Assessment of Anthropogenic and Climate Variation Effects on River Channel Morphology and Vegetation: Impact of Dry Periods on a European Piedmont River. *Earth Surf. Process. Landf.* **2024**, *49*, 1632–1652. [CrossRef]
29. Brenna, A.; Bizzi, S.; Surian, N. How Multiple Anthropic Pressures May Lead to Unplanned Channel Patterns: Insights from the Evolutionary Trajectory of the Po River (Italy). *CATENA* **2024**, *234*, 107598. [CrossRef]
30. Burshtynska, K.; Hrytskiv, N.; Zayats, I.; Babiy, L.; Fijałkowska, A.; Bakuła, K. Monitoring of Channel Deformations of the Lower Dniester Using Topographic Maps and Satellite Imagery with the Implementation of GIS Technologies. *Water* **2024**, *16*, 2148. [CrossRef]
31. Llena, M.; Vericat, D.; Martínez-Casasnovas, J.A.; Smith, M.W. Geomorphic Adjustments to Multi-Scale Disturbances in a Mountain River: A Century of Observations. *CATENA* **2020**, *192*, 104584. [CrossRef]
32. Baena-Escudero, R.; Rinaldi, M.; García-Martínez, B.; Guerrero-Amador, I.C.; Nardi, L. Channel Adjustments in the Lower Guadalquivir River (Southern Spain) over the Last 250 years. *Geomorphology* **2019**, *337*, 15–30. [CrossRef]
33. Calle, M.; Alho, P.; Benito, G. Channel Dynamics and Geomorphic Resilience in an Ephemeral Mediterranean River Affected by Gravel Mining. *Geomorphology* **2017**, *285*, 333–346. [CrossRef]
34. Fernandes, M.R.; Aguiar, F.C.; Martins, M.J.; Rivaes, R.; Ferreira, M.T. Long-Term Human-Generated Alterations of Tagus River: Effects of Hydrological Regulation and Land-Use Changes in Distinct River Zones. *CATENA* **2020**, *188*, 104466. [CrossRef]
35. Wyżga, B.; Zawiejska, J.; Hajdukiewicz, H. Multi-thread rivers in the Polish Carpathians: Occurrence, decline and possibilities of restoration. *Quat. Int.* **2016**, *415*, 344–356. [CrossRef]
36. Zawiejska, J.; Wyżga, B. Twentieth-Century Channel Change on the Dunajec River, Southern Poland: Patterns, Causes and Controls. *Geomorphology* **2010**, *117*, 234–246. [CrossRef]
37. Hajdukiewicz, H.; Wyżga, B. Analysis of Historical Changes in Planform Geometry of a Mountain River to Inform Design of Erodible River Corridor. *Ecol. Eng.* **2023**, *186*, 106821. [CrossRef]
38. Nádudvari, Á.; Czajka, A.; Wyżga, B.; Zygmunt, M.; Wdowikowski, M. Patterns of Recent Changes in Channel Morphology and Flows in the Upper and Middle Odra River. *Water* **2023**, *15*, 370. [CrossRef]
39. Winterbottom, S.J. Medium and Short-Term Channel Planform Changes on the Rivers Tay and Tummel, Scotland. *Geomorphology* **2000**, *34*, 195–208. [CrossRef]
40. Keesstra, S.D.; van Huissteden, J.; Vandenberghe, J.; Van Dam, O.; de Gier, J.; Pleizier, I.D. Evolution of the Morphology of the River Dragonja (SW Slovenia) Due to Land-Use Changes. *Geomorphology* **2005**, *69*, 191–207. [CrossRef]
41. Kidová, A.; Lehotský, M.; Rusnák, M. Geomorphic Diversity in the Braided-Wandering Belá River, Slovak Carpathians, as a Response to Flood Variability and Environmental Changes. *Geomorphology* **2016**, *272*, 137–149. [CrossRef]
42. Belletti, B.; Dufour, S.; Piégay, H. What Is the Relative Effect of Space and Time to Explain the Braided River Width and Island Patterns at a Regional Scale? *River Res. Appl.* **2015**, *31*, 1–15. [CrossRef]
43. Bulteau, T.; Batalla, R.J.; Chapron, E.; Valette, P.; Piégay, H. Geomorphic Effects of a Run-of-the-River Dam in a Multi-Driver Context: The Case of the Upper Garonne (Central Pyrenees). *Geomorphology* **2022**, *408*, 108243. [CrossRef]

44. Jautzy, T.; Schmitt, L.; Rixhon, G. Historical Geomorphological Adjustments of an Upper Rhine Sub-Tributary over the Two Last Centuries (Bruche River, France). *Géomorphologie Relief Process. Environ.* **2022**, *28*, 53–72. [CrossRef]
45. Heckmann, T.; Haas, F.; Abel, J.; Rimböck, A.; Becht, M. Feeding the Hungry River: Fluvial Morphodynamics and the Entrainment of Artificially Inserted Sediment at the Dammed River Isar, Eastern Alps, Germany. *Geomorphology* **2017**, *291*, 128–142. [CrossRef]
46. Eschbach, D.; Schmitt, L.; Imfeld, G.; May, J.-H.; Payraudeau, S.; Preusser, F.; Trauerstein, M.; Skupinski, G. Long-Term Temporal Trajectories to Enhance Restoration Efficiency and Sustainability on Large Rivers: An Interdisciplinary Study. *Hydrol. Earth Syst. Sci.* **2018**, *22*, 2717–2737. [CrossRef]
47. Alexeevsky, N.I.; Chalov, R.S.; Berkovich, K.M.; Chalov, S.R. Channel Changes in Largest Russian Rivers: Natural and Anthropogenic Effects. *Int. J. River Basin Manag.* **2013**, *11*, 175–191. [CrossRef]
48. Spada, D.; Molinari, P.; Bertoldi, W.; Vitti, A.; Zolezzi, G. Multi-Temporal Image Analysis for Fluvial Morphological Characterization with Application to Albanian Rivers. *ISPRS Int. J. Geo-Inf.* **2018**, *7*, 314. [CrossRef]
49. Pavlek, K.; Faivre, S. Geomorphological Changes of the Cetina River Channels since the End of the Nineteenth Century, Natural vs Anthropogenic Impacts (the Dinarides, Croatia). *Environ. Earth Sci.* **2020**, *79*, 482. [CrossRef]
50. Kiss, T.; Balogh, M.; Fiala, K.; Sipos, G. Morphology of Fluvial Levee Series along a River under Human Influence, Maros River, Hungary. *Geomorphology* **2018**, *303*, 309–321. [CrossRef]
51. Kiss, T.; Fiala, K.; Sipos, G. Alterations of Channel Parameters in Response to River Regulation Works since 1840 on the Lower Tisza River (Hungary). *Geomorphology* **2008**, *98*, 96–110. [CrossRef]
52. Kiss, T.; Blanka, V. River Channel Response to Climate- and Human-Induced Hydrological Changes: Case Study on the Meandering Hernád River, Hungary. *Geomorphology* **2012**, *175–176*, 115–125. [CrossRef]
53. Kiss, T.; Andrási, G. Evolution of the Drava Floodplain in Hungary in the Last 100 Years. In *The Drava River*; Lóczy, D., Ed.; Springer: Cham, Switzerland, 2019; pp. 157–175. [CrossRef]
54. Amissah, G.J.; Kiss, T.; Fiala, K. Morphological Evolution of the Lower Tisza River (Hungary) in the 20th Century in Response to Human Interventions. *Water* **2018**, *10*, 884. [CrossRef]
55. Škarpich, V.; Hradecký, J.; Dušek, R. Complex transformation of the geomorphic regime of channels in the forefield of the Moravskoslezské Beskydy Mts.: Case study of the Morávka River (Czech Republic). *CATENA* **2013**, *111*, 25–40. [CrossRef]
56. Hohensinner, S.; Habersack, H.; Jungwirth, M.; Zauner, G. Reconstruction of the Characteristics of a Natural Alluvial River–Floodplain System and Hydromorphological Changes Following Human Modifications: The Danube River (1812–1991). *River Res. Appl.* **2004**, *20*, 25–41. [CrossRef]
57. Rădoane, M.; Perșoiu, I.; Chiriloaei, F.; Cristea, I.; Robu, D. Styles of Channel Adjustments in the Last 150 Years. In *Landform Dynamics and Evolution in Romania. Springer Geography*; Rădoane, M., Vespremeanu-Stroe, A., Eds.; Springer: Cham, Switzerland, 2017; pp. 489–518. [CrossRef]
58. Bălteanu, D.; Jurchescu, M.; Surdeanu, V.; Ionita, I.; Goran, C.; Urdea, P.; Rădoane, M.; Rădoane, N.; Sima, M. Recent Landform Evolution in the Romanian Carpathians and Pericarpathian Regions. In *Recent Landform Evolution: The Carpatho-Balkan-Dinaric Region*; Lóczy, D., Stankoviansky, M., Kotarba, A., Eds.; Springer: Dordrecht, The Netherlands, 2012; pp. 249–286. [CrossRef]
59. Rădoane, M.; Rădoane, N.; Dumitriu, D. Geomorphological Evolution of Longitudinal River Profiles in the Carpathians. *Geomorphology* **2003**, *50*, 293–306. [CrossRef]
60. Dumitriu, D. Flood Events as Geomorphic Thresholds for Channel Bed Level Change. *Carpathian J. Earth Environ. Sci.* **2021**, *16*, 77–92. [CrossRef]
61. Ioana-Toroimac, G.; Zaharia, L.; Neculau, G.; Minea, G. Impact of Channel Incision on Floods: A Case Study in The South-Eastern Subcarpathians (Romania). *Geogr. Environ. Sustain.* **2020**, *13*, 17–24. [CrossRef]
62. Rădoane, M.; Obreja, F.; Cristea, I.; Mihailă, D. Changes in the Channel-Bed Level of the Eastern Carpathian Rivers: Climatic vs. Human Control over the Last 50 Years. *Geomorphology* **2013**, *193*, 91–111. [CrossRef]
63. Rădoane, M.; Rădoane, N.; Cristea, A.; Dinu, O. Contemporary Changes of the Prut River Channel, Romanian Border. *Rev. Geomorfol.* **2008**, *10*, 57–71.
64. Rădoane, M.; Pandi, G.; Rădoane, N. Contemporary Bed Elevation Changes from the Eastern Carpathians. *Carpathian J. Earth Environ. Sci.* **2010**, *5*, 49–60.
65. Grecu, F.; Ioana-Toroimac, G.; Molin, P.; Dramis, F. River channel dynamics in the contact area between the Romanian Plain and the Curvature Subcarpathians. *Rev. Geomorfol.* **2014**, *16*, 5–12.
66. Ioana-Toroimac, G.; Robert, D.; Grecu, F.; Zaharia, L. Evolution 2D de La Bande Active de La Haute Prahova (Roumanie) Durant Les 150 Dernières Années. *Géomorphologie Relief Process. Environ.* **2010**, *16*, 275–286. [CrossRef]
67. Grecu, F.; Grigore, Ș.; Ioana-Toroimac, G.; Ghiță, C. The Dynamics of the Danube Channel over the Last 150 Years (Giurgiu—Călărași Sector. Preliminary Data). In Proceedings of the Romanian Geomorphology Symposium, 33rd edition, Iași, Romania, 11–14 May 2017.

68. Grecu, F.; Zaharia, L.; Ioana-Toroimac, G.; Armaș, I. Floods and Flash-Floods Related to River Channel Dynamics. In *Landform Dynamics and Evolution in Romania. Springer Geography*; Radoane, M., Vespremeanu-Stroe, A., Eds.; Springer: Cham, Switzerland, 2017; pp. 821–844. [CrossRef]
69. Ioana-Toroimac, G. La Dynamique Hydrogéomorphologique de La Rivière Prahova (Roumanie): Fonctionnement Actuel, Évolution Récente et Conséquences Géographiques. Ph.D. Thesis, University of Bucharest, Bucharest, Romania, 2009.
70. Perşoiu, I.; Rădoane, M. Spatial and Temporal Controls on Historical Channel Responses—Study of an Atypical Case: Someşu Mic River, Romania. *Earth Surf. Process Landf.* **2011**, *36*, 1391–1409. [CrossRef]
71. Feier, I.; Rădoane, M. Dinamica în plan orizontal a albiei minore a râului Somesu Mic, înainte de lucrarile hidrotehnice majore (1870–1968). *Analele Univ. Stefan Cel Mare Suceava Sect. Geogr.* **2007**, *16*, 5–22.
72. Chiriloaei, F.; Rădoane, M. The Metamorphosis of a Braided River Channel during the Last 100 Years: Moldova River as a Case Study. *Rev. Geomorfol.* **2015**, *17*, 19–28.
73. Chiriloaei, F.A. Analiza Cantitativa a Modificarilor Albiei Raului Moldova in Sectorul Extracarpatic. Ph.D. Thesis, " Al. I. Cuza" University, Iași, Romania, 2012.
74. Perșoiu, I.; Hosu, M. Historical geomorphic analysis of Someșul Mic River (NW Romania) channel spatial and temporal behavior, as a background for the assessment of present day morphological state of the river. *Rev. Geomorfol.* **2019**, *21*, 59–76. [CrossRef]
75. Brașovanu, L. Studiu Morfodinamic al Albiei Râului Prahova în Sectoarele Montan și Subcarpatic. Ph.D. Thesis, University of Bucharest, Bucharest, Romania, 2018.
76. Armaș, I.; Gogoașe Nistoran, D.; Osaci-Costache, G.; Brașoveanu, L. Morpho-Dynamic Evolution Patterns of Subcarpathian Prahova River (Romania). *CATENA* **2013**, *100*, 83–99. [CrossRef]
77. Grecu, F.; Ioana-Toroimac, G.; Osaci-Costache, G.; Zaharia, L.; Cocoș, O.; Hachemi, K.; Sălăjan, L. Dynamics of Islands and Danube River Channel Along Vedea-Călărași Sector (1856–2019): Hydrogeomorphological Approach. In *The Lower Danube River. Earth and Environmental Sciences Library*; Negm, A., Zaharia, L., Ioana-Toroimac, G., Eds.; Springer: Cham, Switzerland, 2022; pp. 43–67.
78. Salit, F.; Arnaud-Fassetta, G.; Zaharia, L.; Madelin, M.; Beltrando, G. The influence of river training on channel changes during the 20th century in the Lower Siret River (Romania). *Géomorphologie Relief Process. Environ.* **2015**, *21*, 175–188. [CrossRef]
79. Cristea, I. Studiul Geomorfologic al Văii Putna Vrânceană. Ph.D. Thesis, Ștefan cel Mare University, Suceava, Romania, 2011.
80. Buzău-Ialomița Water Basin Administration Flood Risk Management Plan—Updated. Available online: https://inundatii.ro/wp-content/uploads/2022/10/PMRI_actualizat_ciclul-II_ABA-Buzau-Ialomita_versiune-preliminara-1.pdf (accessed on 23 March 2024).
81. *Geological Map of Romania, Sc. 1:200 000*; Romanian Geological Institute: Bucharest, Romania, 1968.
82. *Geological Map of Romania, Sc. 1:200 000, Explanatory Note of Târgoviște Map Sheet L-35-XXVI*; Romanian Geological Institute: Bucharest, Romania, 1968.
83. Clima României. Available online: https://www.meteoromania.ro/clima/clima-romaniei/ (accessed on 18 March 2025).
84. Dumitrescu, A.; Bojariu, R.; Birsan, M.-V.; Marin, L.; Manea, A. Recent Climatic Changes in Romania from Observational Data (1961–2013). *Theor. Appl. Climatol.* **2015**, *122*, 111–119. [CrossRef]
85. Murărescu, O.; Bretcan, P. Management of Water Resources in Upper Ialomiţa River Basin (Carpathians, Sub-Carpathians, Romania). In Proceedings of the International Symposium on Water Management and Hydraulic Engineering, Faculty of Civil Engineering, Department of Hydraulics, Hydrology and River Engineering, University of Ss. Cyril and Methodius, Skopje, North Macedonia, 1–5 September 2009.
86. Murătoreanu, G.; Murărescu, O.; Pehoiu, G.; Sencovici, M. GIS multitemporal analysis of remote sensing images to assess the silting of anthropogenic dam lakes in the Outer Subcarpathian area. Case study: Pucioasa Accumulation Lake, Romania. *J. Sci. Arts* **2021**, *21*, 1133–1142. [CrossRef]
87. Loghin, V. Modificări antropice în profilul longitudinal al Ialomiței și efecte asupra proceselor de albie (Sectorul Moroeni—Târgoviște). *Analele Univ. Valahia Din Târgovişte Ser. Geogr.* **1997**, *1*, 77–84.
88. Activitățile Cartografice și Geodezice Habsburgice în Vechea Românie. Available online: https://geo-spatial.org/vechi/articole/activitatile-cartografice-si-geodezice-habsburgice-in-vechea-romanie (accessed on 13 August 2024).
89. Timár, G. Habsburg Geodetic and Cartographic Activities in the Old Romania. *Stud. şi Cercet. Ser. Geol.-Geogr.* **2008**, *13*, 93–102.
90. Osaci-Costache, G. *Cartografie*, 2nd ed.; Editura Universitară: Bucharest, Romania, 2008.
91. Planurile Directoare de Tragere. Available online: https://geo-spatial.org/vechi/download/planurile-directoare-de-tragere (accessed on 15 August 2024).
92. Military Topographic Maps Sc. 1:25.000. Available online: https://www.geomil.ro/Produse/HartiTopografice25GK_35# (accessed on 15 August 2024).
93. Năstase, A.; Osaci-Costache, G. *Topografie-Cartografie*, 2nd ed.; Editura Fundaţiei România de Mâine: Bucharest, Romania, 2005.
94. Wallachia—Second Military Survey of the Habsburg Empire | Arcanum Maps. Available online: https://maps.arcanum.com/en/map/secondsurvey-wallachia (accessed on 16 August 2023).

95. Planurile Directoare de Tragere Download Service. Available online: https://www.geo-spatial.org/vechi/maps/download-planuri-tragere.php (accessed on 7 February 2023).
96. Color Orthophoto Map Scale 1:5000—Visualization Service. Available online: https://inspire-geoportal.ec.europa.eu/srv/eng/catalog.search#/metadata/1623228948931r5372581473647313 (accessed on 28 October 2024).
97. Cornes, R.C.; van der Schrier, G.; van den Besselaar, E.J.M.; Jones, P.D. An Ensemble Version of the E-OBS Temperature and Precipitation Data Sets. *J. Geophys. Res. Atmos.* **2018**, *123*, 9391–9409. [CrossRef]
98. Dumitrescu, A.; Birsan, M.-V. ROCADA: A Gridded Daily Climatic Dataset over Romania (1961–2013) for Nine Meteorological Variables. *Nat. Hazards* **2015**, *78*, 1045–1063. [CrossRef]
99. Thiemig, V.; Gomes, G.N.; Skøien, J.O.; Ziese, M.; Rauthe-Schöch, A.; Rustemeier, E.; Rehfeldt, K.; Walawender, J.P.; Kolbe, C.; Pichon, D.; et al. EMO-5: A High-Resolution Multi-Variable Gridded Meteorological Dataset for Europe. *Earth Syst. Sci. Data* **2022**, *14*, 3249–3272. [CrossRef]
100. Hackeloeer, A.; Klasing, K.; Krisp, J.M.; Meng, L. Georeferencing: A Review of Methods and Applications. *Ann. GIS* **2014**, *20*, 61–69. [CrossRef]
101. QGIS User Guide—17.3 Georeferencerer. Available online: https://docs.qgis.org/3.34/en/docs/user_manual/working_with_raster/georeferencer.html (accessed on 25 March 2024).
102. Overview of Georeferencing. Available online: https://pro.arcgis.com/en/pro-app/latest/help/data/imagery/overview-of-georeferencing.htm (accessed on 25 March 2024).
103. Boothroyd, R.J.; Nones, M.; Guerrero, M. Deriving Planform Morphology and Vegetation Coverage From Remote Sensing to Support River Management Applications. *Front. Environ. Sci.* **2021**, *9*, 18. [CrossRef]
104. Rădoane, M.; Perșoiu, I.; Cristea, A.; Chiriloaei, F. River Channel Planform Changes Based on Successive Cartographic Data. A Methodological Approach. *Rev. Geomorfol.* **2013**, *15*, 68–88.
105. Nawfee, S.M.; Dewan, A.; Rashid, T. Integrating Subsurface Stratigraphic Records with Satellite Images to Investigate Channel Change and Bar Evolution: A Case Study of the Padma River, Bangladesh. *Environ. Earth Sci.* **2018**, *77*, 89. [CrossRef]
106. Leopold, L.B.; Wolman, M.G. *River Channel Patterns: Braided, Meandering, and Straight*; U.S. Government Printing Office: Washongton, DC, USA, 1957. [CrossRef]
107. Hijmans, R.J. *R Package*, version 1.6-53; Terra: Spatial Data Analysis; R Core Team: Vienna, Austria, 2023; Available online: https://CRAN.R-project.org/package=terra (accessed on 14 April 2025).
108. Mann, H.B. Nonparametric Tests Against Trend. *Econometrica* **1945**, *13*, 245–259. [CrossRef]
109. Kendall, M.G. *Rank Correlation Methods*, 4th ed.; Charles Griffin: London, UK, 1975.
110. McLeod, A.I. *R Package*, version 2.2.1; Kendall: Kendall Rank Correlation and Mann-Kendall Trend Test; R Core Team: Vienna, Austria, 2005; Available online: https://CRAN.R-project.org/package=Kendall (accessed on 14 April 2025).
111. Killick, R.; Haynes, K.; Eckley, I.A. *R Package*, version 2.3; Changepoint: An R Package for Changepoint Analysis; R Core Team: Vienna, Austria, 2006; Available online: https://CRAN.R-project.org/package=changepoint (accessed on 15 April 2025).
112. R Core Team. *R: A Language and Environment for Statistical Computing*; R Foundation for Statistical Computing: Vienna, Austria, 2022.
113. Rousseeuw, P.J. Silhouettes: A Graphical Aid to the Interpretation and Validation of Cluster Analysis. *J. Comput. Appl. Math.* **1987**, *20*, 53–65. [CrossRef]
114. Buffin-Bélanger, T.; Biron, P.M.; Larocque, M.; Demers, S.; Olsen, T.; Choné, G.; Ouellet, M.-A.; Cloutier, C.-A.; Desjarlais, C.; Eyquem, J. Freedom Space for Rivers: An Economically Viable River Management Concept in a Changing Climate. *Geomorphology* **2015**, *251*, 137–148. [CrossRef]
115. Manea, A.; Birsan, M.-V.; Tudorache, G.; Cărbunaru, F. Changes in the Type of Precipitation and Associated Cloud Types in Eastern Romania (1961–2008). *Atmos. Res.* **2016**, *169*, 357–365. [CrossRef]
116. Busuioc, A.; Birsan, M.; Carbunaru, D.; Baciu, M.; Orzan, A. Changes in the Large-scale Thermodynamic Instability and Connection with Rain Shower Frequency over Romania: Verification of the Clausius–Clapeyron Scaling. *Int. J. Climatol.* **2016**, *36*, 2015–2034. [CrossRef]

water

MDPI

Article

Flood Susceptibility Assessment Using Multi-Tier Feature Selection and Ensemble Boosting Machine Learning Models

Rajendran Shobha Ajin [1,2], Romulus Costache [2,3,4,*], Alina Bărbulescu [2,*], Riccardo Fanti [1] and Samuele Segoni [1]

1 Department of Earth Sciences, University of Florence (UNIFI), Via G. La Pira 4, 50121 Florence, Italy; rajendranshobha.ajin@unifi.it (R.S.A.); riccardo.fanti@unifi.it (R.F.); samuele.segoni@unifi.it (S.S.)
2 Faculty of Civil Engineering, Transilvania University of Brașov (UUNITBV), No. 5, Turnului Str., 500152 Brașov, Romania
3 National Institute of Hydrology and Water Management, București-Ploiești Road, 97E, 1st District, 013686 Bucharest, Romania
4 Danube Delta National Institute for Research and Development, 165 Babadag Street, 820112 Tulcea, Romania
* Correspondence: romuluscostache2000@yahoo.com (R.C.); alina.barbulescu@unitbv.ro (A.B.)

Abstract: Flood susceptibility modeling (FSM) plays a key role in advancing proactive disaster risk reduction and spatial planning. This research developed FSM for the Buzău River catchment in Romania—a region historically vulnerable to recurrent flood events—using four state-of-the-art ensemble boosting algorithms: AdaBoost, CatBoost, LightGBM, and XGBoost. Initially, a comprehensive set of 13 flood conditioning factors was assessed, which was subsequently narrowed down to 9 essential factors through multi-tier feature selection strategies. Analysis of performance via receiver operating characteristic (ROC) andprecision–recall curves showed only marginal differences between the models; however, CatBoost excelled with an area under the ROC curve (AUC) of 0.972 and an average precision (AP) of 0.971, with XGBoost following closely behind. The SHAP (SHapley Additive exPlanations) analysis of the CatBoost model indicated that the Slope, Distance from Rivers, Topographic Wetness Index (TWI), and Land Use/Land Cover (LULC) are the key contributing factors. The novelty of this research is found in its comparative analysis of AdaBoost alongside three gradient boosting algorithms—CatBoost, LightGBM, and XGBoost—while utilizing explainable artificial intelligence (XAI) and a multi-tier feature selection strategy to create FSM that are precise and comprehensible. These strategies deliver robust tools for managing flood risks and reinforce the viability of data-driven modeling in the various catchments of Europe.

Keywords: Buzău River catchment; CatBoost; feature selection; flood susceptibility; gradient boosting; machine learning; Romania; SHAP analysis

1. Introduction

Flooding represents a critical and persistent threat worldwide, threatening human lives and inflicting significant economic damage. It is exacerbated by global warming and increasing urbanization [1–4]. Projections indicate that the design-level flood frequency will rise for about 47%, 55%, 70%, and 74% of watersheds during warming intervals of 1.5, 2.0, 2.5, and 3.0 °C according to the SSP245 scenario [2]. Between 1990 and 2022, 4713 floods were documented worldwide, impacting over 3.2 billion individuals, leading to 218,353 fatalities, and incurring economic losses exceeding 1.3 trillion USD [5]. Europe's flood risks and vulnerabilities are comparable to those observed globally. Liu et al. [5] report that 15.02% of all floods worldwide occurred in Europe, affecting 16,669,245 people and causing 5543 fatalities. Flooding represents one of the most prevalent and costly hazards in Europe [6], resulting in damage that averages over 12 billion Euros each year [7].

The Danube River Basin (DRB) is an international river basin in Europe, having experienced floods throughout its history [8,9]. The research by Leščešen et al. [10] revealed a trend of increasing extreme events in the Danube River projected for both the winter and summer seasons. During the last century, the floodplains of the DRB have experienced substantial human interventions, resulting in notable changes to their hydromorphology; specifically, the size of these floodplains has been reduced by 68%, which has greatly affected the river's inherent capacity to mitigate floods [11]. According to a more recent investigation by Eder et al. [12], the area of Danube floodplains has diminished by roughly 79% due to anthropogenic activities. The nations situated in the eastern part of the DRB, including Romania [13,14], Bulgaria [15], Serbia [16,17], Moldova [18], and Ukraine [19], face severe flooding issues due to intricate hydro-climatic factors and rising land use demands. Nearly 97.8% of the territory of Romania lies within the DRB, which extends across several countries [20]. Romania encompasses roughly 30% of the entire area of the DRB within its administrative boundaries [20]. Consequently, Romania is one of the severely flood-prone countries in the region, regularly facing both riverine and flash flood events [20–22].

Flooding in Romania is a persistent hazard driven by a combination of climatic, geographic, and anthropogenic factors [21,23,24]. The country's multifaceted topography, ranging from the Carpathian Mountains to the Danube Delta, makes it particularly prone to various types of floods. Intense precipitation and rapid snowmelt are major natural factors [25–27], whereas human activities such as deforestation, improper land management, and insufficient drainage systems play a crucial role in exacerbating flooding [20,28,29]. Romania's economy suffers an average annual loss of about 140 million Euros due to floods, with some counties facing losses that surpass 4% of their local GDP [30].

Significant flood disasters in Romania occurred in 1970, 1975, 1983, 1988, 1991 [27,31], 2005, 2006, 2008, 2010, 2012, 2018, and 2021 [14,22,31–33]. In Romania, the 2005 European floods resulted in 60 fatalities and damages amounting to 1.66 billion Euros, the 2006 European floods had a profound impact on the entire Danubian watershed, the 2010 floods led to 6 fatalities and damages of 1 billion USD, and the 2021 European floods battered 37 of the 41 counties, as well as the capital, Bucharest [22,32,34]. The 1897 floods were among the most catastrophic, leading to the overflow of the Danube River and producing extensive damage to Galați and Brăila cities, with infrastructure such as roads, bridges, and railway tracks suffering extensively [35]. The major flood event that occurred in 2018 is considered one of the most devastating flood disasters in central Romania, notably in Brasov County, resulting in damages that surpass 6.5 million Euros [33]. The recent 2024 Central European floods, triggered by Storm Boris, had a devastating effect on Romania, especially the counties of Galați and Vaslui, where floodwaters attained depths of 1.5 to 2 m [36,37]. Given Romania's substantial exposure to flood hazards due to its extensive coverage within the DRB, it is crucial to develop reliable flood susceptibility models (FSMs) to facilitate effective risk management and land use planning.

In recent decades, FSM has gained significant importance as a crucial tool for mitigating disaster risks and promoting sustainable watershed management. Conventional statistical techniques such as Frequency Ratio [38], Index of Entropy [39], Logistic Regression [40], and Weights-of-Evidence [38], along with semi-quantitative approaches like Analytic Hierarchy Process (AHP) [41], Analytic Network Process [42], and Fuzzy-AHP [41], have been extensively utilized for FSM. However, recent developments in AI-based data-driven methods have led to a growing adoption of machine learning (ML) or deep learning models such as Random Forest [43], Decision Trees [43], Support Vector Machines [44], Naïve

Bayes [45], Adaptive Boosting (AdaBoost) [46], eXtreme Gradient Boosting (XGBoost) [43], Light Gradient Boosting Machine (LightGBM) [46], Categorical Boosting (CatBoost) [43], K-Nearest Neighbors [47], Artificial Neural Networks [47], and Convolutional Neural Networks [48] for FSM due to their superior performance in capturing nonlinear relationships and intricate interactions among factors [49].

Ensemble learning is a technique that integrates predictions from multiple base (weak) models to attain enhanced performance [50,51] by reducing bias, enhancing generalizability, and improving predictive accuracy [52]. This concept includes three fundamental approaches: bagging, boosting, and stacking [50]. Boosting serves as a strategy that transforms weak learners into strong classifier by decreasing bias and possible variance [50]. AdaBoost and gradient boosting (GB) algorithms—including CatBoost, LightGBM, and XGBoost—are all ensemble ML techniques; however, they vary in their boosting strategies. AdaBoost integrates several weak classifiers to form a robust classifier through a weighted majority voting mechanism, with the impact of each classifier determined by its accuracy [53,54]. GB algorithms iteratively optimize a loss function through gradient descent by creating new models that address the residual errors from prior models [55–57].

Despite many studies applying ML algorithms, only Aydin and Iban [46] have performed a comprehensive comparative analysis of traditional boosting methods like AdaBoost against GB algorithms such as CatBoost, LightGBM, and XGBoost for FSM. Unexpectedly, AdaBoost surpassed the performance of the other three GB algorithms. Existing literature commonly disregards the fact that the performance of these three GB algorithms can vary considerably depending on the dataset characteristics and the geographical context of the study. While there is a scarcity of studies that directly compare these algorithms in the context of FSM, comparative assessments have been carried out in other fields. In some of these studies, CatBoost excelled compared to other GB algorithms because of its effective management of categorical data [58,59], whereas in others, XGBoost or LightGBM produced superior outcomes [46,60]. This variability emphasizes the importance of conducting comparative studies to ascertain the most suitable algorithm for specific geospatial applications like FSM. Moreover, while most FSM studies primarily emphasize topographic and hydrological factors, new indices obtained from remote sensing, despite their demonstrated effectiveness in measuring imperviousness, vegetation health, and water presence, are still underused, yet they are crucial for precise FSM [61,62]. Additionally, the physical properties of soil are infrequently considered in FSM, even though they significantly influence soil infiltration, water retention, and runoff patterns [63–65].

This investigation tackles the identified gap by rigorously assessing the performance and predictive strength of four ensemble boosting ML models. This comparison is particularly significant as it not only measures the performance of AdaBoost in relation to three prevalent GB algorithms, but also compares these GB algorithms amongst themselves to determine the most efficient one for FSM in a real-world hydrological setting. This modeling is novel as it adopts a multi-tier feature selection strategy that utilizes Variance Inflation Factor (VIF), Condition Index (CI), Mutual Information (MI), and Information Gain (IG) to guarantee the inclusion of only the most pertinent and non-redundant factors. Additionally, the modeling integrated a diverse and innovative set of 13 factors, including lesser-used remote sensing (RS) indices such as the Normalized Difference Impervious Surface Index (NDISI), Normalized Difference Greenness Index (NDGI), Urban Index (UI), and Land Surface Water Index (LSWI). These indices are proficient in capturing aspects of imperviousness, vegetation status, and surface water availability, soil clay content and soil bulk density, which offer critical insights into permeability and surface runoff processes.

This study developed susceptibility models for the Buzău River catchment through the application of AdaBoost and three GB algorithms. It integrates a diverse array of

13 conditioning factors (CFs), blending both traditional topographic and hydrological factors with advanced RS indices and soil physical characteristics. A multi-tier feature selection strategy will be utilized to optimize performance and ensure robustness, incorporating VIF, CI, MI, and IG techniques to determine the most pertinent factors. The model's efficacy will be measured using a range of performance metrics, and the relevance of the factors will be evaluated through SHapley Additive exPlanations (SHAP) values.

2. Materials and Methods

2.1. Study Area: Overview of the Buzău River Catchment

The Buzău River basin (Figure 1) is situated in the south-eastern region of Romania and serves as a left tributary to the Siret River [66]. Originating from the Ciucaș Mountains within the Curvature Carpathians, the Buzău River has an overall length of 302 km [66]. This catchment area encompasses 5264 km^2 [66] and receives an average annual precipitation of about 750 mm/year [67].

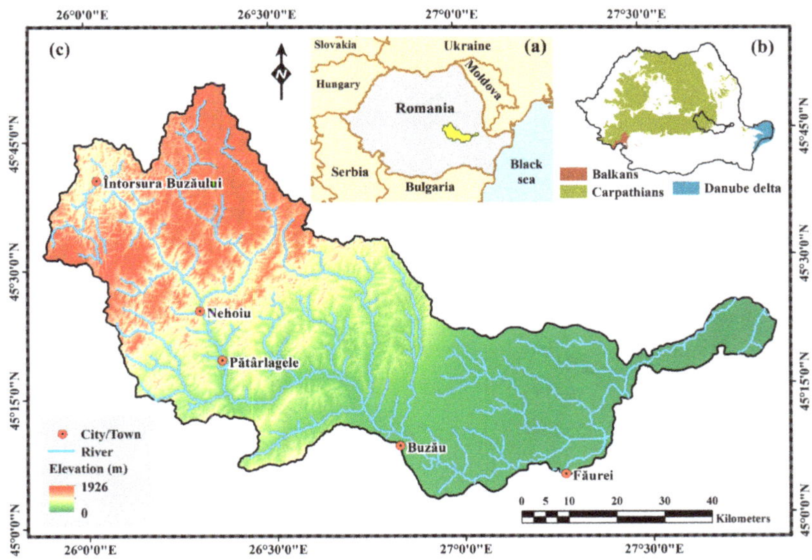

Figure 1. (**a**) Location of the Buzău River catchment in Romania, (**b**) Carpathian and Balkan mountain ranges and the Danube Delta region, and (**c**) Buzău River catchment, highlighting the river network and elevation range.

The catchment region covers five counties: Covasna, Brăila, Brașov, Buzău, and Prahova, along with 116 territorial-administrative units [66]. The catchment showcases a varied topography, from the steep northern Carpathian Mountain slopes to the southern low-lying plains. The basin features thick forests in the upper basin, along with agricultural and urbanized zones downstream, where uncontrolled development and land degradation have amplified the risk of flooding. The Buzău River is among the rivers in Romania that face the greatest risk of flooding, having experienced significant flood events in recent decades [66], making it a crucial region for FSM.

2.2. Methodological Framework for Susceptibility Modeling

The modeling process employed a multi-tier feature selection strategy, followed by the application of four ensemble boosting algorithms and multiple evaluation metrics, as well as explainable artificial intelligence (XAI) techniques such as SHAP to analyze flood susceptibility (Figure 2).

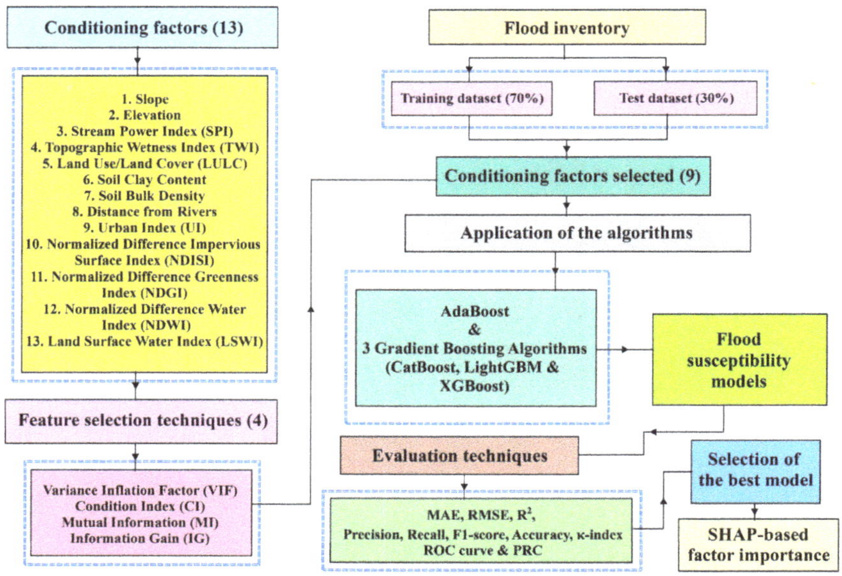

Figure 2. The flowchart of the susceptibility modeling framework.

The modeling was performed on the Kaggle and Google Colab platforms, which provide cloud-based environments with significant computational power. In this analysis, the pixel served as the main spatial and mapping unit, featuring a spatial resolution of 30 m, which guarantees uniform input across all geospatial layers.

2.3. Flood Inventory Dataset and Data Splitting Strategy

This research compiled a total of 205 locations of flood occurrences from earlier studies conducted by Costache et al. [67,68]. The inventory comprises flood events that resulted in socio-economic damage between 1990 and 2020 [67], categorized as the positive class (indicating flood presence) within the FSM framework. An equivalent number of non-flood points (205) were randomly generated to maintain a balanced dataset from regions with no documented flood history. These non-flood sites represent the negative class (indicating flood absence) and were selected carefully to avoid overlap with flood-affected regions, thus maintaining a clear distinction. The dataset, comprising 410 spatial points (205 flood and 205 non-flood locations), was randomly partitioned into two subsets: 70% of the data (287 points) was designated for model training, and the remaining 30% (123 points) was allocated for model validation (Figure 3). The 70:30 split ratio is commonly employed because it facilitates effective model training and precise performance assessment [69,70].

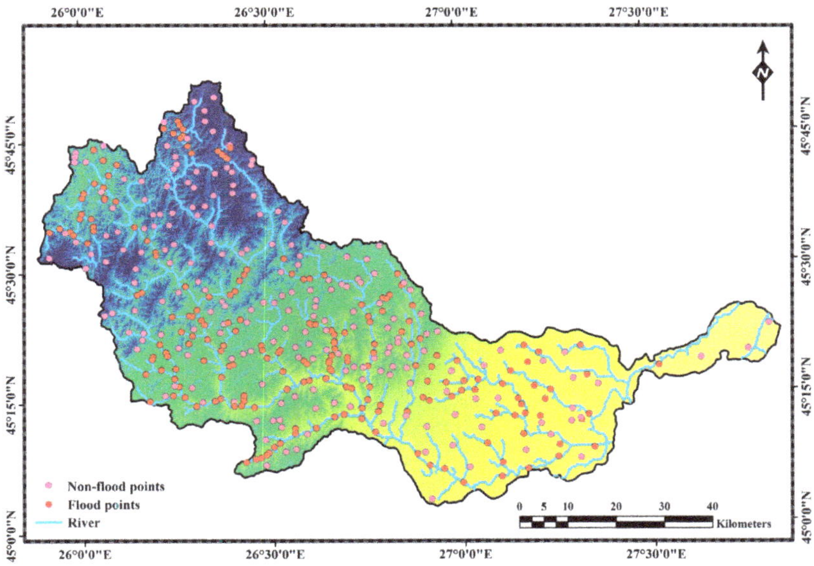

Figure 3. Spatial distribution of flood and non-flood points within the Buzău River catchment.

2.4. Derivation of Conditioning Factors

Based on earlier studies [41,71–73], the modeling process selected 13 CFs, which are presented in Table 1.

Table 1. Overview of the datasets and conditioning factors derived for the modeling, including their sources and spatial resolutions.

Dataset	Source	Conditioning Factor	Scale/Spatial Resolution
SRTM DEM	https://earthexplorer.usgs.gov/ (accessed on 20 April 2025)	Slope Elevation Stream Power Index (SPI) Topographic Wetness Index (TWI)	30 m
CORINE Land Cover	https://land.copernicus.eu/en/products/corine-land-cover (accessed on 20 April 2025)	Land Use/Land Cover (LULC)	100 m
SoilGrids	https://soilgrids.org/ (accessed on 20 April 2025)	Soil Clay Content Soil Bulk Density	250 m
HydroSHEDS	https://www.hydrosheds.org/products/hydrorivers (accessed on 20 April 2025)	Distance from Rivers	-
Landsat 8 and 9 imagery	https://earthexplorer.usgs.gov/ (accessed on 20 April 2025)	Normalized Difference Impervious Surface Index (NDISI)	100 m
Sentinel-2 imagery	https://browser.dataspace.copernicus.eu/ (accessed on 20 April 2025)	Urban Index (UI) Normalized Difference Greenness Index (NDGI) Normalized Difference Water Index (NDWI) Land Surface Water Index (LSWI)	20 m 10 m 10 m 20 m

Slope, elevation, SPI, and TWI were obtained from the DEM utilizing SAGA GIS version 9.5.1 (Institute of Geography at the University of Hamburg). SPI and TWI values were determined based on Equations (1) and (2) [72]. The LULC data were sourced from the CORINE Land Cover portal, whereas the soil clay content and bulk density were acquired from the SoilGrids portal. River networks were obtained from the HydroSHEDS portal (https://www.hydrosheds.org/products), and the distance to these rivers was computed utilizing the Euclidean Distance tool in ArcGIS 10.8 (ESRI, Romania). The five-year (2020–2024) mean UI, NDGI, NDWI, and LSWI were computed using Sentinel-2 surface reflectance data, while NDISI was derived from Landsat 8 and 9 (NASA) data. All five indices were derived using the Google Earth Engine platform, which provides efficient access to multi-temporal satellite imagery and facilitates large-scale geospatial analysis. The selection of a five-year mean aims to mitigate the effects of seasonal and interannual fluctuations, thus providing a more consistent and accurate estimate of land surface conditions. The UI, NDISI, NDGI, NDWI, and LSWI were calculated utilizing Equations (3)–(7) [74–78], respectively.

$$\text{SPI} = \alpha \times tan\beta \quad (1)$$

$$\text{TWI} = ln\left(\frac{\alpha}{tan\beta}\right) \quad (2)$$

where α = catchment area and β = slope angle.

$$\text{UI} = \frac{SWIR2 - NIR}{SWIR2 + NIR} \quad (3)$$

$$\text{NDISI} = \frac{T_b - (MNDWI + NIR + SWIR1)/3}{T_b + (MNDWI + NIR + SWIR1)/3} \quad (4)$$

$$\text{NDGI} = \frac{\alpha \times Green + (1-\alpha) \times NIR - Red}{\alpha \times Green + (1-\alpha) \times NIR + Red} \quad (5)$$

$$\text{NDWI} = \frac{Green - NIR}{Green + NIR} \quad (6)$$

$$\text{LSWI} = \frac{NIR - SWIR}{NIR + SWIR} \quad (7)$$

where T_b = brightness temperature, α = weighted parameter, *SWIR* = Short-Wave Infrared band, *NIR* = Near-Infrared band, *Green* = Green band, *Red* = Red band, and *MNDWI* = Modified Normalized Difference Water Index [41].

2.5. Feature Selection Techniques

In this modeling, the feature selection process includes the evaluation of multicollinearity and the implementation of feature selection algorithms to discard multicollinear and non-relevant factors. The key challenge posed by multicollinearity is its tendency to inflate standard errors, resulting in unstable estimates and unreliable interpretations [79]. Irrelevant and redundant factors may negatively influence algorithms' complexity and functionality, leading to suboptimal results or performance [80]. The feature selection process involves removing irrelevant and redundant factors to boost the performance of algorithms and the accuracy of model outputs [80,81].

2.5.1. Variance Inflation Factor (VIF)

VIF is a statistical indicator utilized to measure the degree of multicollinearity among CFs [70]. A VIF score greater than 10 signifies multicollinearity, which was determined through Equation (8) [82]. Typically, researchers discard all factors that have VIF scores exceeding 10 in one step. However, this study implemented a step-wise analysis of multicollinearity and factor removal, which allowed for a more accurate identification and retention of impactful factors. This approach guarantees that only the most problematic factors are omitted, without compromising important predictive data.

$$\text{VIF} = \frac{1}{1 - R_j^2} \tag{8}$$

where R_j^2 = coefficient of determination (R^2) for the jth factor.

2.5.2. Condition Index (CI)

CI is the square root of the ratio of the maximum eigenvalue to each eigenvalue, as outlined in Equation (9) [83]. Multicollinearity is deemed absent if the CI is 10 or lower, moderate if it ranges from 10 to 30, and severe if it is 30 or higher [84].

$$\text{CI} = \sqrt{\frac{\lambda_{max}}{\lambda_i}} \tag{9}$$

where λ_{max} = maximum eigenvalue and λ_i = ith eigenvalue.

2.5.3. Mutual Information (MI)

MI is a filter-based approach that measures the degree of interdependence among variables, effectively capturing linear and nonlinear associations [85]. MI is a benchmark for choosing appropriate feature subsets by assessing the quantity of information, feature offers concerning the target variable [86]. MI(X;Y) was computed by applying Equation (10) [85,87].

$$\text{MI}(X;Y) = \sum_{x \in X} \sum_{y \in Y} p(x,y)\, log\left(\frac{p(x,y)}{p(x)\, p(y)}\right) \tag{10}$$

where X and Y = two random variables, $p(x,y)$ = joint probability distribution, and $p(x)$ and $p(y)$ = marginal distributions.

2.5.4. Information Gain (IG)

IG is an entropy-based approach that evaluates the information supplied by a feature [81]. The gain (y, A), calculated from the output data categorized by feature A, was computed according to Equation (11) [88].

$$\text{gain}(y, A) = entropy(y) - \sum\nolimits_{C \in vals(A)} \frac{y_c}{y} entropy\ (y_c) \tag{11}$$

with $val(A)$ = possible values of feature A, y = number of samples, and y_c = subset of y.

2.6. *Machine Learning (ML) Algorithms*

2.6.1. Adaptive Boosting (AdaBoost)

AdaBoost is an algorithm based on decision trees that creates a collection of stumps, which are basic trees consisting of a single node and two leaves, typically only one level deep [89,90]. It adopts an iterative approach to learn from these stumps and integrates them into an ensemble [90]. AdaBoost operates by minimizing an exponential loss function, making it sensitive to data noise and outliers [91]. Despite this challenge, it effectively decreases bias and variance, improving overall performance [91].

2.6.2. Gradient Boosting Algorithms

Categorical Boosting (CatBoost): CatBoost can efficiently address both categorical and numerical factors without requiring preprocessing steps such as one-hot encoding or label encoding, instead relying on its inherent 'ordered boosting' approach to manage categorical data [92]. Each model is trained on fresh data by utilizing ordered boosting, which helps alleviate the biases commonly linked to standard GB algorithms [91]. CatBoost employs 'oblivious decision trees' (Figure 4a), which maintain the same splitting criterion at every tree level, creating balanced structures that are less likely to overfit [93,94].

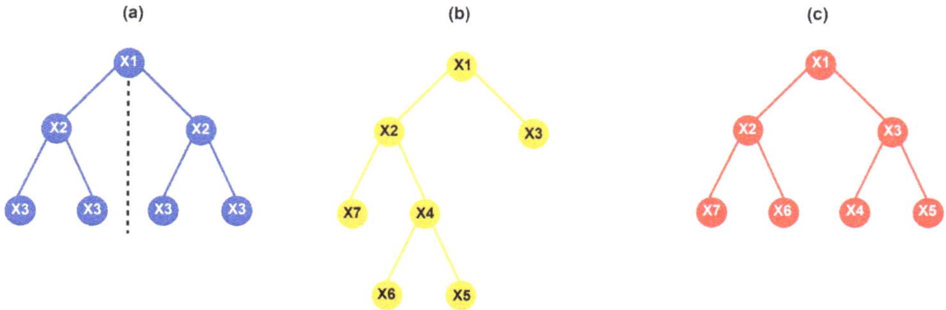

Figure 4. Tree structures and split indexes: (**a**) CatBoost, (**b**) LightGBM, and (**c**) XGBoost.

Light Gradient Boosting Machine (LightGBM): LightGBM presents three novel strategies designed to enhance the efficiency of training: a histogram-based approach for split finding, Exclusive Feature Bundling (EFB), and Gradient-Based One-Side Sampling (GOSS) [92,93]. The histogram-based split finding technique accelerates the training and reduces memory requirements by binning continuous feature values before identifying the best splits [92]. EFB applies heuristics to identify and merge groups of mutually exclusive features, decreasing the dataset's dimensionality [93]. GOSS utilizes gradients to sample the most critical instances of the dataset during each iteration, ensuring the training set distribution remains unchanged [93]. LightGBM builds trees leaf-wise (Figure 4b), resulting in quicker convergence and greater accuracy [91].

eXtreme Gradient Boosting (XGBoost): XGBoost builds additive models sequentially, allowing for the optimization of any differentiable loss function [91]. It incorporates regularization techniques (L1 and L2) to reduce overfitting, thereby enhancing the model's ability to generalize [91]. Additionally, XGBoost applies second-order Taylor series approximations of the loss function to improve both accuracy and computational efficiency [59,91]. It supports parallel processing and internally addresses missing values [91]. XGBoost mainly follows a level-wise tree growth strategy (Figure 4c), where all nodes at a specific depth are split before advancing deeper, which helps control overfitting and maintain balanced trees [94].

2.7. Performance Evaluation Techniques

2.7.1. Mean Absolute Error (MAE) and Root Mean Square Error (RMSE)

MAE and RMSE serve as key indicators of absolute error, primarily applied in model fitting, validation, selection, and comparison [95]. The MAE and RMSE values were computed based on Equations (12) and (13) [51].

$$\text{MAE} = \frac{1}{n}\sum_{i=1}^{n}\left|Y_i - \tilde{Y}_i\right| \tag{12}$$

$$\text{RMSE} = \sqrt{\frac{1}{n}\sum_{i=1}^{n}\left(Y_i - \tilde{Y}_i\right)^2} \tag{13}$$

where Y_i = actual value, $\tilde{Y}_i$ = predicted value, and n = number of observations.

2.7.2. R-Squared (R^2)

The R-squared (R^2) serves as a quantitative measure of how effectively a model captures the variability of the dependent factor as influenced by the independent factors [96]. It represents the proportion of the variation explained by the model out of the overall variation [51], with possible values spanning from 0 (indicating a lack of fit) to 1 (indicating a perfect fit) [97]. R^2 was derived through the application of Equation (14) [51].

$$R^2 = 1 - \frac{\sum_1^n \left(D_{act} - D_{pre}\right)^2}{\sum_1^n \left(D_{act} - \overline{D}_{act}\right)^2} \quad (14)$$

where D_{act} = actual value, D_{pre} = predicted value, $\overline{D}_{act}$ = mean, and n = number of observations.

2.7.3. Accuracy, Precision, Recall, and F1-Score

Accuracy, precision, recall, and F1-score, key performance metrics ranging from 0 (denoting poor performance) to 1 (denoting perfect performance), were computed using Equations (15)–(18) [82,98], respectively.

$$\text{Accuracy} = \frac{TP + TN}{TP + FP + FN + TN} \quad (15)$$

$$\text{Precision} = \frac{TP}{TP + FP} \quad (16)$$

$$\text{Recall} = \frac{TP}{TP + FN} \quad (17)$$

$$\text{F1-score} = 2 \times \frac{Precision \times Recall}{Precision + Recall} \quad (18)$$

where TP (TN) = True Positives (Negatives), FP (FN) = False Positives (Negatives).

2.7.4. Kappa Index (κ-Index)

Cohen's Kappa index (κ-index) is a statistical indicator that evaluates the extent of agreement, with values spanning −1 to +1 [82,99]. A score of −1 reflects total disagreement, 0 indicates a lack of agreement beyond random chance, and +1 signifies total agreement [82,99]. The κ-index was derived based on Equation (19) [82].

$$\kappa\text{-index} = \frac{P_{obs} - P_{exp}}{1 - P_{exp}} \quad (19)$$

where P_{obs} = observed agreement and P_{exp} = expected agreement.

2.7.5. Receiver Operating Characteristic (ROC) Curve

The ROC curve represents a graph that plots the True Positive Rate (TPR) on the Y-axis against the False Positive Rate (FPR) on the X-axis, serving as a measure of overall performance [100,101]. The area under the ROC Curve (AUC) is applied to quantify this performance [100], with AUC values ranging from 0.5 (random performance) to 1 (perfect performance) [102].

2.7.6. Precision Recall Curve (PRC)

The PRC, featuring recall on the X-axis and precision on the Y-axis, is regarded as a more informative tool than the ROC curve for evaluating performance in datasets with class imbalance [103]. Higher average precision (AP) scores signify superior model performance, with a score of 1 indicating perfect performance and 0 denoting poor performance [104].

2.8. Factor Importance Evaluation: SHapley Additive exPlanations (SHAP)

SHAP elucidates a prediction by illustrating the contribution of each feature to the variation from the model's baseline value [105,106]. It is grounded in coalitional game theory [105] and employs a linear explanatory model (Equation (20)) to approximate the original prediction model [107,108]. SHAP will quantify and visualize the impact of each conditioning factor on the predictions made by the model. This will facilitate a more profound comprehension of the relative significance of each factor in relation to flood susceptibility, thus improving the interpretability of the model and aiding in more informed decision-making.

$$f(x) = g(x') = \varnothing_0 + \sum_{i=1}^{M} \varnothing_i x'_i \tag{20}$$

where $f(x)$ = original model, $g(x')$ = explanation model, M = number of input features, $\varnothing_0$ = base value, $\varnothing_i$ = SHAP value, and $x'_i \in \{0,1\}$ = presence (1) or absence (0) of the i-th feature.

The computation of the SHAP value was performed based on Equation (21) [107].

$$\varnothing_i = \sum_{z' \subseteq x' \setminus \{i\}} \frac{|z'|! \cdot (M - |z'| - 1)!}{M!} \left[f_x(z' \cup \{i\}) - f_x(z') \right] \tag{21}$$

where $z' \subseteq x' \setminus \{i\}$ = potential subsets of the simplified input, without feature i, $|z'|$ = number of features in the subset z', $\frac{|z'|! \cdot (M - |z'| - 1)!}{M!}$ = SHAP weight, $f_x(z' \cup \{i\})$ = output of the model when utilizing subset z' with feature i, and $f_x(z')$ = model output when utilizing solely the subset z'.

3. Results

3.1. Conditioning Factors Selected Through Various Feature Selection Methods

An analysis of the multicollinearity for the 13 CFs was performed, identifying three factors with VIF scores surpassing 10 (Table A1). To mitigate this issue, a two-tier selection approach was implemented. During the first stage, the factor with the highest VIF score, LSWI, was discarded, and the multicollinearity was re-evaluated, indicating that two CFs continued to display multicollinearity (Table A2). At the second stage, the subsequent factor with the highest VIF score (UI) was removed, and the multicollinearity was re-evaluated (Table 2).

Table 2. Variance Inflation Factor (VIF) scores of the conditioning factors.

Sl. No.	Conditioning Factor	VIF Score
1	Soil Clay Content	7.844
2	Soil Bulk Density	6.548
3	NDGI	6.330
4	NDISI	5.566
5	TWI	4.888
6	Slope	3.539
7	SPI	3.072
8	Elevation	2.872
9	NDWI	1.763
10	Distance from Rivers	1.486
11	LULC	1.360

Thus, it was established that all remaining 11 CFs maintained VIF scores beneath the threshold of 10. In addition, the CI for all 11 CFs was determined to be beneath the critical threshold of 30 (Table 3), thereby validating that all multicollinear factors have been effectively eliminated.

Table 3. Condition index (CI) of the conditioning factors used in the modeling.

Sl. No.	Conditioning Factor	CI
1	TWI	7.550
2	Distance from Rivers	6.490
3	SPI	5.810
4	Slope	3.400
5	NDWI	2.960
6	NDISI	2.460
7	NDGI	2.360
8	LULC	1.940
9	Elevation	1.570
10	Soil Bulk Density	1.310
11	Soil Clay Content	1.000

To ensure the exclusion of irrelevant factors, the MI scores for the 11 CFs were computed. The analysis revealed that Slope (0.452), TWI (0.403), and Distance from Rivers (0.348) had the highest MI scores, indicating their significant relevance for inclusion, while Soil Clay Content (0.042) and NDWI (0.012) displayed the lowest MI scores, implying minimal relevance (Table 4). Despite the variation in scores, none were zero, and therefore all 11 CFs were retained for subsequent analysis.

Table 4. Conditioning factors and their corresponding Mutual Information (MI) scores.

Sl. No.	Conditioning Factor	MI Score
1	Slope	0.452
2	TWI	0.403
3	Distance from Rivers	0.348
4	SPI	0.211
5	LULC	0.187
6	Elevation	0.146
7	Soil Bulk Density	0.141
8	NDGI	0.107
9	NDISI	0.100
10	Soil Clay Content	0.042
11	NDWI	0.012

Subsequently, IG-based feature selection was implemented due to its effectiveness with tree-based ML algorithms. The analysis demonstrated that SPI (0.000) and NDWI (0.001) exhibited negligible IG scores (Table 5), reflecting their limited relevance; thus, these two factors were discarded. Consequently, the multi-tier feature selection approach led to the identification of nine relevant factors (Figures 5 and 6) and the elimination of four irrelevant factors (Figure A1).

Table 5. Information Gain (IG) scores of the conditioning factors.

Sl. No.	Conditioning Factor	IG Score
1	Slope	0.451
2	Distance from Rivers	0.400
3	TWI	0.375
4	LULC	0.326
5	NDGI	0.116
6	NDISI	0.110
7	Elevation	0.099
8	Soil Bulk Density	0.088
9	Soil Clay Content	0.007
10	NDWI	0.001
11	SPI	0.000

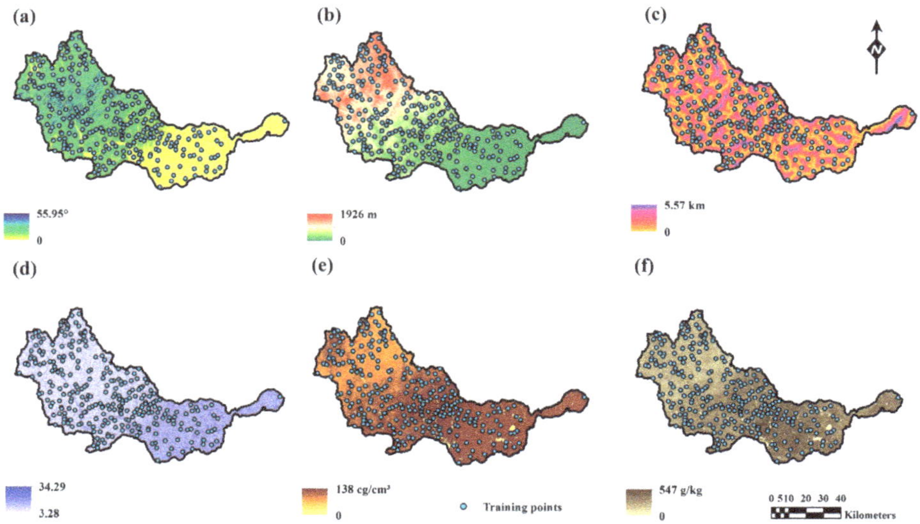

Figure 5. Conditioning factors: (**a**) Slope, (**b**) Elevation, (**c**) Distance from Rivers, (**d**) Topographic Wetness Index (TWI), (**e**) Soil Bulk Density, and (**f**) Soil Clay Content.

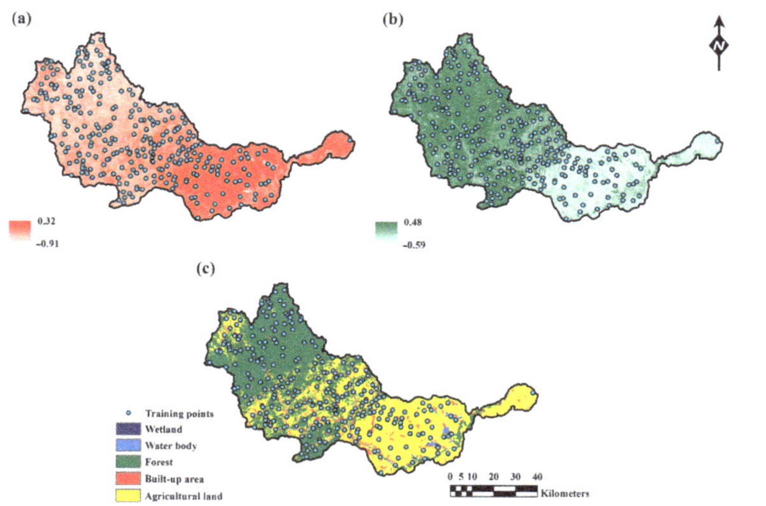

Figure 6. Conditioning factors: (**a**) Normalized Difference Impervious Surface Index (NDISI), (**b**) Normalized Difference Greenness Index (NDGI), and (**c**) Land Use/Land Cover (LULC).

3.2. Flood Susceptibility Models and Their Performance

The susceptibility maps were produced utilizing four ML models—AdaBoost (Figure 7a), CatBoost (Figure 7b), LightGBM (Figure 7c), and XGBoost (Figure 7d)—based on nine CFs. All maps pinpointed the river network and low-lying regions as areas of significant susceptibility. Table 6 illustrates the models' performance as measured by MAE, RMSE, and R^2 for both datasets. CatBoost exhibited the most superior performance, recording the lowest MAE (0.074) and RMSE (0.146) on the training set, as well as the lowest MAE (0.097) and RMSE (0.182) on the testing set. It also achieved the highest R^2 scores of 0.919 and 0.838 for the training and testing sets, respectively.

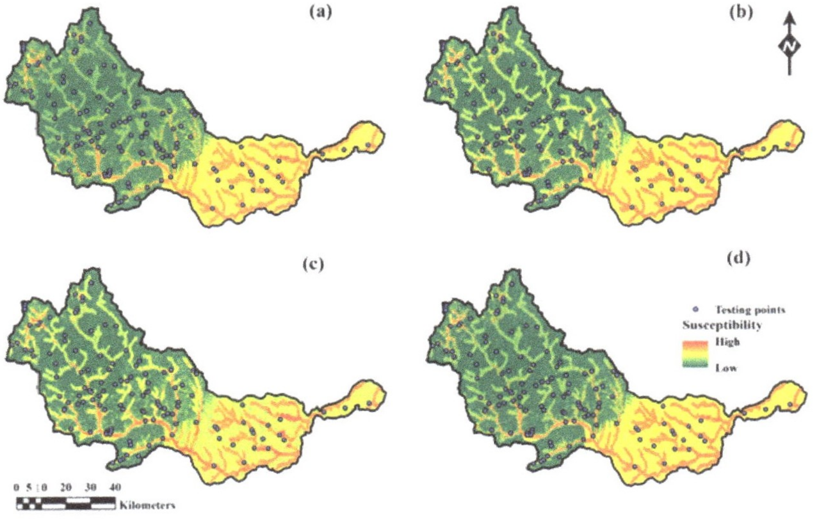

Figure 7. Susceptibility maps: (**a**) AdaBoost, (**b**) CatBoost, (**c**) LightGBM, and (**d**) XGBoost.

Table 6. Performance evaluation of the models using MAE, RMSE, and R^2 metrics on both training and testing datasets.

Model	MAE		RMSE		R^2	
	Train	Test	Train	Test	Train	Test
AdaBoost	0.088	0.117	0.173	0.229	0.880	0.787
CatBoost	0.074	0.097	0.146	0.182	0.919	0.838
LightGBM	0.082	0.111	0.164	0.211	0.892	0.804
XGBoost	0.079	0.102	0.151	0.192	0.914	0.817

3.3. *Results of Susceptibility Model Evaluation Using Various Performance Metrics*

Among the assessed models, CatBoost excelled with a precision of 0.928, recall of 0.917, F1-score of 0.913, accuracy of 0.912, and a κ-index of 0.841, as indicated in Table 7. It consistently achieved superior results compared to AdaBoost, LightGBM, and XGBoost.

Table 7. Performance comparison of flood susceptibility models using various evaluation metrics.

	AdaBoost	CatBoost	LightGBM	XGBoost
Precision	0.904	0.928	0.906	0.916
Recall	0.887	0.917	0.885	0.909
F1-Score	0.885	0.913	0.894	0.908
Accuracy	0.886	0.912	0.894	0.908
κ-index	0.782	0.841	0.801	0.825

In addition, CatBoost achieved the best overall performance, reaching the highest ROC-AUC score of 0.972, trailed by XGBoost with 0.971, LightGBM with 0.967, and AdaBoost with 0.964 (Figure 8).

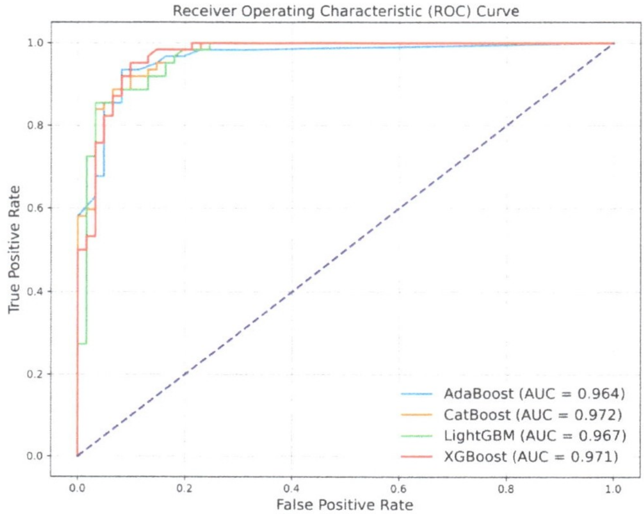

Figure 8. Comparison of ROC curves for the four susceptibility models to assess their predictive performance.

The PRC-AP was similarly the highest for CatBoost (0.971), followed by XGBoost (0.967), AdaBoost (0.963), and LightGBM (0.961) (Figure 9). These findings underscore the enhanced capability of GB models—CatBoost, XGBoost, and LightGBM—compared to AdaBoost.

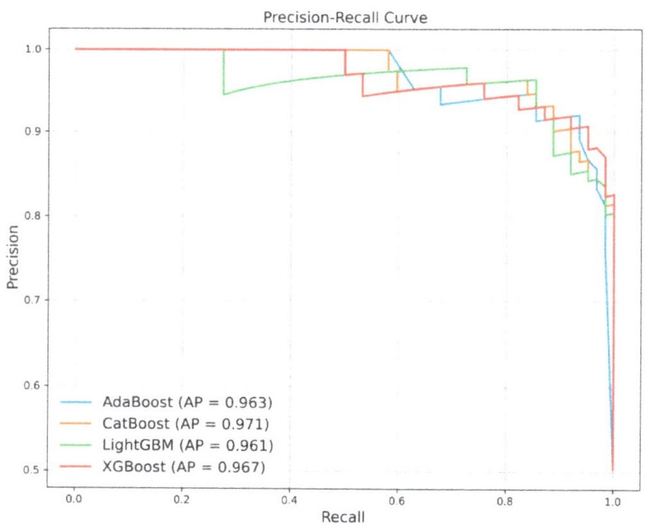

Figure 9. Precision–recall curves of the four susceptibility models: AdaBoost, CatBoost, LightGBM, and XGBoost.

4. Discussion

4.1. Role and Importance of Conditioning Factors

The SHAP-driven factor importance assessment for the CatBoost model, identified as the top performer, revealed that Slope (0.232) and Distance from Rivers (0.155) were the most influential CFs (Table 8). The SHAP summary plot (Figure 10) visually represents how each factor influences the model's predictions, emphasizing that Slope and Distance from Rivers contributed most significantly and consistently across the dataset. Many studies [61,71,109] have identified Slope, Distance from Rivers, and LULC as key contributing factors. Specifically, Costache et al. [68] reported these three as the most influential CFs in the Buzău River catchment.

Table 8. SHAP-based importance of conditioning factors (mean absolute SHAP).

Sl. No.	Conditioning Factor	Mean_Abs_SHAP
1	Slope	0.232
2	Distance from Rivers	0.155
3	TWI	0.061
4	LULC	0.034
5	NDISI	0.026
6	Soil Bulk Density	0.022
7	Elevation	0.019
8	NDGI	0.017
9	Soil Clay Content	0.015

Figure 10. SHAP summary plot showing factor importance and impact on model predictions (SD = Distance from Rivers, SBD = Soil Bulk Density, and SCC = Soil Clay Content).

Slope, with a mean SHAP value of 0.232, stands out as the most critical factor, likely due to its strong effect on surface runoff and water accumulation processes. Steeper slopes facilitate greater surface runoff and limit infiltration, while flatter terrains tend to gather water, thus amplifying flood risks [110,111]. Ranking second (0.155), the Distance from Rivers illustrates the vulnerability of areas adjacent to river channels. Areas near rivers face higher risk due to the immediate impact of river overflow during periods of high discharge; during heavy rainfall, rising river levels tend to inundate the surrounding low-lying regions first [112,113]. TWI (0.061) exhibited a notable impact, emphasizing the crucial role of terrain morphology in retaining water and directing flow. The TWI indicates possible water accumulation within the landscape; elevated values signify areas at risk of saturation and runoff concentration, which can lead to flooding [72,114].

LULC has a moderate contribution of 0.034, suggesting that land management practices and surface conditions play a significant role in flooding. LULC impacts flooding by altering land–rainfall interactions, where urban and agricultural regions enhance surface runoff due to their reduced capacity for infiltration [41,115]. The compaction of soil on agricultural land, caused by heavy machinery and livestock, can facilitate flooding by reducing permeability and hindering water infiltration [116,117]. Moreover, repeated tilling can result in the loss of organic matter and degradation of soil structure [118], consequently heightening surface runoff during heavy rainfall events. Furthermore, the transformation of natural vegetation into agricultural land removes deep-rooted plants that play a crucial role in absorbing and regulating rainwater. Irrigation methods and inadequate drainage can lead to soil saturation, which reduces its ability to absorb further rainfall and raises the risk of flooding [119].

Among the moderately important factors, NDISI (0.026) and Soil Bulk Density (0.022) exhibited limited but relevant influences, indicating secondary contributions to flood occurrence. Higher NDISI values reflect impervious surfaces such as roads and buildings [120], which hinder infiltration and amplify surface runoff, consequently raising the chance of flooding [121,122]. Prior investigations [4,123] emphasized the impact of soil sealing on flooding, illustrating how alterations to the natural hydraulic network can lead to increased flood risk. An increase in bulk density results in reduced soil porosity and infiltration [124], which consequently heightens the probability of surface runoff and flooding.

Elevation plays a crucial role by dictating the natural pathways for water flow; regions at lower elevations are more likely to gather runoff and are at a higher risk of inundation, especially in the event of significant rainfall or river surges [111,125]. NDGI assesses the density and health of vegetation, reflecting trends of degradation and regeneration [126]. Areas with lower NDGI, which denote sparse or unhealthy vegetation [126,127], can heighten the probability of flooding due to the diminished ability of the soil to absorb water, thereby facilitating quicker surface runoff [128]. Clay-dominant soils possess lower permeability, which hinders infiltration and allows for water to remain for longer durations, consequently promoting flooding [41,129].

The SHAP analysis indicates that elevation (0.019), NDGI (0.017), and Soil Clay Content (0.015) have a relatively minor significance, implying their limited effect in the Buzău catchment. Although elevation typically influences hydrological flow and retention behavior, the Buzău catchment may feature localized flooding events that are more concentrated in topographically low areas, irrespective of their absolute elevation. Consequently, local variations in elevation—such as depressions or valleys—may be more significant than the overall elevation.

The NDGI, indicative of vegetation greenness, may have a limited impact during intense rainfall events, as vegetation's ability to reduce runoff diminishes. Likewise, while the clay content in soil can play a role in determining infiltration capacity, this effect may be overshadowed by other soil characteristics, including bulk density. This analysis indicates that although these factors are not entirely irrelevant, their influence is relatively limited within this modeling framework.

4.2. Interpretation of Model Performance Outcomes

The modeling process revealed that CatBoost achieved the highest performance, with XGBoost, LightGBM, and AdaBoost ranking next. Several studies [58,59,89,91,94,130–132] have identified CatBoost as the most effective model among GB algorithms. In contrast, other research has found that XGBoost [93] or LightGBM [60,133] may be more effective. This variation in results underscores the significance of dataset characteristics, feature composition, and preprocessing techniques in determining model efficacy. The superior performance of CatBoost can be linked to its ability to effectively handle categorical variables without one-hot encoding, as well as its application of ordered boosting and symmetric (oblivious) trees [51,134]. These characteristics are recognized for boosting model accuracy and decreasing overfitting [51,134]. This is particularly useful in FSM, where both continuous and categorical inputs are commonly utilized.

XGBoost also demonstrated commendable performance, probably owing to its well-optimized training architecture, proficient parallel processing capabilities, and flexibility regarding objective functions [135,136]. Moreover, its advanced regularization techniques contribute to enhanced generalizability [135]. LightGBM employs a leaf-wise tree growth strategy that is efficient, yet it can cause instability in datasets that are smaller or contain noise [137]. This could account for its comparatively lower performance in this modeling, indicating that LightGBM may be less appropriate for FSM tasks that involve spatial heterogeneity or class imbalance. In this comparison, AdaBoost demonstrated the lowest performance. This is due to its vulnerability to noisy data and outliers, as it amplifies the weight of misclassified instances, which may diminish the overall performance when compared to GB algorithms [138]. Overall, the results bolster the prevailing view that CatBoost is a robust and effective option, especially when dealing with categorical factors. These conclusions are in line with recent literature and stress the need for model selection to be adapted to the specific characteristics of the dataset and the goals of the modeling process.

5. Conclusions

This research emphasizes the efficacy of ensemble boosting algorithms, especially CatBoost, for FSM in the Buzău River catchment, an ecologically and hydrologically sensitive region within the Danube River Basin. Through the integration of multi-tier feature selection and SHAP-based interpretability, this research not only boosts model accuracy but also improves transparency in recognizing essential factors driving floods. The study reveals that Slope, Distance from Rivers, TWI, and LULC are key contributors to flood susceptibility in this catchment. From a policy viewpoint, the conclusions endorse the necessity for targeted land-use planning measures, including more rigorous zoning regulations in critical areas and the integration of sustainable practices in watershed management. Authorities must prioritize the protection and surveillance of low-lying zones and riverine areas while also incorporating topographic and hydrological information into regional disaster management plans. The methodological framework outlined in this research can be adapted to other flood-prone locales and underscores the critical need for science-driven policies to foster climate-resilient communities.

One notable limitation of this study is the lack of an in-depth analysis of model uncertainty and sensitivity testing. Although performance metrics were employed to assess the model, they do not completely encompass the range of uncertainty present in the input data, model parameters, or algorithmic framework, nor do they evaluate the model's sensitivity to variations in individual input factors. Moreover, the analysis incorporated CORINE Land Cover data from 2018, which is the most current dataset that is available. Despite the application of more recent RS-based indices to reflect current surface conditions, the reliance on 2018 land cover data may restrict the accurate representation of the most recent LULC conditions.

Author Contributions: Conceptualization, R.C., A.B. and R.S.A.; methodology, A.B., R.C. and R.S.A.; software, R.S.A. and R.C.; validation, R.S.A. and R.C.; formal analysis, R.S.A., R.C. and A.B.; investigation, R.S.A. and R.C.; resources, R.S.A. and R.C.; data curation, R.S.A. and R.C.; writing—original draft preparation, R.S.A., R.C. and A.B.; writing—review and editing, A.B., S.S., R.C. and R.F.; visualization, R.S.A.; supervision, A.B.; project administration, R.F.; funding acquisition, A.B. All authors have read and agreed to the published version of the manuscript.

Funding: This study was carried out by the first author within the RETURN Extended Partnership and received funding from the European Union Next-GenerationEU (National Recovery and Resilience Plan—NRRP, Mission 4, Component 2, Investment 1.3—D.D. 1243 2/8/2022, PE0000005). The work was supported for the second author by the Ministry of Research, Innovation and Digitization, Romania, CNCS/CCCDI—UEFISCDI, project number PN-IV-P8-8.1-PRE-HE-ORG-2023-0135, within PNCDI IV.

Data Availability Statement: The raw data supporting the conclusions of this article will be made available by the authors on request.

Acknowledgments: The authors express their sincere gratitude to the Head of Department of Civil Engineering and ERASMUS responsible of Faculty of Civil Engineering, Transilvania University of Brașov (UNITBV), for providing the necessary facilities and institutional support to the first author during the three-month visiting Ph.D. student mobility period at UNITBV, where this article was developed.

Conflicts of Interest: The authors declare no conflicts of interest.

Appendix A

Table A1. Initial VIF scores of all conditioning factors before removal of multicollinear factors.

Sl. No.	Conditioning Factor	VIF Score
1	LSWI	138.576
2	UI	68.691
3	NDISI	33.641
4	Soil Clay Content	8.081
5	NDGI	7.756
6	Soil Bulk Density	6.694
7	TWI	4.996
8	Elevation	3.594
9	Slope	3.592
10	SPI	3.234
11	NDWI	2.570
12	Distance from Rivers	1.487
13	LULC	1.378

Table A2. VIF scores after the first iteration of removing the most collinear conditioning factor.

Sl. No.	Conditioning Factor	VIF Score
1	UI	23.294
2	NDISI	20.292
3	Soil Clay Content	7.857
4	NDGI	6.603
5	Soil Bulk Density	6.569
6	TWI	4.949
7	Slope	3.583
8	SPI	3.170
9	Elevation	3.048
10	NDWI	2.309
11	Distance from Rivers	1.487
12	LULC	1.373

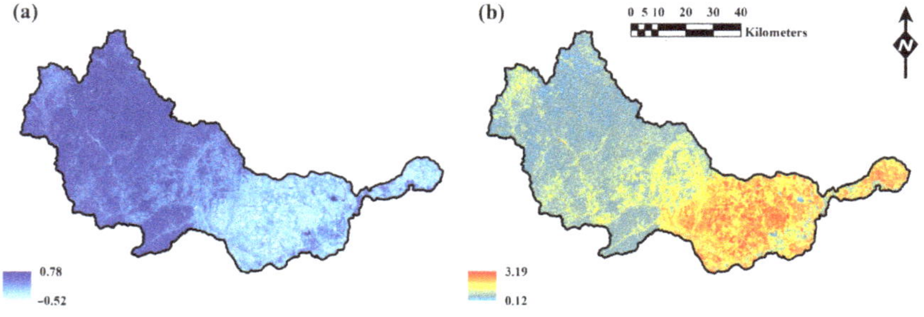

Figure A1. *Cont.*

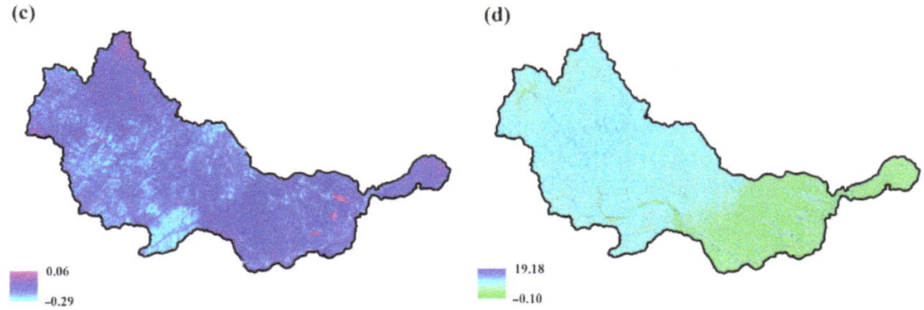

Figure A1. Discarded factors: (**a**) Land Surface Water Index (LSWI), (**b**) Urban Index (UI), (**c**) Normalized Difference Water Index (NDWI), and (**d**) Stream Power Index (SPI).

References

1. Petrucci, O.; Aceto, L.; Bianchi, C.; Bigot, V.; Brázdil, R.; Pereira, S.; Kahraman, A.; Kılıç, Ö.; Kotroni, V.; Llasat, M.C.; et al. Flood Fatalities in Europe, 1980–2018: Variability, Features, and Lessons to Learn. *Water* **2019**, *11*, 1682. [CrossRef]
2. Chen, J.; Shi, X.; Gu, L.; Wu, G.; Su, T.; Wang, H.M.; Kim, J.S.; Zhang, L.; Xiong, L. Impacts of Climate Warming on Global Floods and Their Implication to Current Flood Defense Standards. *J. Hydrol.* **2023**, *618*, 129236. [CrossRef]
3. Furtak, K.; Wolińska, A. The Impact of Extreme Weather Events as a Consequence of Climate Change on the Soil Moisture and on the Quality of the Soil Environment and Agriculture—A Review. *CATENA* **2023**, *231*, 107378. [CrossRef]
4. Gatto, A.; Martellozzo, F.; Clo', S.; Ciulla, L.; Segoni, S. The Downward Spiral Entangling Soil Sealing and Hydrogeological Disasters. *Environ. Res. Lett.* **2024**, *19*, 084023. [CrossRef]
5. Liu, Q.; Du, M.; Wang, Y.; Deng, J.; Yan, W.; Qin, C.; Liu, M.; Liu, J. Global, Regional and National Trends and Impacts of Natural Floods, 1990–2022. *Bull. World Health Organ.* **2024**, *102*, 410–420. [CrossRef]
6. Paprotny, D.; Terefenko, P.; Śledziowski, J. HANZE v2.1: An Improved Database of Flood Impacts in Europe from 1870 to 2020. *Earth Syst. Sci. Data* **2024**, *16*, 5145–5170. [CrossRef]
7. Economic Losses from Weather- and Climate-Related Extremes in Europe. Available online: https://www.eea.europa.eu/en/analysis/indicators/economic-losses-from-climate-related (accessed on 13 May 2025).
8. Morlot, M.; Brilly, M.; Šraj, M. Characterisation of the Floods in the Danube River Basin through Flood Frequency and Seasonality Analysis. *Acta Hydrotech.* **2019**, *32*, 73–89. [CrossRef]
9. Bezak, N.; Petan, S.; Kobold, M.; Brilly, M.; Bálint, Z.; Balabanova, S.; Cazac, V.; Csík, A.; Godina, R.; Janál, P.; et al. A Catalogue of the Flood Forecasting Practices in the Danube River Basin. *River Res. Appl.* **2021**, *37*, 909–918. [CrossRef]
10. Leščešen, I.; Basarin, B.; Pavić, D.; Mudelsee, M.; Pekarova, P.; Mesaroš, M. Are Extreme Floods on the Danube River Becoming More Frequent? A Case Study of Bratislava Station. *J. Water Clim. Chang.* **2024**, *15*, 1300–1312. [CrossRef]
11. Hein, T.; Schwarz, U.; Habersack, H.; Nichersu, I.; Preiner, S.; Willby, N.; Weigelhofer, G. Current Status and Restoration Options for Floodplains along the Danube River. *Sci. Total Environ.* **2016**, *543*, 778–790. [CrossRef]
12. Eder, M.; Perosa, F.; Hohensinner, S.; Tritthart, M.; Scheuer, S.; Gelhaus, M.; Cyffka, B.; Kiss, T.; Van Leeuwen, B.; Tobak, Z.; et al. How Can We Identify Active, Former, and Potential Floodplains? Methods and Lessons Learned from the Danube River. *Water* **2022**, *14*, 2295. [CrossRef]
13. Romanescu, G.; Stoleriu, C. Causes and Effects of the Catastrophic Flooding on the Siret River (Romania) in July–August 2008. *Nat. Hazards* **2013**, *69*, 1351–1367. [CrossRef]
14. Romanescu, G.; Cimpianu, C.I.; Mihu-Pintilie, A.; Stoleriu, C.C. Historic Flood Events in NE Romania (Post-1990). *J. Maps* **2017**, *13*, 787–798. [CrossRef]
15. Sekulova, F.; van den Bergh, J.C.J.M. Floods and Happiness: Empirical Evidence from Bulgaria. *Ecol. Econ.* **2016**, *126*, 51–57. [CrossRef]
16. Momčilović Petronijević, A.; Petronijević, P. Floods and Their Impact on Cultural Heritage—A Case Study of Southern and Eastern Serbia. *Sustainability* **2022**, *14*, 14680. [CrossRef]
17. Petrović, A.M.; Leščešen, I.; Radevski, I. Unveiling Torrential Flood Dynamics: A Comprehensive Study of Spatio-Temporal Patterns in the Šumadija Region, Serbia. *Water* **2024**, *16*, 991. [CrossRef]
18. Ana, J. Assessment of Pluvial Floods Potential on the Rivers of the Republic of Moldova. *Present Environ. Sustain. Dev.* **2018**, *12*, 121–133. [CrossRef]
19. Agayar, E.; Armon, M.; Wernli, H. The Catastrophic Floods in 2008, 2010 and 2020 in Western Ukraine: Hydrometeorological Processes and the Role of Upper-Level Dynamics. In Proceedings of the EGU General Assembly 2025, Vienna, Austria, 27 April–2 May 2025. [CrossRef]

20. Albano, R.; Samela, C.; Crăciun, I.; Manfreda, S.; Adamowski, J.; Sole, A.; Sivertun, Å.; Ozunu, A. Large Scale Flood Risk Mapping in Data Scarce Environments: An Application for Romania. *Water* **2020**, *12*, 1834. [CrossRef]
21. Albulescu, A.C. Exploring the Links between Flood Events and the COVID-19 Infection Cases in Romania in the New Multi-Hazard-Prone Era. *Nat. Hazards* **2023**, *117*, 1611–1631. [CrossRef]
22. Armaş, I.; Dobre, D.; Fekete, A.; Rufat, S.; Albulescu, A.C. Hinging on the Preparedness of First Responders. A Case Study on the 2021 Flood Operations in Romania. *Int. J. Disaster Risk Red.* **2025**, *116*, 105008. [CrossRef]
23. Costache, R.; Crăciun, A.; Ciobotaru, N.; Bărbulescu, A. Intelligent Methods for Estimating the Flood Susceptibility in the Danube Delta, Romania. *Water* **2024**, *16*, 3511. [CrossRef]
24. Popescu, N.C.; Bărbulescu, A. A Practical Approach on Reducing the Flood Impact: A Case Study from Romania. *Appl. Sci.* **2024**, *14*, 10378. [CrossRef]
25. Popa, M.C.; Simion, A.G.; Peptenatu, D.; Dima, C.; Draghici, C.C.; Florescu, M.; Dobrea, C.R.; Diaconu, D.C. Spatial Assessment of Flash-flood Vulnerability in the Moldova River Catchment (N Romania) Using the FFPI. *J. Flood Risk Manag.* **2020**, *13*, e12624. [CrossRef]
26. Costache, R.; Arabameri, A.; Costache, I.; Crăciun, A.; Md Towfiqul Islam, A.R.; Abba, S.I.; Sahana, M.; Pham, B.T. Flood Susceptibility Evaluation through Deep Learning Optimizer Ensembles and GIS Techniques. *J. Environ. Manag.* **2022**, *316*, 115316. [CrossRef]
27. Ionescu, C.S.; Gogoașe-Nistoran, D.E.; Baciu, C.A.; Cozma, A.; Motovilnic, I.; Brașovanu, L. The Impact of a Clay-Core Embankment Dam Break on the Flood Wave Characteristics. *Hydrology* **2025**, *12*, 56. [CrossRef]
28. Constantin-Horia, B.; Simona, S.; Gabriela, P.; Adrian, S. Human Factors in the Floods of Romania. In Proceedings of the Threats to Global Water Security; Jones, J.A.A., Vardanian, T.G., Hakopian, C., Eds.; Springer: Dordrecht, The Netherlands, 2009; pp. 187–192. [CrossRef]
29. Peptenatu, D.; Grecu, A.; Simion, A.G.; Gruia, K.A.; Andronache, I.; Draghici, C.C.; Diaconu, D.C. Deforestation and Frequency of Floods in Romania. In *Water Resources Management in Romania*; Negm, A.M., Romanescu, G., Zeleňáková, M., Eds.; Springer International Publishing: Cham, Switzerland, 2020; pp. 279–306. [CrossRef]
30. World Bank. *Romania Water Diagnostic Report: Moving Toward EU Compliance, Inclusion, and Water Security*; World Bank: Washington, DC, USA, 2018.
31. Cojoc, G.M.; Romanescu, G.; Tirnovan, A. Exceptional Floods on a Developed River: Case Study for the Bistrita River from the Eastern Carpathians (Romania). *Nat. Hazards* **2015**, *77*, 1421–1451. [CrossRef]
32. Stancalie, G.; Craciunescu, V.; Irimescu, A. Development of a Downstream Emergency Response Service for Flood and Related Risks in Romania Based on Satellite Data. *E3S Web Conf.* **2016**, *7*, 17007. [CrossRef]
33. Tudose, N.C.; Ungurean, C.; Davidescu, Ș.; Clinciu, I.; Marin, M.; Nita, M.D.; Adorjani, A.; Davidescu, A. Torrential Flood Risk Assessment and Environmentally Friendly Solutions for Small Catchments Located in the Romania Natura 2000 Sites Ciucas, Postavaru and Piatra Mare. *Sci. Total Environ.* **2020**, *698*, 134271. [CrossRef]
34. Romanescu, G.; Stoleriu, C.C. Exceptional Floods in the Prut Basin, Romania, in the Context of Heavy Rains in the Summer of 2010. *Nat. Hazard. Earth Syst. Sci.* **2017**, *17*, 381–396. [CrossRef]
35. Ionita, M.; Nagavciuc, V. Shedding Light on the Devastating Floods in June 1897 in Romania: Early Instrumental Observations and Synoptic Analysis. *J. Hydrometeorol.* **2024**, *25*, 1729–1745. [CrossRef]
36. Romania: Floods-DREF Operation N° MDRRO006—Romania | ReliefWeb. Available online: https://reliefweb.int/report/romania/romania-floods-dref-operation-ndeg-mdrro006 (accessed on 26 May 2025).
37. Floods in Central-Eastern Europe—September 2024. Available online: http://emergency.copernicus.eu/news/floods-in-central-eastern-europe-september-2024/ (accessed on 26 May 2025).
38. Rahmati, O.; Pourghasemi, H.R.; Zeinivand, H. Flood Susceptibility Mapping Using Frequency Ratio and Weights-of-Evidence Models in the Golastan Province, Iran. *Geocarto Int.* **2016**, *31*, 42–70. [CrossRef]
39. Sharma, A.; Poonia, M.; Rai, A.; Biniwale, R.B.; Tügel, F.; Holzbecher, E.; Hinkelmann, R. Flood Susceptibility Mapping Using GIS-Based Frequency Ratio and Shannon's Entropy Index Bivariate Statistical Models: A Case Study of Chandrapur District, India. *ISPRS Int. J. Geo-Inf.* **2024**, *13*, 297. [CrossRef]
40. Edamo, M.L.; Ayele, E.G.; Ukumo, T.Y.; Kassaye, A.A.; Haile, A.P. Capability of Logistic Regression in Identifying Flood-Susceptible Areas in a Small Watershed. *H2Open J.* **2024**, *7*, 351–374. [CrossRef]
41. Senan, C.P.C.; Ajin, R.S.; Danumah, J.H.; Costache, R.; Arabameri, A.; Rajaneesh, A.; Sajinkumar, K.S.; Kuriakose, S.L. Flood Vulnerability of a Few Areas in the Foothills of the Western Ghats: A Comparison of AHP and F-AHP Models. *Stoch. Environ. Res. Risk Assess.* **2023**, *37*, 527–556. [CrossRef]
42. Yariyan, P.; Avand, M.; Abbaspour, R.A.; Torabi Haghighi, A.; Costache, R.; Ghorbanzadeh, O.; Janizadeh, S.; Blaschke, T. Flood Susceptibility Mapping Using an Improved Analytic Network Process with Statistical Models. *Geomat. Nat. Hazard. Risk* **2020**, *11*, 2282–2314. [CrossRef]

43. Lyu, H.M.; Yin, Z.Y. Flood Susceptibility Prediction Using Tree-Based Machine Learning Models in the GBA. *Sustain. Cities Soc.* **2023**, *97*, 104744. [CrossRef]
44. Tehrany, M.S.; Pradhan, B.; Mansor, S.; Ahmad, N. Flood Susceptibility Assessment Using GIS-Based Support Vector Machine Model with Different Kernel Types. *CATENA* **2015**, *125*, 91–101. [CrossRef]
45. Chen, W.; Li, Y.; Xue, W.; Shahabi, H.; Li, S.; Hong, H.; Wang, X.; Bian, H.; Zhang, S.; Pradhan, B.; et al. Modeling Flood Susceptibility Using Data-Driven Approaches of Naïve Bayes Tree, Alternating Decision Tree, and Random Forest Methods. *Sci. Total Environ.* **2020**, *701*, 134979. [CrossRef]
46. Aydin, H.E.; Iban, M.C. Predicting and Analyzing Flood Susceptibility Using Boosting-Based Ensemble Machine Learning Algorithms with SHapley Additive ExPlanations. *Nat. Hazards* **2023**, *116*, 2957–2991. [CrossRef]
47. Yu, H.; Luo, Z.; Wang, L.; Ding, X.; Wang, S. Improving the Accuracy of Flood Susceptibility Prediction by Combining Machine Learning Models and the Expanded Flood Inventory Data. *Remote Sens.* **2023**, *15*, 3601. [CrossRef]
48. Wang, Y.; Fang, Z.; Hong, H.; Peng, L. Flood Susceptibility Mapping Using Convolutional Neural Network Frameworks. *J. Hydrol.* **2020**, *582*, 124482. [CrossRef]
49. Chen, C.; Wang, J.; Li, D.; Sun, X.; Zhang, J.; Yang, C.; Zhang, B. Unraveling Nonlinear Effects of Environment Features on Green View Index Using Multiple Data Sources and Explainable Machine Learning. *Sci. Rep.* **2024**, *14*, 30189. [CrossRef] [PubMed]
50. Mienye, I.D.; Sun, Y. A Survey of Ensemble Learning: Concepts, Algorithms, Applications, and Prospects. *IEEE Access* **2022**, *10*, 99129–99149. [CrossRef]
51. Ajin, R.S.; Segoni, S.; Fanti, R. Optimization of SVR and CatBoost Models Using Metaheuristic Algorithms to Assess Landslide Susceptibility. *Sci. Rep.* **2024**, *14*, 24851. [CrossRef]
52. Roy, D.K.; Sarkar, T.K.; Munmun, T.H.; Paul, C.R.; Datta, B. A Review on the Applications of Machine Learning and Deep Learning to Groundwater Salinity Modeling: Present Status, Challenges, and Future Directions. *Discov. Water* **2025**, *5*, 16. [CrossRef]
53. Ding, Y.; Zhu, H.; Chen, R.; Li, R. An Efficient AdaBoost Algorithm with the Multiple Thresholds Classification. *Appl. Sci.* **2022**, *12*, 5872. [CrossRef]
54. Hussain, S.S.; Zaidi, S.S.H. AdaBoost Ensemble Approach with Weak Classifiers for Gear Fault Diagnosis and Prognosis in DC Motors. *Appl. Sci.* **2024**, *14*, 3105. [CrossRef]
55. Sarker, I.H. Machine Learning: Algorithms, Real-World Applications and Research Directions. *SN Comput. Sci.* **2021**, *2*, 160. [CrossRef]
56. Ganie, S.M.; Pramanik, P.K.D.; Zhao, Z. Ensemble Learning with Explainable AI for Improved Heart Disease Prediction Based on Multiple Datasets. *Sci. Rep.* **2025**, *15*, 13912. [CrossRef]
57. Rizkallah, L.W. Enhancing the Performance of Gradient Boosting Trees on Regression Problems. *J. Big Data* **2025**, *12*, 35. [CrossRef]
58. Abujayyab, S.K.M.; Kassem, M.M.; Khan, A.A.; Wazirali, R.; Coşkun, M.; Taşoğlu, E.; Öztürk, A.; Toprak, F. Wildfire Susceptibility Mapping Using Five Boosting Machine Learning Algorithms: The Case Study of the Mediterranean Region of Turkey. *Adv. Civ. Eng.* **2022**, *2022*, 3959150. [CrossRef]
59. Deng, J.; Ji, W.; Liu, H.; Li, L.; Wang, Z.; Hu, Y.; Wang, Y.; Zhou, Y. Development and Validation of a Machine Learning-Based Framework for Assessing Metabolic-Associated Fatty Liver Disease Risk. *BMC Public Health* **2024**, *24*, 2545. [CrossRef] [PubMed]
60. Nguyen, N.; Ngo, D. Comparative Analysis of Boosting Algorithms for Predicting Personal Default. *Cogent Econ. Fin.* **2025**, *13*, 2465971. [CrossRef]
61. Hajji, S.; Krimissa, S.; Abdelrahman, K.; Boudhar, A.; Elaloui, A.; Ismaili, M.; El Bouzekraoui, M.; Chikh Essbiti, M.; Kahal, A.Y.; Mondal, B.K.; et al. Enhancing Flood Prediction through Remote Sensing, Machine Learning, and Google Earth Engine. *Front. Water* **2025**, *7*, 1514047. [CrossRef]
62. Tian, J.; Chen, Y.; Yang, L.; Li, D.; Liu, L.; Li, J.; Tang, X. Enhancing Urban Flood Susceptibility Assessment by Capturing the Features of the Urban Environment. *Remote Sens.* **2025**, *17*, 1347. [CrossRef]
63. Shi, S.; Zhao, F.; Ren, X.; Meng, Z.; Dang, X.; Wu, X. Soil Infiltration Properties Are Affected by Typical Plant Communities in a Semi-Arid Desert Grassland in China. *Water* **2022**, *14*, 3301. [CrossRef]
64. Zhang, H.; Liu, Q.; Liu, S.; Li, J.; Geng, J.; Wang, L. Key Soil Properties Influencing Infiltration Capacity after Long-Term Straw Incorporation in a Wheat (*Triticum Aestivum L.*)—Maize (*Zea Mays L.*) Rotation System. *Agric. Ecosyst. Environ.* **2023**, *344*, 108301. [CrossRef]
65. Wang, D.; Chen, J.; Tang, Z.; Zhang, Y. Effects of Soil Physical Properties on Soil Infiltration in Forest Ecosystems of Southeast China. *Forests* **2024**, *15*, 1470. [CrossRef]
66. Popa, M.C.; Peptenatu, D.; Drăghici, C.C.; Diaconu, D.C. Flood Hazard Mapping Using the Flood and Flash-Flood Potential Index in the Buzău River Catchment, Romania. *Water* **2019**, *11*, 2116. [CrossRef]
67. Costache, R.; Arabameri, A.; Elkhrachy, I.; Ghorbanzadeh, O.; Pham, Q.B. Detection of Areas Prone to Flood Risk Using State-of-the-Art Machine Learning Models. *Geomat. Nat. Hazards Risk* **2021**, *12*, 1488–1507. [CrossRef]

68. Costache, R.; Popa, M.C.; Tien Bui, D.; Diaconu, D.C.; Ciubotaru, N.; Minea, G.; Pham, Q.B. Spatial Predicting of Flood Potential Areas Using Novel Hybridizations of Fuzzy Decision-Making, Bivariate Statistics, and Machine Learning. *J. Hydrol.* **2020**, *585*, 124808. [CrossRef]
69. Nguyen, Q.H.; Ly, H.-B.; Ho, L.S.; Al-Ansari, N.; Le, H.V.; Tran, V.Q.; Prakash, I.; Pham, B.T. Influence of Data Splitting on Performance of Machine Learning Models in Prediction of Shear Strength of Soil. *Math. Probl. Eng.* **2021**, *2021*, 4832864. [CrossRef]
70. Segoni, S.; Ajin, R.S.; Nocentini, N.; Fanti, R. Insights Gained from the Review of Landslide Susceptibility Assessment Studies in Italy. *Remote Sens.* **2024**, *16*, 4491. [CrossRef]
71. Kaya, C.M.; Derin, L. Parameters and Methods Used in Flood Susceptibility Mapping: A Review. *J. Water Clim. Chang.* **2023**, *14*, 1935–1960. [CrossRef]
72. Mabdeh, A.N.; Ajin, R.S.; Razavi-Termeh, S.V.; Ahmadlou, M.; Al-Fugara, A. Enhancing the Performance of Machine Learning and Deep Learning-Based Flood Susceptibility Models by Integrating Grey Wolf Optimizer (GWO) Algorithm. *Remote Sens.* **2024**, *16*, 2595. [CrossRef]
73. Islam, T.; Zeleke, E.B.; Afroz, M.; Melesse, A.M. A Systematic Review of Urban Flood Susceptibility Mapping: Remote Sensing, Machine Learning, and Other Modeling Approaches. *Remote Sens.* **2025**, *17*, 524. [CrossRef]
74. Suharyadi, R.; Umarhadi, D.A.; Awanda, D.; Widyatmanti, W. Exploring Built-Up Indices and Machine Learning Regressions for Multi-Temporal Building Density Monitoring Based on Landsat Series. *Sensors* **2022**, *22*, 4716. [CrossRef]
75. Oñate-Valdivieso, F.; Oñate-Paladines, A.; Collaguazo, M. Spatiotemporal Dynamics of Soil Impermeability and Its Impact on the Hydrology of an Urban Basin. *Land* **2022**, *11*, 250. [CrossRef]
76. Cao, R.; Feng, Y.; Liu, X.; Shen, M.; Zhou, J. Uncertainty of Vegetation Green-Up Date Estimated from Vegetation Indices Due to Snowmelt at Northern Middle and High Latitudes. *Remote Sens.* **2020**, *12*, 190. [CrossRef]
77. Laonamsai, J.; Julphunthong, P.; Saprathet, T.; Kimmany, B.; Ganchanasuragit, T.; Chomcheawchan, P.; Tomun, N. Utilizing NDWI, MNDWI, SAVI, WRI, and AWEI for Estimating Erosion and Deposition in Ping River in Thailand. *Hydrology* **2023**, *10*, 70. [CrossRef]
78. Xiang, K.; Yuan, W.; Wang, L.; Deng, Y. An LSWI-Based Method for Mapping Irrigated Areas in China Using Moderate-Resolution Satellite Data. *Remote Sens.* **2020**, *12*, 4181. [CrossRef]
79. Chan, J.Y.L.; Leow, S.M.H.; Bea, K.T.; Cheng, W.K.; Phoong, S.W.; Hong, Z.W.; Chen, Y.L. Mitigating the Multicollinearity Problem and Its Machine Learning Approach: A Review. *Mathematics* **2022**, *10*, 1283. [CrossRef]
80. Odhiambo Omuya, E.; Onyango Okeyo, G.; Waema Kimwele, M. Feature Selection for Classification Using Principal Component Analysis and Information Gain. *Expert Syst. Appl.* **2021**, *174*, 114765. [CrossRef]
81. Qu, K.; Xu, J.; Hou, Q.; Qu, K.; Sun, Y. Feature Selection Using Information Gain and Decision Information in Neighborhood Decision System. *Appl. Soft Comp.* **2023**, *136*, 110100. [CrossRef]
82. Sinha, A.; Nikhil, S.; Ajin, R.S.; Danumah, J.H.; Saha, S.; Costache, R.; Rajaneesh, A.; Sajinkumar, K.S.; Amrutha, K.; Johny, A.; et al. Wildfire Risk Zone Mapping in Contrasting Climatic Conditions: An Approach Employing AHP and F-AHP Models. *Fire* **2023**, *6*, 44. [CrossRef]
83. Kim, J.H. Multicollinearity and Misleading Statistical Results. *Korean J. Anesthesiol.* **2019**, *72*, 558–569. [CrossRef]
84. Dar, I.S.; Chand, S.; Shabbir, M.; Kibria, B.M.G. Condition-Index Based New Ridge Regression Estimator for Linear Regression Model with Multicollinearity. *Kuwait J. Sci.* **2023**, *50*, 91–96. [CrossRef]
85. Huang, L.; Zhou, X.; Shi, L.; Gong, L. Time Series Feature Selection Method Based on Mutual Information. *Appl. Sci.* **2024**, *14*, 1960. [CrossRef]
86. Sulaiman, M.A.; Labadin, J. Feature Selection with Mutual Information for Regression Problems. In Proceedings of the 2015 9th International Conference on IT in Asia (CITA), Sarawak, Malaysia, 4–5 August 2015; pp. 1–6. [CrossRef]
87. Vergara, J.R.; Estévez, P.A. A Review of Feature Selection Methods Based on Mutual Information. *Neural Comput. Appl.* **2014**, *24*, 175–186. [CrossRef]
88. Prasetiyowati, M.I.; Maulidevi, N.U.; Surendro, K. Determining Threshold Value on Information Gain Feature Selection to Increase Speed and Prediction Accuracy of Random Forest. *J. Big Data* **2021**, *8*, 84. [CrossRef]
89. Omer, Z.M.; Shareef, H. Comparison of Decision Tree Based Ensemble Methods for Prediction of Photovoltaic Maximum Current. *Energy Convers. Manag. X* **2022**, *16*, 100333. [CrossRef]
90. Khan, A.A.; Chaudhari, O.; Chandra, R. A Review of Ensemble Learning and Data Augmentation Models for Class Imbalanced Problems: Combination, Implementation and Evaluation. *Expert Syst. Appl.* **2024**, *244*, 122778. [CrossRef]
91. Levent, İ.; Şahin, G.; Işık, G.; van Sark, W.G.J.H.M. Comparative Analysis of Advanced Machine Learning Regression Models with Advanced Artificial Intelligence Techniques to Predict Rooftop PV Solar Power Plant Efficiency Using Indoor Solar Panel Parameters. *Appl. Sci.* **2025**, *15*, 3320. [CrossRef]
92. Shen, F.; Jha, I.; Isleem, H.F.; Almoghayer, W.J.K.; Khishe, M.; Elshaarawy, M.K. Advanced Predictive Machine and Deep Learning Models for Round-Ended CFST Column. *Sci. Rep.* **2025**, *15*, 6194. [CrossRef]

93. Boldini, D.; Grisoni, F.; Kuhn, D.; Friedrich, L.; Sieber, S.A. Practical Guidelines for the Use of Gradient Boosting for Molecular Property Prediction. *J. Cheminform.* **2023**, *15*, 73. [CrossRef]
94. So, B. Enhanced Gradient Boosting for Zero-Inflated Insurance Claims and Comparative Analysis of CatBoost, XGBoost, and LightGBM. *Scandinav. Actuar. J.* **2024**, *10*, 1013–1035. [CrossRef]
95. Karunasingha, D.S.K. Root Mean Square Error or Mean Absolute Error? Use Their Ratio as Well. *Inform. Sci.* **2022**, *585*, 609–629. [CrossRef]
96. Romeo, G. Chapter 13—Data Analysis for Business and Economics. In *Elements of Numerical Mathematical Economics with Excel*; Romeo, G., Ed.; Academic Press: Boston, MA, USA, 2020; pp. 695–761. [CrossRef]
97. Ross, S.M. Chapter 12—Linear Regression. In *Introductory Statistics*, 3rd ed.; Ross, S.M., Ed.; Academic Press: Boston, MA, USA, 2010; pp. 537–604. [CrossRef]
98. AlZoman, R.M.; Alenazi, M.J.F. A Comparative Study of Traffic Classification Techniques for Smart City Networks. *Sensors* **2021**, *21*, 4677. [CrossRef]
99. Feizizadeh, B.; Darabi, S.; Blaschke, T.; Lakes, T. QADI as a New Method and Alternative to Kappa for Accuracy Assessment of Remote Sensing-Based Image Classification. *Sensors* **2022**, *22*, 4506. [CrossRef]
100. Nayak, R.; Pati, U.C.; Das, S.K. A Comprehensive Review on Deep Learning-Based Methods for Video Anomaly Detection. *Image Vis. Comput.* **2021**, *106*, 104078. [CrossRef]
101. Nahm, F.S. Receiver Operating Characteristic Curve: Overview and Practical Use for Clinicians. *Korean J. Anesthesiol.* **2022**, *75*, 25–36. [CrossRef] [PubMed]
102. Melo, F. Area under the ROC Curve. In *Encyclopedia of Systems Biology*; Dubitzky, W., Wolkenhauer, O., Cho, K.-H., Yokota, H., Eds.; Springer: New York, NY, USA, 2013; pp. 38–39. [CrossRef]
103. Fu, G.-H.; Xu, F.; Zhang, B.-Y.; Yi, L.-Z. Stable Variable Selection of Class-Imbalanced Data with Precision-Recall Criterion. *Chemom. Intell. Lab. Syst.* **2017**, *171*, 241–250. [CrossRef]
104. Saito, T.; Rehmsmeier, M. The Precision-Recall Plot Is More Informative than the ROC Plot When Evaluating Binary Classifiers on Imbalanced Datasets. *PLoS ONE* **2015**, *10*, e0118432. [CrossRef]
105. Rasheed, K.; Qayyum, A.; Ghaly, M.; Al-Fuqaha, A.; Razi, A.; Qadir, J. Explainable, Trustworthy, and Ethical Machine Learning for Healthcare: A Survey. *Comput. Biol. Med.* **2022**, *149*, 106043. [CrossRef]
106. Keleko, A.T.; Kamsu-Foguem, B.; Ngouna, R.H.; Tongne, A. Health Condition Monitoring of a Complex Hydraulic System Using Deep Neural Network and DeepSHAP Explainable XAI. *Adv. Eng. Softw.* **2023**, *175*, 103339. [CrossRef]
107. Huang, X.; Kroening, D.; Ruan, W.; Sharp, J.; Sun, Y.; Thamo, E.; Wu, M.; Yi, X. A Survey of Safety and Trustworthiness of Deep Neural Networks: Verification, Testing, Adversarial Attack and Defence, and Interpretability. *Comput. Sci. Rev.* **2020**, *37*, 100270. [CrossRef]
108. Mangalathu, S.; Hwang, S.-H.; Jeon, J.-S. Failure Mode and Effects Analysis of RC Members Based on Machine-Learning-Based SHapley Additive ExPlanations (SHAP) Approach. *Eng. Struct.* **2020**, *219*, 110927. [CrossRef]
109. Khodaei, H.; Nasiri Saleh, F.; Nobakht Dalir, A.; Zarei, E. Future Flood Susceptibility Mapping under Climate and Land Use Change. *Sci. Rep.* **2025**, *15*, 12394. [CrossRef]
110. Tariq, A.; Yan, J.; Ghaffar, B.; Qin, S.; Mousa, B.G.; Sharifi, A.; Huq, M.E.; Aslam, M. Flash Flood Susceptibility Assessment and Zonation by Integrating Analytic Hierarchy Process and Frequency Ratio Model with Diverse Spatial Data. *Water* **2022**, *14*, 3069. [CrossRef]
111. Al-Kindi, K.M.; Alabri, Z. Investigating the Role of the Key Conditioning Factors in Flood Susceptibility Mapping Through Machine Learning Approaches. *Earth Syst. Environ.* **2024**, *8*, 63–81. [CrossRef]
112. Suwanno, P.; Yaibok, C.; Pornbunyanon, T.; Kanjanakul, C.; Buathongkhue, C.; Tsumita, N.; Fukuda, A. GIS-Based Identification and Analysis of Suitable Evacuation Areas and Routes in Flood-Prone Zones of Nakhon Si Thammarat Municipality. *IATSS Resear.* **2023**, *47*, 416–431. [CrossRef]
113. Wang, Z.; Chen, X.; Qi, Z.; Cui, C. Flood Sensitivity Assessment of Super Cities. *Sci. Rep.* **2023**, *13*, 5582. [CrossRef] [PubMed]
114. Agarwal, D.S.; Bharat, A. Nature-Based Solutions for Flood–Drought Mitigation Using a Composite Framework: A Case-Based Approach. *J. Water Clim. Chang.* **2023**, *14*, 778–795. [CrossRef]
115. Ma, S.; Wang, L.-J.; Jiang, J.; Zhao, Y.-G. Land Use/Land Cover Change and Soil Property Variation Increased Flood Risk in the Black Soil Region, China, in the Last 40 Years. *Environ. Impact Assess. Rev.* **2024**, *104*, 107314. [CrossRef]
116. Saco, P.M.; McDonough, K.R.; Rodriguez, J.F.; Rivera-Zayas, J.; Sandi, S.G. The Role of Soils in the Regulation of Hazards and Extreme Events. *Philos. Trans. R. Soc. B* **2021**, *376*, 20200178. [CrossRef]
117. Mileusnić, Z.I.; Saljnikov, E.; Radojević, R.L.; Petrović, D.V. Soil Compaction Due to Agricultural Machinery Impact. *J. Terramech.* **2022**, *100*, 51–60. [CrossRef]
118. Singh, O.; Shahi, U.P.; Dutta, D.; Shivangi; Rajput, V.D.; Singh, A. Strategic Tillage for Improved Soil Health and Nutrient Dynamics. In *Strategic Tillage and Soil Management—New Perspectives*; de Sousa, R.N., Ed.; IntechOpen: London, UK, 2024. [CrossRef]

119. Szejba, D. Importance of the Influence of Drained Clay Soil Retention Properties on Flood Risk Reduction. *Water* **2020**, *12*, 1315. [CrossRef]
120. Zhang, F.; Gao, Y. Composite Extraction Index to Enhance Impervious Surface Information in Remotely Sensed Imagery. *Egypt. J. Remote Sens. Sp. Sci.* **2023**, *26*, 141–150. [CrossRef]
121. Öztürk, Ş.; Yılmaz, K.; Dinçer, A.E.; Kalpakcı, V. Effect of Urbanization on Surface Runoff and Performance of Green Roofs and Permeable Pavement for Mitigating Urban Floods. *Nat. Hazard* **2024**, *120*, 12375–12399. [CrossRef]
122. Shrestha, S.; Dahal, D.; Poudel, B.; Banjara, M.; Kalra, A. Flood Susceptibility Analysis with Integrated Geographic Information System and Analytical Hierarchy Process: A Multi-Criteria Framework for Risk Assessment and Mitigation. *Water* **2025**, *17*, 937. [CrossRef]
123. Pistocchi, A.; Calzolari, C.; Malucelli, F.; Ungaro, F. Soil Sealing and Flood Risks in the Plains of Emilia-Romagna, Italy. *J. Hydrol. Reg. Stud.* **2015**, *4*, 398–409. [CrossRef]
124. Frene, J.P.; Pandey, B.K.; Castrillo, G. Under Pressure: Elucidating Soil Compaction and Its Effect on Soil Functions. *Plant Soil* **2024**, *502*, 267–278. [CrossRef]
125. Ashfaq, S.; Tufail, M.; Niaz, A.; Muhammad, S.; Alzahrani, H.; Tariq, A. Flood Susceptibility Assessment and Mapping Using GIS-Based Analytical Hierarchy Process and Frequency Ratio Models. *Glob. Planet. Chang.* **2025**, *251*, 104831. [CrossRef]
126. Nedkov, R. Normalized Differential Greenness Index for Vegetation Dynamics Assessment. *Comptes Rendus De L'academie Bulg. Des Sci.* **2017**, *70*, 1143–1146.
127. Xu, J.; Tang, Y.; Xu, J.; Chen, J.; Bai, K.; Shu, S.; Yu, B.; Wu, J.; Huang, Y. Evaluation of Vegetation Indexes and Green-Up Date Extraction Methods on the Tibetan Plateau. *Remote Sens.* **2022**, *14*, 3160. [CrossRef]
128. Hidayat, M.; Djufri, D.; Basri, H.; Ismail, N.; Idroes, R.; Ikhwali, M.F. Influence of Vegetation Type on Infiltration Rate and Capacity at Ie Jue Geothermal Manifestation, Mount Seulawah Agam, Indonesia. *Heliyon* **2024**, *10*, e25783. [CrossRef]
129. Firoozi, A.A.; Firoozi, A.A. Water Erosion Processes: Mechanisms, Impact, and Management Strategies. *Result. Eng.* **2024**, *24*, 103237. [CrossRef]
130. Sahin, E.K. Comparative Analysis of Gradient Boosting Algorithms for Landslide Susceptibility Mapping. *Geocarto Int.* **2022**, *37*, 2441–2465. [CrossRef]
131. Szczepanek, R. Daily Streamflow Forecasting in Mountainous Catchment Using XGBoost, LightGBM and CatBoost. *Hydrology* **2022**, *9*, 226. [CrossRef]
132. Yavuz Ozalp, A.; Akinci, H.; Zeybek, M. Comparative Analysis of Tree-Based Ensemble Learning Algorithms for Landslide Susceptibility Mapping: A Case Study in Rize, Turkey. *Water* **2023**, *15*, 2661. [CrossRef]
133. Heddam, S. Explainability of Machine Learning Using Shapley Additive ExPlanations (SHAP): CatBoost, XGBoost and LightGBM for Total Dissolved Gas Prediction. In *Machine Learning and Granular Computing: A Synergistic Design Environment*; Pedrycz, W., Chen, S.-M., Eds.; Springer: Cham, Switzerland, 2024; pp. 1–25. [CrossRef]
134. Hancock, J.T.; Khoshgoftaar, T.M. CatBoost for Big Data: An Interdisciplinary Review. *J. Big Data* **2020**, *7*, 94. [CrossRef] [PubMed]
135. Kharazi Esfahani, P.; Peiro Ahmady Langeroudy, K.; Khorsand Movaghar, M.R. Enhanced Machine Learning—Ensemble Method for Estimation of Oil Formation Volume Factor at Reservoir Conditions. *Sci. Rep.* **2023**, *13*, 15199. [CrossRef] [PubMed]
136. Onyelowe, K.C.; Kamchoom, V.; Hanandeh, S.; Ebid, A.M.; Llamuca Llamuca, J.L.; Cayán Martínez, J.C.; Rose, E.; Awoyera, P.; Avudaiappan, S. Predicting the Strengths of Basalt Fiber Reinforced Concrete Mixed with Fly Ash Using AML and Hoffman and Gardener Techniques. *Sci. Rep.* **2025**, *15*, 12074. [CrossRef] [PubMed]
137. Ileri, K. Comparative Analysis of CatBoost, LightGBM, XGBoost, RF, and DT Methods Optimised with PSO to Estimate the Number of k-Barriers for Intrusion Detection in Wireless Sensor Networks. *Int. J. Mach. Learn. Cyber.* **2025**. [CrossRef]
138. Chakraborty, D.; Ghosh, A.; Saha, S. Chapter 2—A Survey on Internet-of-Thing Applications Using Electroencephalogram. In *Emergence of Pharmaceutical Industry Growth with Industrial IoT Approach*; Balas, V.E., Solanki, V.K., Kumar, R., Eds.; Academic Press: Boston, MA, USA, 2020; pp. 21–47. [CrossRef]

water

MDPI

Article

Evolution of Nuntași-Tuzla Lake Chemistry in the Context of Human Intervention

Gabriel Dobrica [1], Carmen Elena Maftei [2,*], Ionela Carazeanu Popovici [3,*] and Naliana Lupascu [4]

1 National Administration "Romanian Waters"-Dobrogea Littoral, 900592 Constanta, Romania; gabriel.dobrica88@gmail.com
2 Civil Engineering Faculty, Transilvania University of Brasov, 500036 Brasov, Romania
3 Applied Science and Engineering Faculty, Ovidius University of Constanta, 900527 Constanta, Romania
4 National Institute for Research and Development of Marine Geology and Geoecology, 900581 Constanta, Romania; naliana.lupascu@geoecomar.ro
* Correspondence: cemaftei@gmail.com (C.E.M.); ionela.popovici@univ-ovidius.ro (I.C.P.)

Abstract: This paper analyzes the chemical evolution of Nuntași-Tuzla Lake (Romania) in the context of human intervention. Situated on the shore of the Black Sea, approximately 35 km north of Constanța, Nuntași-Tuzla Lake is part of the Razim–Sinoe Lake complex and a component of the Danube Delta Biosphere Reserve. This area has undergone significant transformations over the past 120 years: canalization of the connecting channels with the St. George arm, construction of polders for agriculture, closure of the connections to the Black Sea, and construction of the Razim–Sinoe irrigation system. After the irrigation system stopped working (around 2000), due to the isolation of the lake and the low flow coming from the two rivers that supply the lake with fresh water, it completely dried up in 2020. All these interventions have led to the ecological, hydrological, and chemical deterioration of the lake's water. The main effects are (i) a decrease in water salinity and (ii) reduction in the production of sapropelic mud as the salinity decreases due to the influx of fresh water.

Keywords: Nuntași-Tuzla Lake (Romania); human intervention; hydrochemical characteristics; water quality index

1. Introduction

The Water Framework Directive (WFD), formally known as Directive 2000/60/EC, is a key piece of European Union legislation that establishes a framework for protecting and managing water resources across the EU. The directive aims to achieve "good status" for all EU waters by setting out a comprehensive, integrated, and coordinated approach to water management [1,2]. The WFD aims to bring all water bodies, including rivers, lakes, coastal waters, and groundwater, to a "good status" by a set deadline (up to 2027). The directive mandates the prevention of any further deterioration in the status of water bodies, ensuring that their ecological and chemical quality is maintained or even improved [2,3]. As a result, it is necessary to (i) prevent any further deterioration of the state of water bodies, ensuring that their ecological and chemical quality is maintained or improved; (ii) encourage sustainable practices in using water to achieve a balance between the needs of the population, protection of the environment, and the development of economic activities; (iii) protect and improve aquatic ecosystems and, where necessary, improve the water quality through specific action plans; (iv) promote transparency by involving the public in water management decisions; and (v) achieve water management at the level of the hydrographic basin, crossing political borders, in order to manage water resources holistically [2,3].

Romania, as a member of the European Union, has integrated the WFD into its national legislation and established a monitoring network to regularly determine the chemical, ecological, and quantitative status of its surface and groundwater bodies. Romania's implementation of the Water Framework Directive represents a significant step towards sustainable water management and environmental protection. While progress has been made in aligning national legislation with EU standards, ongoing efforts are needed to address challenges related to infrastructure, enforcement, and the impacts of climate change. Public participation and continuous monitoring remain critical components of Romania's strategy to meet the WFD's objectives and protect its water resources for future generations [3].

After the second report of Basin Management Plan (BMP) required by the WFD [2], the European Environment Agency (EEA) presented a report in which the authors [1] drew attention to the fact that only 40% of surface waters have reached a good ecological condition. After the third BMP, 55.9% of surface waters of Europe have reached a good ecological condition (WISE, 2024), which means an increase of 16%, approximately [4].

There are 1988 water bodies, from a total of 2014 water bodies, that have an unchanged status. From the second to the third BMP report, there were not too many changes regarding the ecological condition of the lakes (188 lakes from a total of 189 maintain the same status).

Nuntași-Tuzla Lake is a significant brackish water body located in the Dobrogea region of Romania, near the Black Sea coast. Over the centuries, the lake has undergone various natural and anthropogenic changes that have shaped its current state. Understanding the history of Nuntași-Tuzla Lake reveals an insight into its ecological importance, the challenges faced, and the efforts undertaken for its preservation and restoration [5]. Nuntași-Tuzla Lake is part of 32% of Romanian lakes that have not achieved "good" status recently [4].

The main objective of this paper is to determine the water quality using both the water quality index (WQI) and principles of the WFD. The specific objectives are (i) investigation of the limits and advantages of the Water Frame Directive (WFD), (ii) review of the evolution of the WQI, and (iii) review of human intervention in the Nuntași-Tuzla Lake basin. Reviewing the literature related to WHD implementation in Romania gives the possibility to identify gaps in the knowledge and can bring some novelty in research by using other methodologies, such as WQI.

2. Materials and Methods

2.1. Methodology and Methods

The flowchart of the methodology proposed in this paper is represented in Figure 1. The qualitative methods consist of (i) a discussion based on the specific literature of the Water Framework Directive (WFD) implementation into Romanian legislation related to water quality, its limits, and advantages, and (ii) the water quality index (WQI) method. The results are presented in the Results (Section 3.1). The quantitative methods consist of data collection and analysis, and the determination of water quality is conducted through two methods, WQI index calculation and WFD principle application. The two methods are presented in Section 2.2 Methods. The results obtained will be compared.

2.1.1. Study Area

The presented methodology will be applied to Nuntași-Tuzla Lake. Part of the Razim–Sinoe lacustrine system, Nuntași-Tuzla Lake is situated in the Dobrogea region on the Black Sea coast, 35 km north of Constanta. In ancient times, the Halmyris Bay was located here, on the shore of which a series of fortresses was built. The area has an important cultural and spiritual value due to the existence of the Histria citadel (which the ancient Milesian colony created in VII BC century) and the discovery of artifacts from the late Roman era in the Nuntași and Săcele areas. Nuntași-Tuzla Lake was once well known as the second therapeutic lake, after the Techirghiol Lake, for its sapropelic mud.

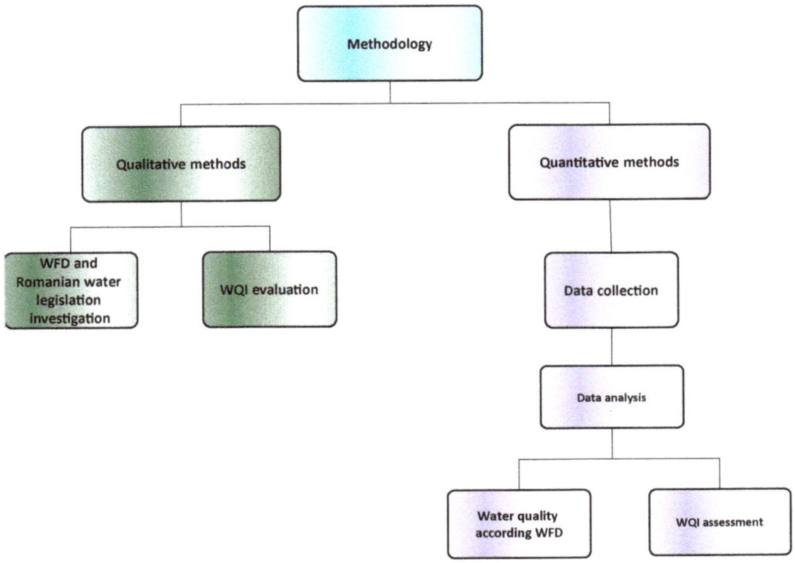

Figure 1. Methodology flowchart used in this paper.

The climate is a temperate continental one with maritime influences. During the cold period of the year, influenced by the Black Sea, the temperature remains positive [6–9]. During the warm period of the year, the climate is influenced by the sea breeze. According to Dobrica et al. [6], for the 1965–2021 period, the multiannual precipitation was 458 mm, and the multiannual average of temperatures at the Jurilovca meteorological station was 11.3 °C.

The area studied is part of the Central Dobrogea Plateau, a subunit of the Istrian Plateau, and has a coastal plain aspect (Prispa Hamangia) developed between the Casimcea Plateau and the Black Sea, being a combination of sea plains (Sacele and Chituc), sandy barriers, and shallow lakes (Sinoe, Histria, and Nuntași). The DEM (digital elevation model) is presented in Figure 2.

The watershed of the lake is located at low altitudes (average altitude is 100.6 m, with a maximum of 218 m). The relief is stepped. The spaces between the valleys have a form of a high plain whose slopes have an eastern and southeastern direction. The average terrain slope is 3.7%. The watershed analyzed is a "mature" one, the relief is somewhat stable, the soil erosion or landslide processes are completed, and the material has been completely evacuated outside the basin [6].

From a hydrological point of view, the Dobrogea region is divided into two large basins: BH Littoral (71%) and BH Danube (29%). Nuntași-Tuzla Lake basin (code: ROLN05) is part of a littoral basin, located at the confluence of two rivers, Nutași and Săcele. According to Dobrica et al. [6], the annual Nuntași River discharge is 0.348 m^3/s, and that of the Sacele River is 0.082 m^3/s.

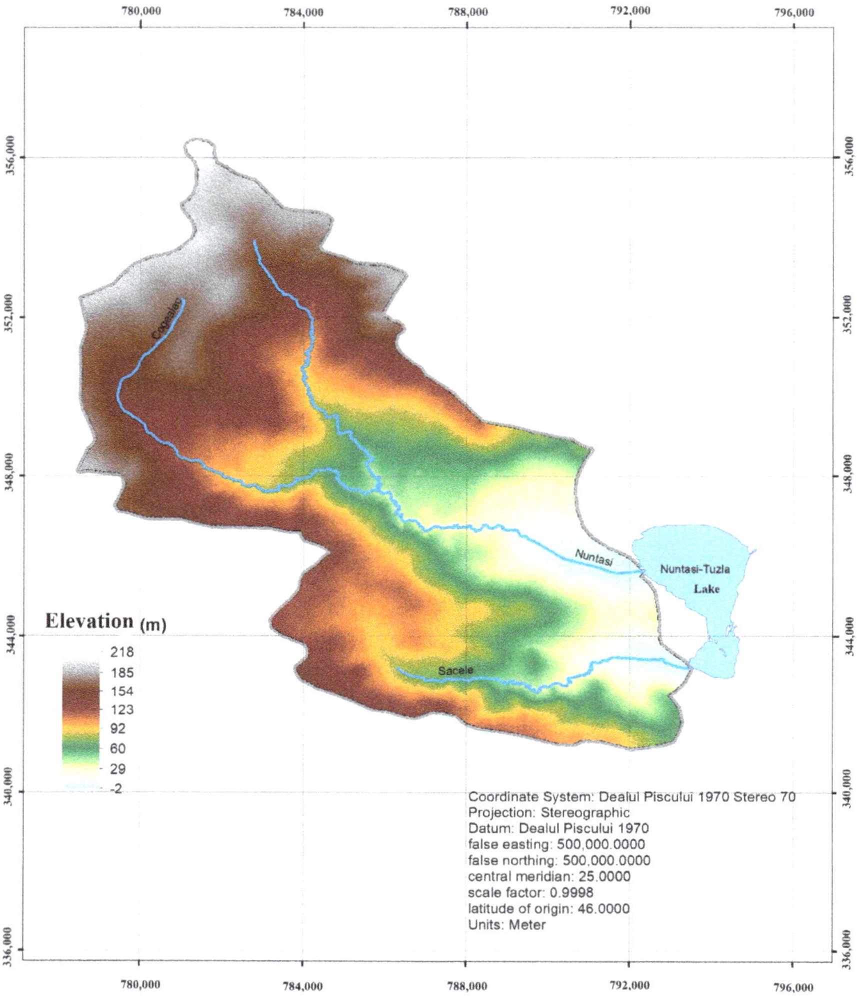

Figure 2. DEM of Nuntași-Tuzla Lake basin [6].

2.1.2. Data Series

According to SMIAR (Romanian Integrated Water Monitoring System), in the Dobrogea region, there are the following monitoring profiles: (i) Danube basin—23 monitoring points for rivers, 11 monitoring points for 9 lakes; (ii) Littoral basin—19 rivers, 11 lakes—33 monitoring stations; and (iii) 41 groundwater monitoring stations [10].

As we already mentioned, Nuntași-Tuzla Lake is situated in the Littoral basin, and starting in 1995, there were three monitoring points (Center Lake, camping baths, and Nuntași River connection). A data series was provided by the Romanian Water National Administration—Dobrogea Littoral Branch (RWNA-DLB). The records cover the 1981–2022 period, with some gaps (1983–1986, 1988–1994) due to unsystematic monitoring of the lakes. After 1998, the monitoring of chemical and biological parameters became a systematic one [11,12].

The following parameters are generally available: (1) physical, thermal, and acidification indicators, (2) oxygenation regimen (dissolved oxygen, CBO_5, CCOMn), (3) nutrients,

(4) salinity condition (only until 2009), (5) toxic pollutants (only since 2010), and (6) indicators of the eutrophication degree. Not all the parameters were measured in all the time periods and for all the sampling points. A discussion on the measured parameters will be presented in the Results section. Initially, samples were taken from 6–7 points of the lake. Since 1995, the number of sampling points has been reduced to three.

Some historical data related to chloride (Cl^-), sulfate (SO_4^{-2}), calcium (Ca^{2+}), magnesium (Mg^{2+}), and bicarbonate (HCO_3^-) are available for different periods (e.g., 1934–1981). These data were collected from different documents [11–15].

2.2. Methods

In this work, two methods of classifying water quality are used: (i) the method stipulated in the WFD and implemented into Romanian legislation through the water law and other subsequent documents, explained in detail in Section 3.1, and (ii) the quality index of water (WQI).

2.2.1. WFD Method

WFD is the principal legislation of water policy in EU and introduces key principles for management and to ensure water quality in all EU countries. In this respect, a series of criteria/indicators is used to assess the status of the water bodies.

In Romania, the WFD requirements are implemented into legislation through the water law and subsequent documents. The ecological and chemical status are divided into 5 categories, from "High" to "Bad", and 5 categories of trophy. Based on Romanian legislation, for the period and available data, the following elements are established: (i) the degree of eutrophication, (ii) the physical and chemical state, (iii) the biological state, and (iv) the ecological state of Nuntași-Tuzla Lake.

The eutrophication establishment degree is based on the use of the following indicators: total phosphorus, total mineral nitrogen, total biomass, and chlorophyll "a" [16–21]. The physical state is established by the pH indicator. The chemical state is established based on physics and chemical elements, namely, oxygen regimen and salinity. The biological state is established based on total phosphorus, total mineral nitrogen, biomass phytoplanktonic (FTK), and chlorophyll "a". Starting in 2010, ABADL has provided a metrical index (IM) calculated for each biological element: phytoplankton, phytobenthos, and macroinvertebrates. The ecological state of Nuntași-Tuzla Lake is based on the "one out–all out" principle. This principle is applied to each indicator to establish the general ecological state. The "one out–all out" principle was imposed by the WFD to assess the ecological status of surface water. This principle expresses that the overall status is selected based on the worst ecological status.

The calculation of the indicators is performed differently, depending on the volume of data selected: if the length of the data series is less than 30, then the evaluation is based on the arithmetic mean of the measured values; if it is greater than 30, the evaluation is based on the 90-percentile value, respectively, 10 percentiles for "dissolved oxygen".

Additionally, for the 1981–2009 period, the state of salinity was determined based on the chloride, sulfate, calcium, and magnesium concentrations. For the 2010–2023 period, the priority substances state was determined.

2.2.2. WQI Method

The water quality index (WQI) is a valuable tool for monitoring and managing water resources, ensuring they remain safe for human use and ecological health. The WQI simplifies complex water quality data into a single number, making it easier for policymakers, scientists, and the public to understand a water body's health. WQI is typically calculated using various water quality parameters, each of which is assigned a weight based on its relative importance to overall water quality [22–25].

For this paper, based on data availability, the WQI selected is the one proposed by Brown [24], namely, "Weighted Arithmetic Water Quality Index Method" (WaWQI). According to the methods proposed, the equation used to calculate WQI is as follows:

$$WQI = \frac{\sum_{i=0}^{n} Q_i \cdot w_i}{\sum_{i=0}^{n} w_i} \quad (1)$$

where n is the number of parameters; Q_i is the evaluation scale; and w_i is the weight of each parameter.

The evaluation scale formula is as follows:

$$Q_i = \frac{V_i - V_0}{S_{ni} - V_0} \quad (2)$$

where V_i is the concentration value measured for each parameter considered; V_o represents the ideal value of concentration in pure water, which is 7 for pH and approximately 0 for all the other parameters; and S_{ni} is the standard concentration value for each parameter selected; in this case, we selected the value corresponding to "good" quality from the guideline.

The standard concentration value for the parameters selected in this paper is presented in Table 1, and the scale value for WQI is presented in Table 2. The determination of weight, w_i, is based on inverse proportionality to the standard concentration.

$$S_{ni} \ (w_i = {}^{k}/_{S_{ni}}) \quad (3)$$

where k_i is the constant of proportionality.

Table 1. Standard concentration (S_{ni}) value for selected parameters, in line with OM no. 161/2006, for "good" conditions.

pH	DO (mg/L)	BOD5 (mg/L)	CCOMn	Ammonium	Nitrate	N total	Phosphate
7	7	5	10	0.8	3	7	0.2
Total Phosphorus	Chloride	Sulfate	Ca	Mg	IM FPK	IM-B	IM-M
0.4	50	120	100	50	0.6	0.6	0.6

Table 2. Categories of WQI.

High/Excellent	Good	Moderate	Poor	Bad
≤25	25–50	50–75	75–100	≥100

3. Results

3.1. WFD and Romanian Legislation Related to Water Quality

The Water Framework Directive (WFD) represents a comprehensive and integrated approach to water management in the European Union. It aims to achieve good water status for all water bodies, promote sustainable use, and protect aquatic ecosystems through River Basin Management Plans, monitoring, and stakeholder involvement. Since it led to significant improvements in water quality and management, ongoing challenges require sustained efforts and adaptations to ensure the objectives are met. The implementation tool of the WFD, regulated by Article 13 and Annex VII, is represented by the management plan of the river basin (BMP). The technical documents, namely, "guidance" documents, are available on the CIRCABC (Communication and Information Resource Centre for Administrations, Businesses and Citizens) website (available online: https://circabc.europa.eu/ (accessed on 15 May 2024)). There are 34 guidance documents, and their role is to assist those involved in the WFD implementation. Figure 3 presents a schematic evolution of WFD application in Romania.

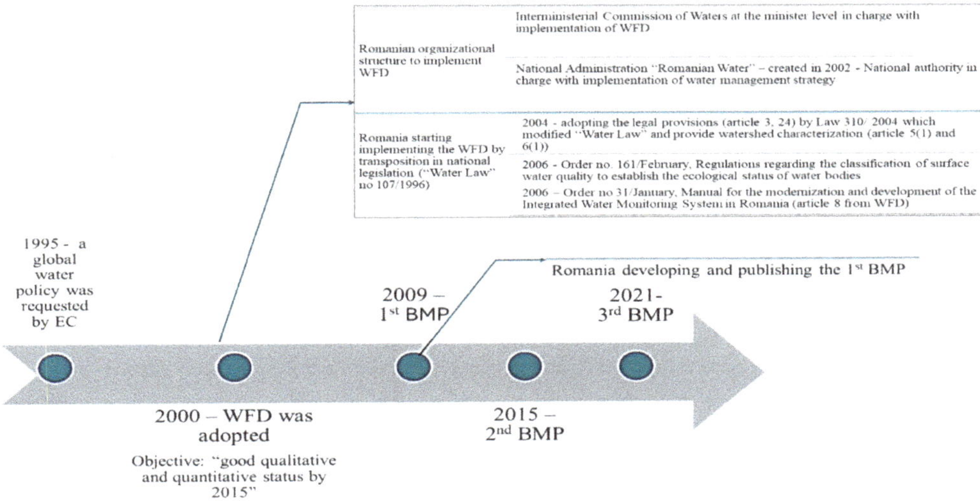

Figure 3. Schematic representation of WFD implementation stages in Romanian legislation. Legend: EC—European Commission, BMP—Basin Management Plan.

After the WFD adoption (2000), the Romanian administration started to modify the Water Law (no. 17/1996) and create the organizational structure to implement the directive. Introduction of the WFD in national legislation started in 2004 by Law 310/2004, which modified Water Law (no. 17/1996) when Romania adopted legal provisions (articles 3 and 24 from the WFD). The law no. 310/2004 also provides the (i) "Scheme of the watershed with their boundaries", (ii) conditions for achieving water and aquatic ecosystem protection objectives for all surface and groundwater bodies, (iii) a list of hazardous substances, (iv) a list of principal pollutants, and (v) economic analysis. In this way, the legal framework for the application of articles 4 (1 a and b) and 5 (1 and 2) was created.

Romania completed the Romanian Basin Management Plan (RBMP) (Article 13) and developed the monitoring program (Article 8) between 2004 and 2009. In the next period, several ministry orders were adopted to introduce (i) regulation regarding the surface water quality classification to establish the ecological status of water bodies and (ii) a manual to develop the Integrated Water Monitoring System in Romania (compliance of article 8). The Romanian Integrated Water Monitoring System (SMIAR) was organized by the water law and restructured according to the WFD.

In January 2006, a guide for the modernization and development of SMIAR was adopted to ensure the implementation of SMIAR through OM No. 31. According to the legislation adopted, two main types of monitoring are used within SMIAR: (i) observation monitoring and (ii) operational monitoring. Six subsystems are monitored: rivers, lakes, transitional and coastal waters, and groundwater. Additionally, in February 2006, a regulation regarding the classification of surface water quality was adopted to ensure the ecological establishment status of water bodies. Observation monitoring has the role of assessing the state of water bodies at the level of all water bodies and is carried out every year based on a management plan identified as not being at risk of not reaching the environmental objectives.

The operational program aims to establish the state of water bodies that present the risk of not meeting the environmental objectives. The quality elements and measurement frequency used to establish the state is well explained in both the Romanian legislation and the Romanian Basin Management Plan (RBMP—chapter 6) and refers to (i) biological elements; (ii) hydro-morphological elements; (iii) physics and chemical elements; (iv) priority substances; (v) priority substance—sediments; and (vi) priority substance—biota. The groups (iv), (v), and (vi) were introduced in line with Annex I of Directive 2008/105/EC, modified by Directive 2013/39/EU, article 3 (6), and article 3 (2), respectively, which were transposed into national legislation by H.G. 570/2016; they are measured only in the case of the existence of pollution sources and/or their identification within the screening analysis.

Based on those indicators, according to OM 161/2006, the ecological state of the water lake is divided into five categories: very good (I), good (II), moderate (III), poor (IV), and bad (V). For lakes, the degree of trophy will be established. OM 161/2006 also establishes five degrees of trophy for lakes, namely, ultraoligotrophic, oligotrophic, mesotrophic, eutrophic, and hypertrophic (article 1 (5)). According to article 3 (4), the elements of chemical and physical–chemical quality in the group "Salinity" represent substances of natural origin and do not indicate pollution.

The general principle for establishing the ecological status of water surfaces stipulated in Appendix 5 of the WFD is "one out–all out", which means that the most unfavorable situation is selected.

Due to the measures taken, the second BMP shows that Romania achieved a good status for 60–70% of the surface water bodies [1]. But 30–40% of the surface water (including lakes) still fail to achieve "good" ecological status. This is also the case with Nuntaşi-Tuzla Lake.

3.2. Human Intervention in Nuntași-Tuzla Lake Basin

The first information about the Nuntași-Tuzla Lake was provided by Ionescu MD [15]. Accordingly, this lake was called Tuzla at that time; the word means "with salt" in Turkish. He also mentioned a value of salinity of 26 mg/L. More than this, he stated that the main use of Lake Tuzla was the production of salt "by capture", given that the lake had salty water. Until 1897, there were five salt deposits on the shores of lake [15]. In 1924, Poruciuc [26] referred to the same lake by naming it the Tuzla–Sinoe or Duingi Lake (today Nuntași). He stated that the lake communicated with the Black Sea through Istria and Golovita Lakes through the mouth of Portița and had 2000 hectares, but it was still isolated from freshwater sources, having no freshwater springs. For this reason, having a small depth and a high salinity (unspecified) leads to the deposition of salt. He declared that a deposit of 200,000 kg of salt was found in the area, left by the Turks to the Romanian authorities, after the annexation of Dobrogea in 1879.

In 1928, Bratescu C [5] mentioned in his work that Tuzla Lake (the name is kept) communicated with the Sinoe Lake through a "mouth/gate" between the Istria hill in the north and the Histria fortress in the south. In his opinion, the surface of the lake was 1600 hectares, and the lake's water was salty. The first scientific information about the morphological characteristics of the Nuntaşi-Tuzla Lake was provided by Breier [11]. In her work, Braier established the morphometric characteristics of Nuntași-Tuzla Lake as follows: the surface is 1050 hectares; the water volume is $9.28{\cdot}106\ m^3$; the length is 6.2 km; the width varies between a minimum of 1.7 and a maximum of 3 km; the maximum elongation coefficient is 2.00; the average depth is 0.6 m, the maximum being 1.0 m; and the slope of the lake bottom is 0.0020 m/m. Breier [12] also provided some information about the water chemistry, classifying the lake as one with polyhaline waters with a salinity in the range of 10–17 g/L.

From this brief historical presentation, we conclude that the studied lake had water with high salinity. The lack of precipitation in the summer and the high temperatures lasting until September led to a negative hydrological balance, which determines the decrease in water in the lake and its salinization. As was already mentioned, the Nuntași-Tuzla Lake is a part of the Razim–Sinoe lacustrine system, which is connected to both the Black Sea and St. Gheorghe Branch.

Gâstescu and Braier [13] stated that the main canals that connect the St. Gheorghe branch are Lipovenilor, Dunavăț, and Dranov. The connection with the Black Sea is made through the so-called "gates" or "periboine". These are Gura Portiţei, in front of Goloviţa Lake; Periteșca gate; Leahova gate, in front of Razim Lake and in front of Sinoe Lake; and Periboina and Edighiol gates. This area has been the subject of major hydrotechnical works over the past 120 years, which can be divided into two great categories: (1) the canalization phase and (2) the development of the Razim–Sinoe irrigation system. The canalization phase was very well described by Pons L.J. [27] and it refers to (i) canalization of the Sulina Branch to ensure its navigability (1889–1902), (ii) canalization of the canals (including Dranov and Dunavat (1903 and 1912, respectively) to ensure the fisheries, connections between smaller canals, connections between the Danube branches (Chilia, Sulina and St. George) and the Razim–Sinoe lacustrine systems (1950–1965), and canalization of the St. Gheorghe Branch to ensure the navigability (by regularization) and fresh water supply of the Razim–Sinoe irrigation system (1965, approximately).

The description of the Razim–Sinoe irrigation system development was described by Grumezea et al. [28]. This irrigation system was first used in 1974–1975; the studies and design activity of it were achieved in the 1969–1971, period and execution of it was carried out during the 1971–1974 (1975) period. The development of this irrigation system supposed the isolation of lakes from the Black Sea. To complete these three major hydraulic projects, the following steps were necessary: (i) closing communication between lagoons (Razim, Golovita, Zmeica, Babadag) and the Black Sea, which is supposed to be nearest to Portita Gate, and creating dammings to consolidate the coastal area (littoral belt) so as to prevent deterioration of lake shorelines; (ii) controlling the communication between Razim and Sinoe Lakes to ensure the discharge of the fresh water surplus of Sinoe Lake (by Edighiol and Peribiona gates) and the navigation on Sinoe and Razim Lakes; (iii) arranging the lagoon area as a great reservoir by the canalization of the Dunavat, Dranov, and Mustaca canals (Figure 4), which ensure the communication between St. Gheorghe Branch and the lagoon area; those canals ensure an average discharge of 80 m^3/s.

In a 1973 publication, the authors drew attention to the elaborated development plan for the Razim–Sinoe irrigation complex that would "sweeten" the waters of the lakes in about two years and would be able to be used for the irrigation of about 123,000 ha in the NE of Dobrogea [13].

Based on salinity data, the evolution of this parameter is presented in Figure 5. Starting with 1981, the water chemistry reached the limits of the necessary amount for irrigation (1 g/L). Therefore, the Nutasi-Tuzla Lake remains isolated, and the canal between Istria Lake and Nuntași-Tuzla Lake was closed, being silted since 1976 [12].

Dobrica et al. [6] stated that the maximum exploitation period of the irrigation system was 1981–1989, and starting with 1989, the system worked only partially and was completely closed in 2006. This also affects the evolution of the Nuntași and Săcele Rivers' flow rates (Table 3).

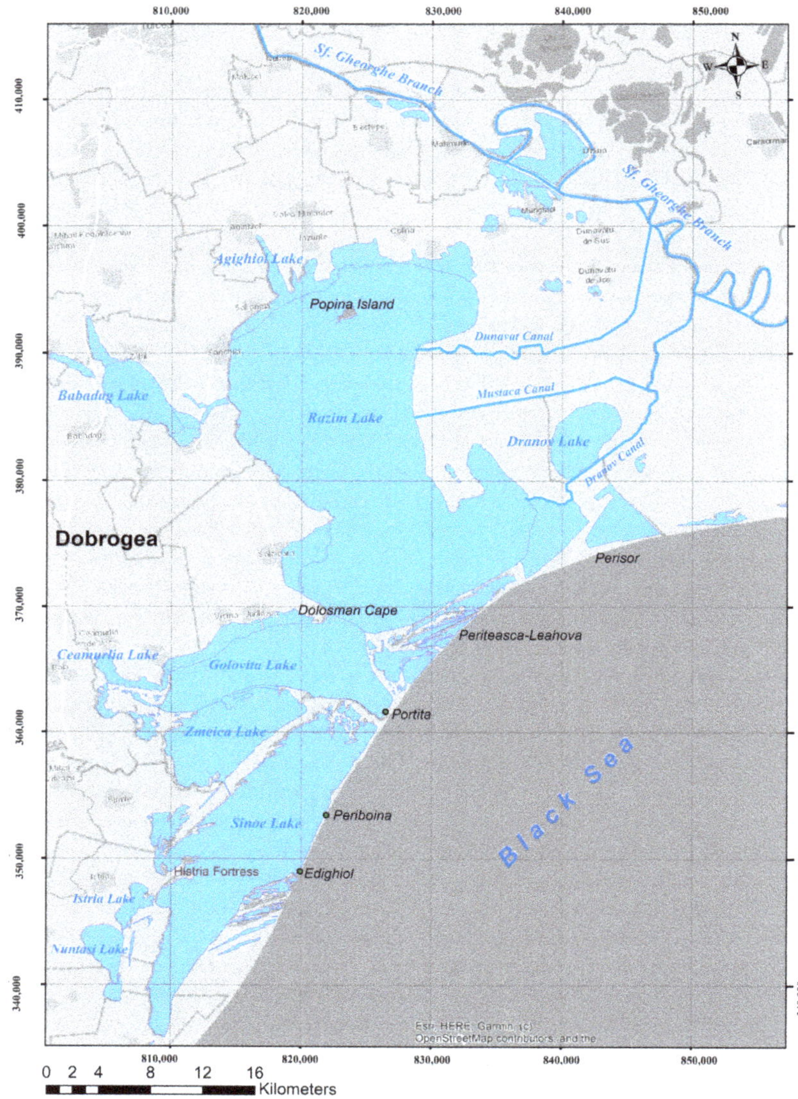

Figure 4. Razim–Sinoe lacustrine system.

Table 3 shows that during the maximum period of the operation of the irrigation system (1981–1989), the average multiannual flow of the Nuntași River is 1.4 times higher than that of the previous period (1965–1980). This increase is due to the infiltration resulting from the water used for irrigation.

After 1989, the multi-year discharge for the period of 1990–1997 returned to the baseline (1965–1980) and continued to decline, reaching a dangerous level (6.75 times lower than the average annual discharge for the period of 1965–1980). After closing the irrigation system and reducing the discharge rates on the two rivers, the Nutasi-Tuzla Lake level began decreasing, which led to a decrease in the lake surface. Research conducted in the past reveals that during the irrigation period, the production of sapropelic mud decreased due to decreases in salinity [14].

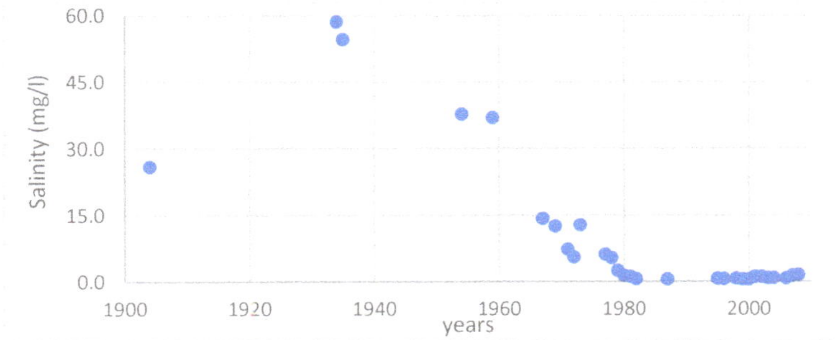

Figure 5. Evolution of salinity in Nuntași-Tuzla Lake.

Table 3. Evolution of discharge on Nuntași and Săcele Rivers [6].

Hydrometric Station	Subseries	Multi Annual Discharge (m^3/s)	Observation
Nuntași	1965–1980	0.467	Before the maximum operation of the irrigation system
	1981–1989	0.638	Maximum operation of irrigation system
	1990–1996	0.409	Partial work
	1997–2006	0.239	Closer
	2007–2020	0.092	
Săcele	1965–2003	0.104	
	2004–2020	0.037	

3.3. Data Analysis

Based on the methodology proposed, four time series of data were established to evaluate the degree of eutrophication, corresponding to the indicators used (total phosphorus [mg/L], biomass [mg/L], mineral nitrogen [mg/L], and chlorophyll "a"; there are only three years with registered values). Generally (86%), the lake is hypertrophic, 7% eutrophic, and 7% mesotrophic.

Physical components refer only to the pH parameter. The pH of the Nuntași-Tuzla Lake varies, depending on several factors, including its salinity, temperature, and the presence of organic material. However, as a saltwater lagoon, the pH is typically expected to be in the range of 6.5 to 8.5 [21]. This range is generally alkaline due to the dissolved salts and the buffering capacity of carbonate ions present in the water.

In recent years, environmental changes, such as droughts, reduced water inflow, and human activities, have impacted the lake, possibly leading to fluctuations in its pH levels. These changes could cause the pH to deviate from the typical range, potentially affecting the local ecosystem. From a total amount of 31 recordings, only 23% are situated in the 6.5–8.5 range; the rest of the values (77%) are outside a value of 8.5.

Oxygen regimen refers to dissolved oxygen (DO) [mg/L], biochemical oxygen demand—BOD5 [mg/L], and chemical oxygen consumption—CCOMn [mg/L]. The level of dissolved oxygen in the Nuntași-Tuzla Lake is a critical indicator of the lake's health and its ability to support aquatic life. Dissolved oxygen levels can vary significantly depending on factors such as temperature, salinity, organic load, and water circulation [17–21,27–31].

In saltwater or brackish environments like Nuntași-Tuzla Lake, dissolved oxygen levels are typically lower than in freshwater due to the higher salinity, which reduces the solubility of oxygen in water. Given these factors, the dissolved oxygen in the Nuntași-Tuzla Lake might be relatively low, particularly in warmer months or during periods of high organic decomposition.

The dynamics of dissolved oxygen are in an interdependent relationship with the intensity of biochemical and chemical oxygen consumption processes in water, which reflects the level of pollution of an aquatic ecosystem with organic substances.

The biochemical oxygen consumption (BOD), expressed by the BOD5 indicator, indirectly indicates the loading of water with biodegradable organic substances. This depends on the quantity and biochemical decomposition capacity of the existing organic substances in the water, under the influence of oxygen consumption. CCOMn is a variant of BOD5 testing that uses manganese as the oxidizing agent to measure the carbonaceous component of organic matter in water. It is particularly useful for assessing the organic carbon load in water bodies and industrial affluents, with applications in water treatment and environmental monitoring [21]. The evolution of all those parameters is presented in Figure 6.

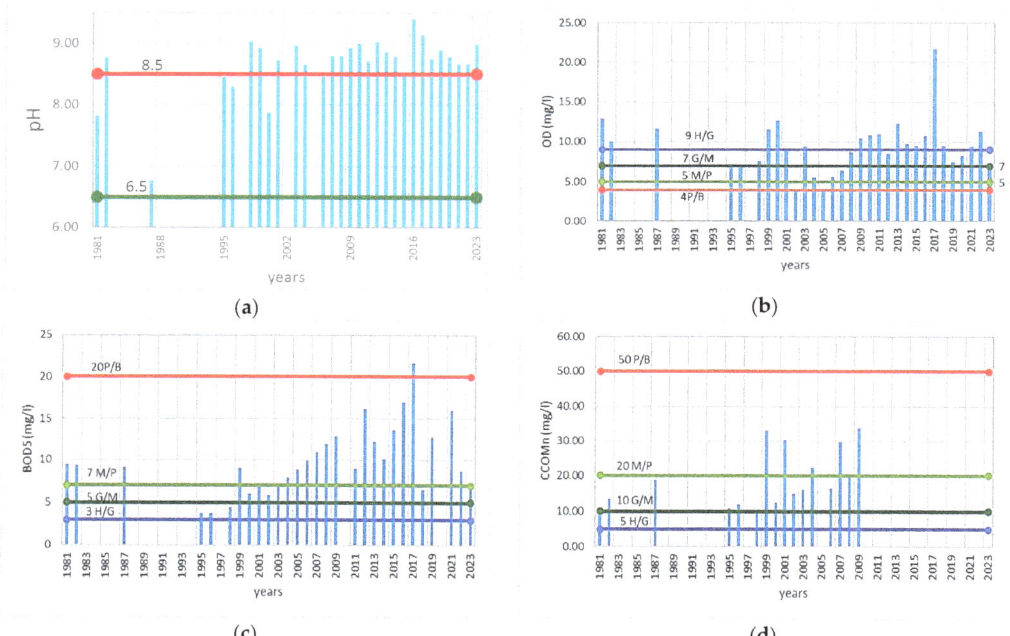

The level's degree colors are as follows:

bad	moderate	good	high

Figure 6. The dynamics in time of (**a**) pH, (**b**) DO, (**c**) BOD5, and (**d**) COD (CCOMn).

As Figure 6 shows, of 31 DO records, 17 (58%) represent a high status, 26% good, 13% moderate, and the rest poor (3%). No bad condition was registered. Concerning BOD5, of 29 records, 21 (72%) represent "poor" conditions, 14% moderate, 13% good conditions, and 3% bad conditions. No high condition was registered. As for CCOMn, of 16 records, 9 (56%) represent moderate status, 31% represent poor status, and 13% represent good status. No high condition was registered.

The nutrient levels in the Nuntași-Tuzla Lake, like in many other coastal lagoons, are important for understanding its ecological health and potential for issues like eutrophication. The key nutrients of interest are typically nitrogen (in forms such as nitrates and ammonium) and phosphorus (primarily in the form of phosphates). Nitrogen can enter the lake from various sources, including agricultural runoff (fertilizers), atmospheric deposition, and the decomposition of organic matter. In Nuntași-Tuzla Lake, nitrogen often presents as nitrates (NO_3^-), nitrites (NO_2^-), and ammonium (NH_4^+). Elevated levels of nitrogen can lead to algal blooms, which, upon decomposition, can decrease oxygen levels, causing hypoxic conditions that are harmful to aquatic life. Phosphorus typically enters the lake from runoff containing fertilizers, detergents, and natural sources, like sediment erosion. In coastal environments, phosphorus is usually present in the form of phosphates (PO_4^{3-}) [21,29–31].

The nutrient regimen of Nuntași-Tuzla Lake is determined based on the following indicators: ammonium (NH_4^+) [mg/L], nitrate (NO_3^-) [mg/L], total nitrogen (N) [mg/L], phosphate (PO_4^{3-}) [mg/L], and total phosphorous (P) [mg/L] (Figure 7). The nutrient classification according to OM161/2006 is presented in Table 4.

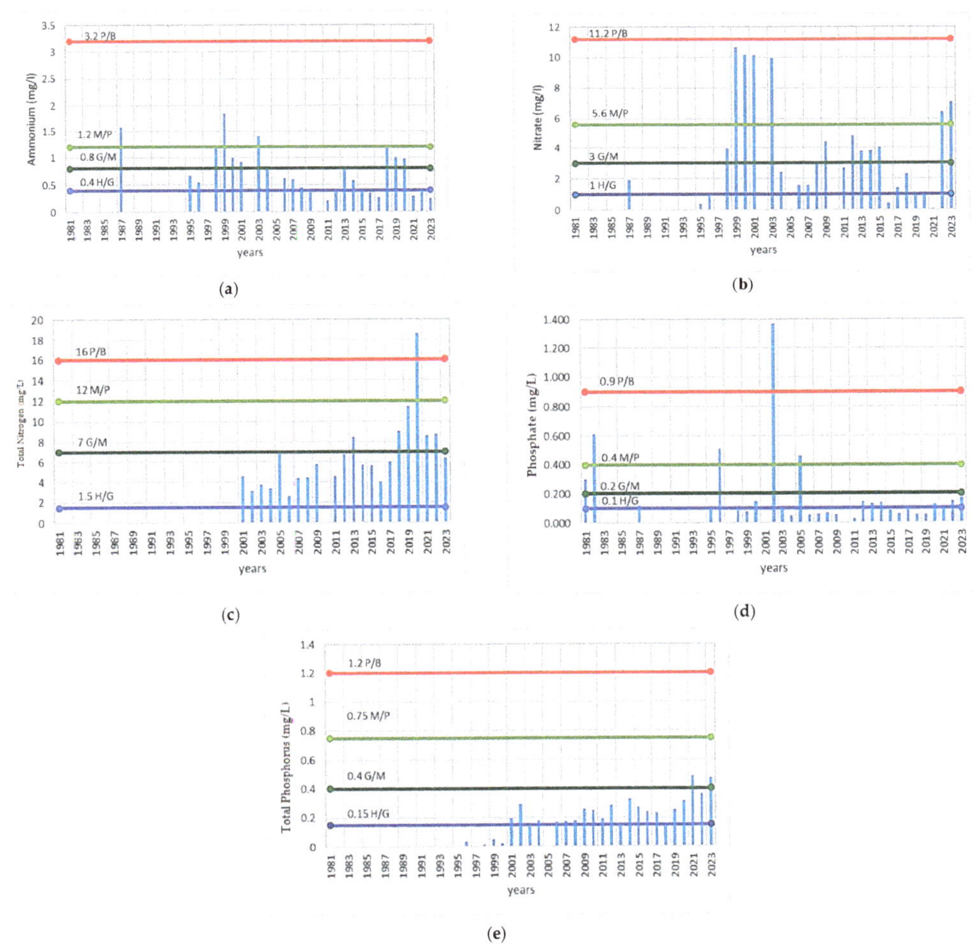

Figure 7. Evolution in time of nutrients. (**a**) Ammonium. (**b**) Nitrate. (**c**) Total Nitrogen. (**d**) Phosphate. (**e**) Total Phosphorus.

Table 4. Classification of nutrient indicators.

Classification	Ammonium		Nitrate		Total Nitrogen *		Orthophosphate		Total Phosphorous	
	Total Records	%	Total Records	%	Total Records	%	Total Records	%	Total Records	%
high	26	31	26	19	22	no	29	41	26	23
good		27		38		73		41		69
moderate		27		23		23		3		8
poor		15		19		no		10		no
bad		no		no		no		3		no

Notes: * Total nitrogen includes all forms of nitrogen (nitrates, nitrites, ammonia, and organically bound nitrogen).

Nuntași-Tuzla Lake's salinity is an important environmental parameter that requires ongoing observation to ensure the ecological stability of this unique coastal lagoon. The salinity of Nuntași-Tuzla Lake varies depending on several environmental factors, but it is generally recognized as a saline lagoon. The salinity indicator is based on the concentration of chlorides (Cl^-), sulfates (SO_4^{2-}), calcium (Ca^{2+}), and magnesium (Mg^{2+}). The values are measured in mg/L [31,32]. The variation of these indicators is presented in Figure 8, and the classification of the salinity indicator is presented in Table 5.

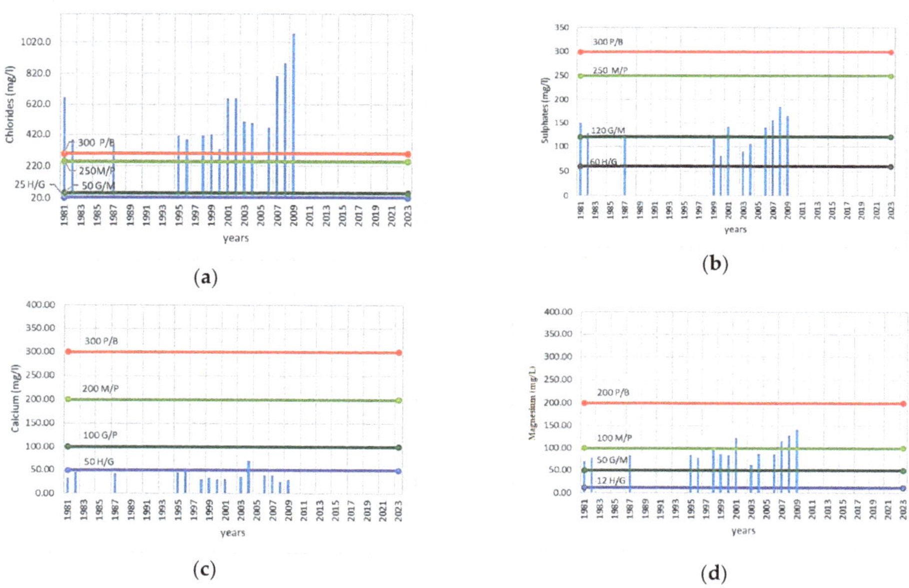

Figure 8. Variations of salinity indicators. (**a**) Chloride. (**b**) Sulfates. (**c**) Calcium. (**d**) Magnesium.

The salinity of Nuntași-Tuzla Lake has experienced significant changes, particularly due to human interventions, such as the construction of dikes and channels, which have altered water flow and exchange with the Black Sea. This has led to periods of both increased and decreased salinity, impacting the lake's flora and fauna.

3.4. *Water Quality Determination According to the WFD*

Nuntași-Tuzla Lake, a part of the larger Razim–Sinoe lagoon complex in Romania, has faced various ecological challenges over the years. The lake is important both ecologically and economically, but its status has been influenced by a combination of natural and anthropogenic factors. The ecological status of Nuntași-Tuzla Lake for the entire study period is presented in Figure 9.

Table 5. Classification function of salinity indicator.

Classification	Chlorides		Sulfate		Calcium		Magnesium	
	Total Records	**%**	**Total Records**	**%**	**Total Records**	**%**	**Total Records**	**%**
high	16	no	12	no	15	93	15	No
good		no		33		7		no
moderate		no		67		no		67%
poor		no		no		no		33%
bad		100		no		no		no

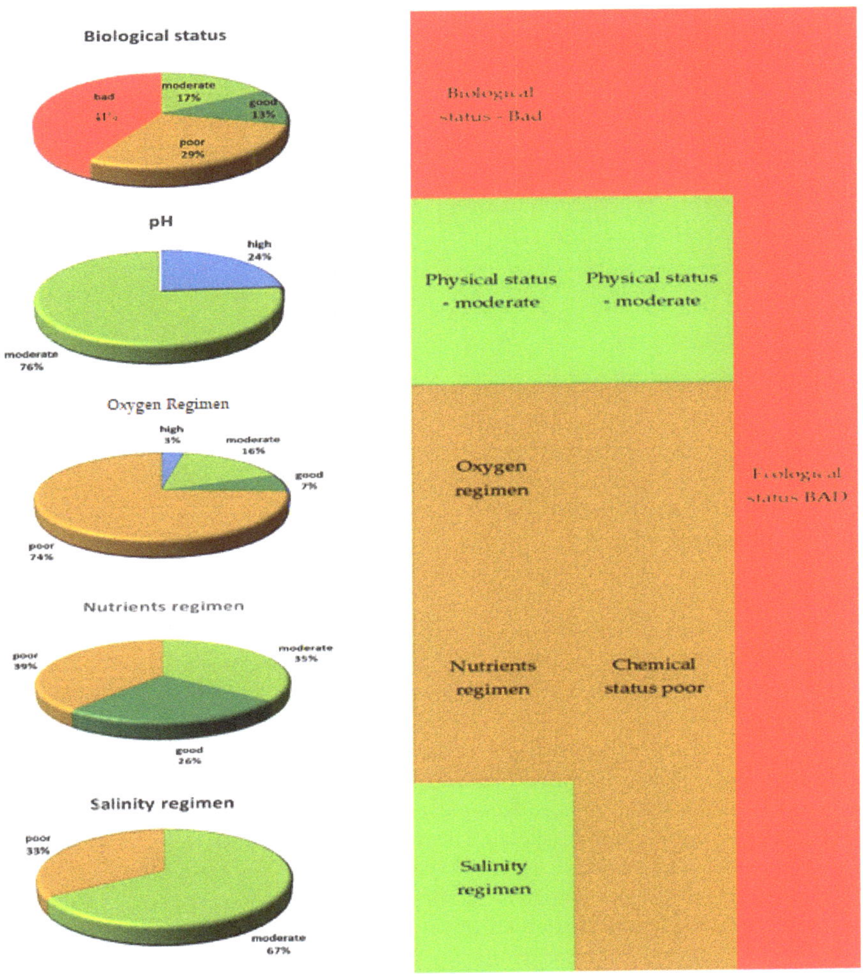

Figure 9. Ecological status of Nuntași-Tuzla Lake for 1981–2023 period.

Generally, throughout this period, the lake had a biological status of "bad" (42%) This classification result could be divided into two periods: (i) 1981–2009, when the biological status was generally bad (63%), and (ii) 2010–2023, when the biological status was generally "moderate" (31%). Otherwise, the ecological status percentages are equal: 23% for each category: "good"—2014–2016, "poor" (2010, 2021–2022), and "bad" (2018–2020). For the year 2023, the biological status could not be determined, and the ecological status, which is "poor", was based on physical–chemical elements.

3.5. Water Quality Index Determination

Based on the equations presented in the Methods section, the variation of the WQI is presented in Figure 10.

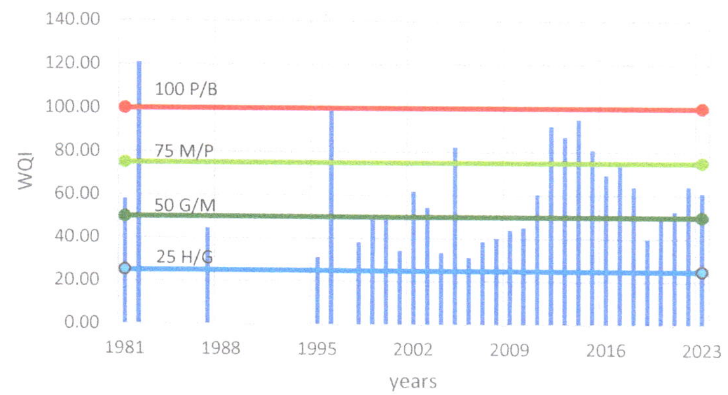

Figure 10. Evolution of WQI of Nuntași-Tuzla Lake.

Figure 10 shows that only in 1982, the WQI was greater than 100, which means "bad" conditions. Otherwise, out of 31 WQI values, 39% represent "good" conditions, 35% represent "moderate" conditions, and 23% represent "poor" conditions. It should be remembered that the series representing the salinity indicators is for the period 1981–2009 (with the gaps already indicated) and the one relating to the biological elements is for the period 2010–2023.

4. Discussion

The WFD implementation has had a great influence on the water management systems of EU countries by introducing the concept of watershed management [33]. This was not the case in Romania, which had this concept introduced in the old Water Law no. 8 of March 1974. To assess the ecological state of water, the WFD introduced an integrative method in the evaluation of water quality by establishing its quality based on several biological indicators, at the expense of chemical ones. Appendix V of the WFD outlines a main group as "quality elements", the biological one and two supporting ones, hydro-morphological and physico-chemical. This principle was implemented in Romanian legislation, as it is explained in the Results section. The use of the principle "one out–all out" leads to a restrictive ecological status classification system in relation to the definition of environmental objectives. It means that a body of water cannot achieve good ecological status if any element has a value that deviates from undisturbed conditions. The big advantage is that if an indicator is missing from a group of indicators, the status could be determined based on the others. For example, as is shown in Figure 6, the DO data series has 31 values registered, BOD5 has 29 values registered, and the CCOMn data series presents some gaps (no values registered for the 2010–2023 period); in this condition, the oxygen regimen status is determined based on DO and BOD5 conditions.

In the early years of application of the WFD methodology, a major obstacle was the fact that no consistent biological datasets were available for lakes, in particular. This impediment was resolved by the creation of geographical groups whose aim was to harmonize the methodologies used by the member states. In line with this, Romania participated in an "intercalibration exercise" with Bulgaria and Hungary and established a harmonized methodology to evaluate the phytoplankton [34] by the new Decision of European Commission [3].

Regarding the monitoring systems, through the WFD, the existing gaps in the monitoring systems were addressed by introducing the two types of monitoring programs: surveillance and operational. Unfortunately, there are no national websites dedicated to the monitoring system. For example, the Romanian system, SMIAR, is not as transparent as the National Network for Monitoring Air Quality (RNMCA), which limits the follow-up by the public of the general water condition. Discussing Nuntași-Tuzla Lake's ecological condition, based on data provided by RWNA-DLB Constanta, the ecological status for the entire period is "bad" (Figure 10). This period could be divided into two distinguished periods: 1981–2009 and 2010–2023. The period of 1981–2009 is based on a large lack of records; only in 1982 and 2007–2009 are there records regarding the types of phytoplankton that would lead to the possibility of calculating the metric index. In the rest of the periods, there are only records with biomass. There are no determinations regarding the other biological elements (phytobenthos and macroinvertebrates). For 2023, there are no registrations, so the ecological status is based on physical–chemical elements. In this condition, we consider that for the 1981–2009 period and the year 2023, the ecological condition is based only on chemical elements, and for the 2010–2022 period, it is based on biological elements. The results obtained show that the ecological status is "bad" for the 1981–2009 period, "moderate" for the period of 2010–2022, and "poor" for 2023. The "bad" ecological status of the 1981–2009 period is due to the chloride conditions, which were "bad" (Figure 8a). But in this case, salinity is not a pollutant. The high salinity values are due to natural conditions. If the ecological conditions determined based on the salinity regime were to be abandoned, then the ecological condition for the period of 1981–2010 would fall into the poor class. Even if the 2010–2022 period is declared "moderate", the oxygen regimen conditions for the 2011–2017, 2019, and 2021–2023 periods are "poor". This poor condition is due to the BOD5 indicator; the values of this indicator are "poor" for the entire period (2004–2022). A rise in BOD5 indicates a high degree of organic contamination in water bodies. So, could the ecological status be declared as "poor"? If the answer is "yes", it means that the lake was in poor condition for the entire period. The decision according to which the ecological status assigned to a water body depends on the most seriously affected quality element can lead to anomalies. In this case, Nuntași-Tuzla Lake could not achieve a "good" ecological status due to the chloride concentration, which, starting in 2001, has been increasing.

The studied lake presents some particularities: (i) it has a small water depth and was isolated from the rest of the lacustrine system (Razim–Sinoe) for a long time (1976–2020), which could lead to degradation of the oxygen regime. (ii) It was strongly affected by anthropogenic measures (the operation of the irrigation system), which led to the change in salinity and the disappearance of the sapropelic mud. Related to sapropelic mud, Bulgareanu showed in 1974 that the bottom deposit in the lake consists of black or black-gray, unctuous, sapropelic mud about 0.6 m thick [35]. These deposits were mostly located in the center of the lake, below the loess deposits on the western shore, and less on the eastern shores. In 1982 (ICPGA, 1982), the research authors stated that Nuntași-Tuzla Lake is the second therapeutic lake in Romania due to its sapropelic mud reserve [14]. Based on data recorded in 2002, Dragan-Bularda et al. [36] considered that the sapropelic mud from Nuntași-Tuzla Lake belongs to less active mud, which diminishes its therapeutic qualities. The anthropic impact together with climate changes (increasing temperatures after 1997 up to 0.8 °C [7]) led to the complete drying of the lake in August 2000 [6,8,10]. Șerban et al. [10] showed that in the summer of 2003, spring of 2007, autumn of 2021, summer of 2012, and 2013, the surface of the lake fell below the recorded values. If measures are taken to improve the hydrological conditions by periodically unclogging the connection channels and removing fishing hooks, then the oxygenation regime will be improved significantly.

The decision to classify the ecological conditions of the lakes from an arid zone, although it can provide general indications, does not give links and explicit explanations regarding the contribution of each element, and should be made based on an analysis of the specificity of the lakes.

The water quality index (WQI) could be an alternative if the WFD methodology is not too restrictive or gives anomalies for specific cases. WQI models use a single value to provide a better understanding of the surface water bodies' overall water quality. In practice, this method involves determining a weight and standardizing the data series of each indicator using, in this example, the standard value corresponding to the "good" class. Both chemical and biological indicators are used in this example. According to the WQI results, the conditions of the lake water are generally "good" to "moderate".

5. Conclusions

Situated in the southern extremity of the Razim–Sinoe complex lake, the Nuntași-Tuzla Lake was well-known in the past (the middle of the XIX century and early XX century) for the salt and fisheries production (especially mullet). Although connected by a network of canals (backwater) with the other component lakes of the lagoon complex, Nuntași-Tuzla Lake sometimes remained isolated from the other lakes (in the sense that it no longer received fresh water). This situation, combined with the climate (extremely dry summers), led to a high salinity of the lake water (for example, in 1912 the salinity was 18‰), which favored the crystallization of salt and the migration of fish from the Black Sea. Salinity also favored the formation of sapropelic mud. As we described in this work, the operation of the irrigation system led to an ecological imbalance that determined the salinity, mud production reduction, and the appearance of freshwater fish. To conclude, Nuntași-Tuzla Lake, with its unique hydrochemical and ecological characteristics, plays a vital role in the local environment and water management. Understanding and managing its water quality helps to ensure that the lake remains a healthy and sustainable ecosystem. Regular monitoring, effective management strategies, and public involvement represent the key to achieving and maintaining good water status for the lake.

Through both the WQI and the WFD methodology, the parameters used are determined and compared with the standard limits prescribed by national agencies. The first major conclusion of this study is that salinity is not a pollutant but is a natural characteristic due to the lake's position within the lake complex and the region's climate. We propose that for similar areas, both in terms of climatic conditions and other natural conditions (location, etc.), the indicator "salinity" should not be considered when analyzing the ecological status of the water. The second major conclusion is that the Water Framework Directive's heavy focus on ecology rather than chemical water quality leads, in this case, to certain anomalies. An ecological status was never intended as a substitute for water quality measurement, nor is it applicable in the same way. Despite certain limitations (e.g., non-consideration of pollutants), the current study serves as a basis for future research on developing new tools to investigate the quality of lake water situated in arid zones.

Author Contributions: Conceptualization, C.E.M.; methodology, C.E.M.; software, G.D.; validation, C.E.M., I.C.P. and N.L.; formal analysis, N.L.; resources, G.D.; data curation, C.E.M., I.C.P. and G.D.; writing—original draft preparation, C.E.M.; writing—review and editing, I.C.P. All authors have read and agreed to the published version of the manuscript.

Funding: This research received no external funding.

Data Availability Statement: The raw data on which this study was based belong to the Romanian National Water Administration, Dobrogea Litoral, and are available at https://dobrogea-litoral.rowater.ro/ (accessed on 10 May 2025), with their permission.

Conflicts of Interest: The authors declare no conflicts of interest.

References

1. Kristensen, P.; Whalley, C.; Nery, F.; Christiansen, T.; Schmedtje, U.; Solheim, A.; Austnes, K.; Kampa, E.; Rouillard, J.; Prchalová, H.; et al. 2018 EEA European Waters Assessment. 2018. Available online: https://www.researchgate.net/publication/329482267_2018_EEA_European_waters_assessment (accessed on 15 May 2024).
2. Directive 2000/60/EC of the European Parliament and of the Council of 23 October 2000 Establishing a Framework for Community Action in the Field of Water Policy, 2000/60/EC, EP, CONSIL, 327 OJ L (2000). Available online: http://data.europa.eu/eli/dir/2000/60/oj/eng (accessed on 20 May 2024).
3. European Comission. Decizia (UE) 2024/721 a Comisiei din 27 Februarie 2024 de Stabilire, în Temeiul Directivei 2000/60/CE a Parlamentului European și a Consiliului, a Valorilor Pentru Clasificările Sistemelor de Monitorizare ale Statelor Membre ca Rezultat al Exercițiului de Intercalibrare și de Abrogare a Deciziei (UE) 2018/229 a Comisiei [Notificată cu Numărul C(2024) 1113] (Decizia (UE) 2024/72 C(2024) 1113). EU Comission. 2024. Available online: https://eur-lex.europa.eu/legal-content/RO/TXT/PDF/?uri=OJ:L_202400721 (accessed on 15 April 2024).
4. Water Information System for Europe, Fp. Surface Water Bodies: Ecological Status or Potential (Group) [Table]. 23 February 2024. Available online: https://water.europa.eu/freshwater/resources/metadata/draft-wfd-dashboards/surface-water-bodies-ecological-status-or-potential-group-table (accessed on 15 April 2024).
5. Bratescu, C. *Pământul Dobrogei (Dobrogea Land)*, 1st ed.; Cultura Națională: Bucuresti, Romania, 1928.
6. Dobrica, G. *Evaluarea Secetei Hidrologice în Bazinul Hidrografic al Lacului Nuntași-Tuzla, Județul Constanța (Assessment of the Hydrological Drought in the Nuntași-Tuzla Lake Watershed, Constanța County)*; Ovidius University of Constanta: Constanta, Romania, 2023; *unpublished*.
7. Maftei, C.; Barbulescu, A. Statistical analysis of climate evolution in Dobrudja region. In Proceedings of the World Congress on Engineering, London, UK, 2–4 July 2008; Volume II.
8. Maftei, C.; Barbulescu, A. Statistical analysis of precipitation time series in the Dobrudja region. *Mausam* **2012**, *63*, 553–564. [CrossRef]
9. Paltineanu, C.; Chitu, E.; Tanasescu, N.; Apostol, G.; Pufu, M.N. Irrigation water requirements for some fruit trees specific to the Arges-Vedea river basin, Romania. *Acta Hortic.* **2000**, *537*, 113–119. [CrossRef]
10. Șerban, C.; Maftei, C.; Dobrică, G. Surface Water Change Detection via Water Indices and Predictive Modeling Using Remote Sensing Imagery: A Case Study of Nuntași-Tuzla Lake, Romania. *Water* **2022**, *14*, 556. [CrossRef]
11. Breier, A. Raportul dintre caracteristicile mor/ogralice morfometrice ale lacurilor de pe litoralul românesc al Marii Negre. *Stud. Cercet. Geol. Geofiz. Geogr.* **1970**, *XVII*, 187–194.
12. Breier, A. *Lacurile de pe Litoralul Romanesc al Marii Negre. Studiu Hidrogeografic*; Academiei Republicii Socialiste România: Bucuresti, Romania, 1976.
13. Gastescu, P.; Breier, A. Complexul Razim-Sinoe-Geneză, morfometrie şi regim hidric. *Peuce* **1973**, *III*, 49–76.
14. Institutul National de Hidrologie si Gospodarire a Apelor (National Institute of Hydrologie and Water Management). *Stabilirea Dinamicii Calității Apei Lacurilor Techirghiol, Nuntași, Istria, Amara și a Măsurilor de Protecție Necesare (Establishing the Water Quality Dynamics of Techirghiol, Nuntași, Istria, Amara Lakes and the Necessary Protective Measures) [Contract no 310]*; Institutul National de Hidrologie si Gospodarire a Apelor: Bucuresti, Romania, 1982.
15. Ionescu, M.D. *Dobrogia în Pragul Veacului al XX-lea: Geografia Matematică, Fisică, Politică Economică și Militară (Dobrogia at the Threshold of the 20th Century: Mathematical Geography, Physics, Economic and Military Policy, Graphic Workshops—In English)*, Atelierele Grafice I. V. Socecu: Bucuresti, Romania, 1904. Available online: https://www.ziuaconstanta.ro/images/stories/2017/11/carti-online/Dobrogia-in-pragul-veacului-al-XX-lea-low.pdf (accessed on 15 February 2024).
16. Dumitran, G.E.; Vuta, L.I.; Popa, B. Overview of the Eutrophication in Romanian Lakes and Reservoirs. *Limnol. Rev.* **2024**, *24*, 76–104. [CrossRef]

17. Chaves Fortes, A.C.; Guimarães Barrocas, P.R.; Kligerman, D.C. Water quality indices: Construction, potential, and limitations. *Ecol. Indic.* **2023**, *157*, 111187. [CrossRef]
18. Khan, M.H.R.B.; Ahsan, A.; Imteaz, M.; Shafiquzzaman, M.; Al-Ansari, N. Evaluation of the surface water quality using global water quality index (WQI) models: Perspective of river water pollution. *Sci. Rep.* **2023**, *13*, 20454. [CrossRef] [PubMed]
19. Jin, Y.; Zhu, B.; Wang, F.; Sun, S.; Wang, P.; Liu, X. Analysis of Water Chemistry Characteristics and Main Ion Controlling Factors of Lakes in the Nagqu Area of the Qinghai–Tibet Plateau in Summer. *Water* **2023**, *15*, 2900. [CrossRef]
20. Gao, Z.; Chen, J.; Wang, G.; Ren, S.; Fang, L.; Yinglan, A.; Wang, Q. A novel multivariate time series prediction of crucial water quality parameters with Long Short-Term Memory (LSTM) networks. *J. Contam. Hydrol.* **2023**, *259*, 104262. [CrossRef] [PubMed]
21. Vasistha, P.; Ganguly, R. Water quality assessment of natural lakes and its importance: An overview. *Mater. Today Proc.* **2020**, *32*, 544–552. [CrossRef]
22. Lumb, A.; Sharma, T.C.; Bibeault, J.F. A Review of Genesis and Evolution of Water Quality Index (WQI) and Some Future Directions. *Water Qual. Expo. Health* **2011**, *3*, 11–24. [CrossRef]
23. Gibrilla, A.; Bam, E.K.P.; Adomako, D.; Ganyaglo, S.; Osae, S.; Akiti, T.T.; Kebede, S.; Achoribo, E.; Ahialey, E.; Ayanu, G.; et al. Application of Water Quality Index (WQI) and Multivariate Analysis for Groundwater Quality Assessment of the Birimian and Cape Coast Granitoid Complex: Densu River Basin of Ghana. *Water Qual. Expo. Health* **2011**, *3*, 63–78. [CrossRef]
24. Brown, R.M.; McClelland, N.I.; Deininger, R.A.; O'Connor, M.F. *A Water Quality Index—Crashing the Psychological Barrier*; Thomas, W.A., Ed.; Springer USA: New York, NY, USA, 1972; pp. 173–182. [CrossRef]
25. Horvath, C. *Studiul Lacurilor de Acumulare din Bazinul Superior al Crisului Repede/Study of the Storage Lakes in the Upper Basin of the Crişul Repede River*; Casa Cărţii de Ştiinţă: Cluj-Napoca, Romania, 2008.
26. Porucic, T. Lacurile sarate din sudul Basarabiei. Studiu geografico-geologic si economic cu privire la industria extragerii sarii de mare. *Stud. Limnol.* **1924**, *I*, 1–144.
27. Pons, L.J. Natural resources. In *Conservation Status of the Danube Delta*; IUCN: Gland, Switzerland, 1992; Volume 4, pp. 23–36.
28. Grumezea, N.; Kleps, C.; Tusa, C. *Evolutia Nivelului si Chimismului Apei Freatice din Amenajarile de Irigatii in Interrelatie cu Mediul Inconjurator*; Intreprinderea poligrafica Oltenia: Craiova, Romania, 1990.
29. Hammami Abidi, J.; Eldin Elzain, H.; Sabarathinam, C.; Selmane, T.; Selvam, S.; Farhat, B.; Mammou, A.B.; Senapathi, V. Evaluation of groundwater quality indices using multi-criteria decision-making techniques and a fuzzy logic model in an irrigated area. *Groundw. Sustain. Dev.* **2024**, *25*, 101122. [CrossRef]
30. Sylvie, N.M.E.; Armel Zacharie, E.B.; Ayissi Mbomo, R.E.; Akono, J.V.; Bachirou, D.; Kamtchouing, P. Evaluation of surface water contamination and its impacts on health in the mining districts of Kambele and Betare-Oya (Eastern-Cameroon). *Heliyon* **2024**, *10*, e29189. [CrossRef] [PubMed]
31. Balcerowska-Czerniak, G.; Gorczyca, B. Rapid assessment of surface water quality using statistical multivariate analysis approach: Oder River system case study. *Sci. Total Environ.* **2024**, *912*, 168754. [CrossRef] [PubMed]
32. Keci, E. Ecological Status of Water Quality. In *Hydrology-Current Research and Future Directions*; IntechOpen: London, UK, 2024. [CrossRef]
33. Hering, D.; Borja, A.; Carstensen, J.; Carvalho, L.; Elliott, M.; Feld, C.K.; Heiskanen, A.-S.; Johnson, R.K.; Moe, J.; Pont, D.; et al. The European Water Framework Directive at the age of 10: A critical review of the achievements with recommendations for the future. *Sci. Total Environ.* **2010**, *408*, 4007–4019. [CrossRef] [PubMed]
34. Joint Research Centre (European Commission); Borics, G.; Poikane, S.; Wolfram, G.; Chiriac, G.; Belkinova, D.; Donabaum, K. *Intercalibration of the National Classifications of Ecological Status for Eastern Continental Lakes: Biological Quality Element Phytoplankton*; Publications Office of the European Union: Luxembourg, 2018; ISBN 978-92-79-92972-4. [CrossRef]
35. Bulgăreanu, V.; Momea, G.; Momea, L. Unele consideraţii limnogeologice asupra cuvetei Nuntaşi-Tuzla (complexul Sinoe). *Stud. Geol. Cuaternarului* **1974**, *5*, 123–159.
36. Dragan-Bularda, M.; Grigore, C.E.; Tura, D. Utilizarea indicatorului enzimatic al calităţii nămolului în scopul valorificării şi protecţiei lacurilor saline (Utilisation of the Enzymatic Indicator of Mud Quality for Exploitation and Protection of Salt Lake). *Stud. Univ. Babeş-Bolyai Biol.* **2004**, *XLIX*. Available online: http://studia.ubbcluj.ro/download/pdf/18.pdf (accessed on 12 May 2024).

water

MDPI

Article

GIS-Based Accessibility Analysis for Emergency Response in Hazard-Prone Mountain Catchments: A Case Study of Vărbilău, Romania

Cristian Popescu [1] and Alina Bărbulescu [2,*]

[1] Doctoral School, Technical University of Civil Engineering of Bucharest, 122-124 Bd. Lacul Tei, 020396 Bucharest, Romania; cristiannicolae.popescu@gmail.com
[2] Department of Civil Engineering, Transilvania University of Brasov, 5 Turnului Str., 900152 Brașov, Romania
* Correspondence: alina.barbulescu@unitbv.ro

Abstract: The intensification of extreme hydrologic events, such as flash floods and landslides, has amplified the challenges of ensuring timely and effective emergency response. A key factor in the efficiency of such interventions is the accessibility of affected areas, which often becomes compromised during hazard events. In this context, the present study focuses on the Vărbilău River catchment in Romania, a region highly exposed to frequent flash floods and terrain instability. The research evaluates the spatial accessibility of emergency intervention units. Four major intervention centers were assessed under both normal and constrained scenarios. Accessibility was quantified through travel-time thresholds, incorporating variables such as road quality, network density, topography, and hazard-induced disruptions. Findings indicate that southern localities enjoy relatively short intervention times (less than 10 or between 10 and 20 min) due to favorable terrain and proximity to well-equipped centers. In such cases, the speed on main roads is 50–60 km/h, while the accessibility index is 5. Conversely, northern areas and villages like Lutu Roşu face elevated isolation risks, as single-road access and weak connectivity heighten their vulnerability during floods or landslides. In such cases, speeds reduce to 10 km/h and accessibility is very low, with the accessibility index of 1. Scenario modeling further demonstrated that the loss of key hubs (e.g., Ploieşti or Văleni) severely undermines coverage efficiency, particularly in high-risk zones, where the access times increases over 40 min. These results emphasize the need for dynamic intervention planning, infrastructure reinforcement, and the systematic integration of hazard-prone areas into emergency response strategies. Moreover, the methodological framework developed here can be adapted to other regions exposed to hydrologic hazards.

Keywords: accessibility metrics; emergency; GIS; road; network; discontinuity; flood risks; landslide risks

1. Introduction

Natural hazards significantly affect both the natural and human environment, often resulting in destructive consequences and substantial material losses. They encompass a wide range of phenomena, including earthquakes, volcanic eruptions, and gravitational processes such as landslides, rockfalls, and avalanches, as well as hydrological hazards like flash floods, river floods, forest fires, desertification, storms, and hurricanes [1–3]. The drivers of extreme events are diverse, ranging from large-scale processes such as the differential heating of continents and oceans that generate monsoons, to local factors like the orographic and altimetric influence of relief forms (e.g., vertical climatic zoning, orographic barriers). In temperate zones, a sudden rise in spring temperatures can accelerate snowmelt, triggering floods.

Areas prone to flooding and flash flooding require targeted measures to reduce impacts when hazards occur. In addition to hydrotechnical constructions designed to mitigate flood damage, the spatial distribution of emergency units and the structure of the road network play a crucial role in shaping the effectiveness of response efforts [4,5].

Travel accessibility is heavily conditioned by road quality, network density, and proximity to emergency centers managed by the Inspectorate for Emergency Situations, (called ISU units). Disruptions such as inundated roads, damaged bridges, or landslides introduce discontinuities that hinder rapid intervention. Accessibility in the context of emergencies has been the subject of several studies. Borowska-Stefańska and Wiśniewski [6] emphasized that while floods cannot be fully prevented, preparedness through optimal evacuation planning is essential. Full accessibility simulations are critical to identifying potential bottlenecks in transport networks prior to an event, thereby facilitating the development of efficient evacuation strategies. Similarly, Papilloud and Keiler [7] analyzed evacuation routes in relation to residential distribution, showing that vulnerability can persist even in well-connected areas.

Other contributions have emphasized access to essential services during hazard conditions. Gangwal et al. [8] investigated the impact of flood-related road disruptions on hospital access for vulnerable communities, demonstrating that reduced travel speeds and network failures significantly increase travel times. Their work underscores the need for targeted investments in critical infrastructure to enhance community resilience. Florath et al. [9] advanced this approach by developing a multi-scale framework to assess road accessibility under hazard conditions. Their methodology integrates the identification of damaged road segments with network analyses using metrics such as inter-center connectivity, central node proximity, free-flow indices, and rerouting efficiency. Results revealed substantial shifts in accessibility when roads were impaired, confirming the cascading effects of hazard-induced disruptions.

Recent research has further refined methodological approaches to accessibility analysis. Loreti et al. [10] introduced accessibility metrics for flood-affected areas that move beyond percolation models, focusing instead on how local towns maintain (or lose) access to critical services during extreme events. Morelli and Cunha [11], in turn, developed resilience metrics to evaluate urban road networks under flood conditions, proposing network continuity and alternative efficiency as key measures of functional capacity. Both approaches demonstrate the importance of quantifying accessibility not only spatially but also dynamically, in relation to hazard-induced disruptions.

Transport accessibility also plays a broader role in urban development. Ford et al. [12] demonstrated that accessibility is not only central to emergency response but also a driver of sustainable urban growth. Using a GIS-based methodology applied to London, they showed how accessibility can be measured across multiple transport modes, with implications for reducing environmental impacts. Similarly, Ertugay and Duzgun [13] emphasized the relationship between accessibility, equity, and service distribution, arguing that accessibility analyses should guide infrastructure planning and national or regional policy strategies.

In the Romanian context, hazard research has highlighted both the challenges and opportunities of integrating accessibility analyses into emergency management. Stăncalie et al. [14] used GIS and Earth observation technologies to support flood management after the 2000 floods, while other works on various river basins in Romania utilized flood potential indices and machine learning to improve hazard mapping [15–18]. Chelariu et al. [19] proposed GIS-based modeling of shelter placement and pedestrian evacuation in rural mountain catchments, providing insights highly relevant to communities in steep Carpathian basins. Puie and Mihai [20] extended this line of research by analyzing the dual impacts of rockfalls and floods on transport corridors, showing how multi-hazard interactions compromise both road and rail access. Complementary efforts, such as landslide-hazard mapping project [21,22], also underscore the necessity of integrating slope instability data into national risk management strategies, especially in the Carpathian region.

River basin-level research on road accessibility is insufficient in Romania. There are no detailed maps regarding emergency interventions, comparative scenarios of travel at different speeds, or studies simulating how this distribution was due to the presence of hydrological hazards.

Flood response operations in Romania are coordinated by the ISU, with intervention teams stationed at specialized points distributed according to population density. Response times, however, are highly dependent on road slope, quality, and density, as well as the number and spatial distribution of ISU centers. Although a greater number of centers generally improve coverage, multiple obstacles persist. Flooding, debris, or landslides may render key roads impassable, while bridges can be destroyed by strong currents, completely isolating localities when they represent the sole access route. Road foundation undermining further exacerbates risks, leading to collapses. These interruptions delay intervention, often compounding damage. Moreover, effective planning is hampered by outdated road maps, insufficiently detailed hazard data, and incomplete knowledge of flood frequencies or landslide susceptibilities.

Against this backdrop, the present study aims to address the accessibility of emergency intervention teams in the Vărbilău Catchment, Romania, under flood and landslide scenarios. This region has experienced multiple events requiring urgent evacuation and emergency support, yet no dedicated accessibility analysis has been carried out. Our approach builds on GIS-based methods to evaluate road accessibility, update road maps with field-verified information, and model intervention scenarios. The findings are designed to inform decision-makers responsible for transport planning and hazard response, offering insights for strengthening resilience through improved accessibility modeling.

2. Study Area and Methodology

2.1. Study Area

The Vărbilău catchment is located near the 45° north parallel, approximately 100 km north of Bucharest, in Prahova County, Romania (Figure 1). The area's relief, vegetation, and river network strongly influence the layout of roads and human settlements. Population density is highest in the central and southern valleys, while mountainous areas are sparsely inhabited, with densities approaching 0 inhabitants per km^2 [23]. The total population of the study area is approximately 30,000 [24]. Most towns are situated close to rivers, making them particularly susceptible to flooding. Localities such as Vărbilău, Dumbrăvești, and Aluniș (Figure 2) exhibit higher road densities, while Slănic—the only town in the catchment—has a very high road density [24]. Bertea and Ștefești show average road density, whereas Lutu Roșu, in the northern part of the catchment, has relatively low road density. The primary road in the catchment is County Road DJ 102, which connects Slănic with Dumbrăvești in the south and provides the main route to Ploiești, the nearest major city.

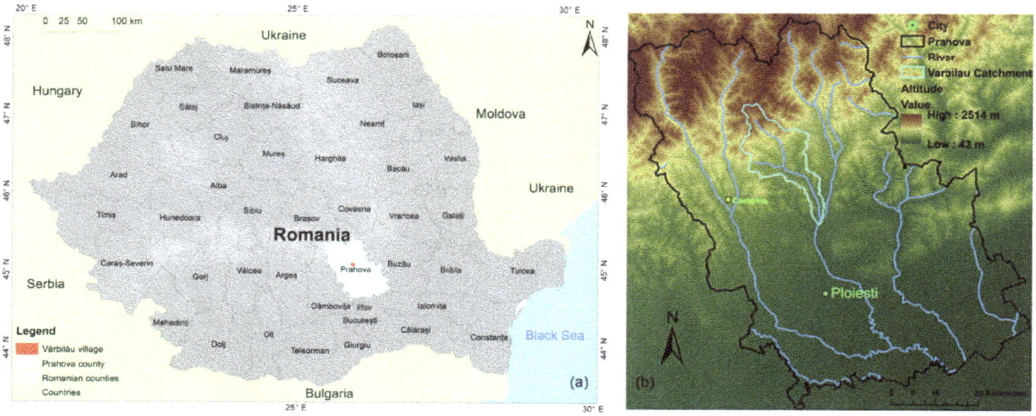

Figure 1. Location of the Vărbilău catchment (**a**) in Prahova County, Romania, and (**b**) detailed view.

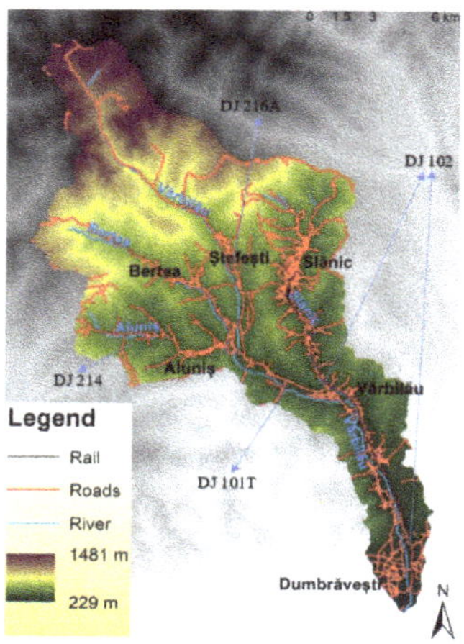

Figure 2. Road network and settlements in the Vărbilău Catchment.

Other significant roads include DJ 101T, linking Bertea and Vărbilău; DJ 216A, connecting Ștefești to DJ 101T; and DJ 214, which provides access between Aluniș and the western part of the catchment. Overall, the road network allows easier travel from north to south in the eastern and southern areas, while in the central and western regions, movement is more accessible along a northwest–southeast direction.

Figure 3a shows the layout of Vărbilău commune, situated at the confluence of two rivers—a position that increases its exposure to flood risk. Given this vulnerability, assessing the locality's accessibility is essential for effective emergency response and evacuation planning. Preliminary observations suggest that although the commune is intersected by key roads, including DJ 101T, some peripheral areas may be underserved. During floods, low-lying roads near the rivers are particularly susceptible to water coverage or structural damage, potentially causing delays in intervention. Understanding the road configuration and identifying critical points within the network is vital for improving the resilience of this locality.

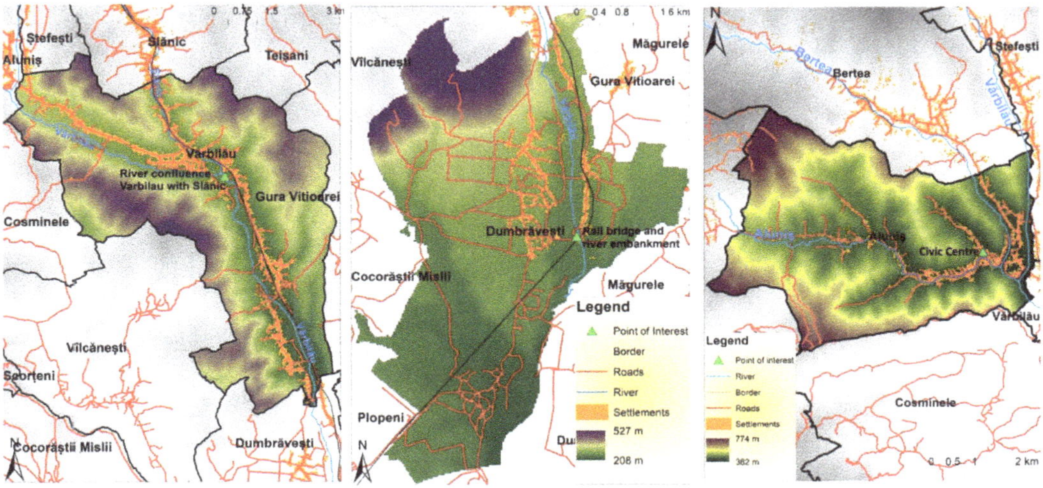

(a) (b) (c)

Figure 3. (a) The Vărbilău commune, (b) Dumbrăvești, and (c) Aluniş village.

Figure 3b illustrates the Dumbrăveşti commune, where the nearby rail embankment can influence flooding [25] and impact accessibility—a topic that will be discussed further in the Section 3. Figure 3c presents Aluniș village, highlighting the Strâmba neighborhood north of the Civic Center. This area is especially vulnerable to flooding, making adequate coverage by emergency response units critical.

Emergency intervention in the region is provided by several ISU centers located in Ploieşti (which hosts two ISU units), Câmpina, and Vălenii de Munte (Figure 4).

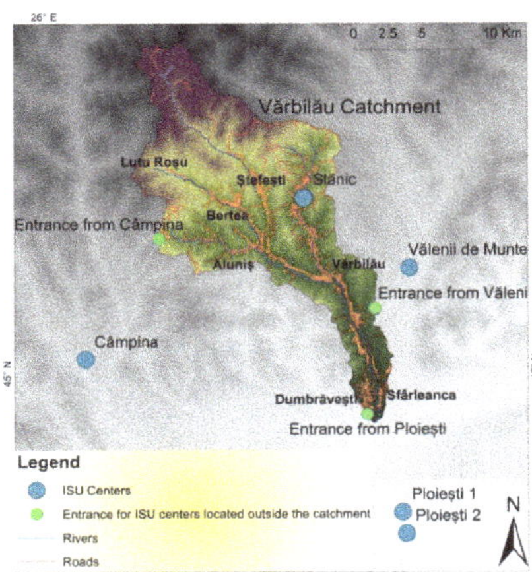

Figure 4. Location of ISU centers and access points in the Vărbilău Catchment.

Within the hydrographic basin, the only intervention center is a small unit based in Slănic. In addition to the road network, the area is also served by a railway line connecting Ploieşti to Slănic, which may provide alternative logistical options in the event of road disruptions.

The Vărbilău Catchment has experienced several exceptional events in recent years, underlining the importance of analyzing road accessibility in the context of recurrent natural hazards, as follows:

- June 2021—Aluniş village: Heavy flooding occurred after a tributary of the Aluniş River overflowed its banks, inundating nearby households. Following its confluence with the Aluniş, additional households and the main road were flooded. The localities of Vărbilău and Ştefeşti were also affected.
- June 2019—Slănic Prahova town: Torrential rains triggered flash floods that damaged roads and households.
- July 2018—Vărbilău commune: Rapid floods originating from surrounding hills covered roads, independent of flows from the Slănic or Vărbilău rivers.
- Landslide at Lutu Roşu (monitored since 1978): The county road DJ 101T, which connects Lutu Roşu with Bertea, has repeatedly been blocked by active sliding. Restoration works are ongoing.
- April 2024—Slănic town center: Land subsidence affected an area of several dozen square meters, with a breach depth of around 2 m. Dozens of residents were evacuated.

These events highlight the hydrological vulnerability of the basin and demonstrate the critical importance of ensuring reliable road accessibility for both local communities and emergency response.

The following subchapters will present a detailed analysis of how these emergency centers serve the Vărbilău Catchment. Specific intervention scenarios will be examined, with a focus on two localities (Lutu Roşu and Dumbrăveşti-Sfârleanca) where landslides and flooding frequently disrupt road access, posing significant challenges to timely emergency response.

2.2. Methodology

Identifying accessibility within a region is a fundamental step in emergency planning. It supports the design of travel routes from intervention centers to affected zones and facilitates the evacuation of populations to designated shelter areas. Moreover, understanding accessibility is essential for evaluating the capacity and effectiveness of emergency response authorities [26]. The findings of this study, which present the results of an accessibility analysis within the Vărbilău Catchment, provide a significant contribution to broader research aimed at defining optimal intervention strategies under various hazard scenarios.

To achieve this objective, both field observations and GIS-based analysis were employed. Fieldwork focused on identifying road types, the location of ISU centers, and access routes used by emergency services operating from outside the Vărbilău Catchment. GIS analysis, performed using ArcMap 10.2.2, facilitated the creation of maps that present current accessibility conditions within the catchment. In addition to general accessibility for emergency units, several potential intervention scenarios are examined.

Field surveys covered primary, secondary, and forest roads to approximate travel speeds between intersections. These speeds were then integrated with a GIS-based road database. Supplementary information was also collected, including road classification (county, communal, forest), road surface quality, network density, access points into the catchment, and distances from intervention centers to the areas they serve during emergencies.

Table 1 summarizes the categories, types, formats, and sources of the data used. The road network data are stored in shapefile (shp) format, specifically as polyline vector data compatible with GIS software.

Table 1. Data sources and format (the links were accessed last time on 15 May 2025).

Category	Data Source	Data Type	Format
Roads	Geofabrik.de (https://download.geofabrik.de/europe/romania.html) (accessed on 15 September 2025)	polyline	shp
Rivers	Geofabrik.de (https://download.geofabrik.de/europe/romania.html) (accessed on 15 September 2025)	Polyline	shp
River catchment	Romanian Water Cadaster (https://data.gov.ro/dataset/hidrografie) (accessed on 15 September 2025)	polygon	shp
DEM Romania	SRTM Elevation Data (https://portal.opentopography.org/raster?opentopoID=OTSRTM.082015.4326.1) (accessed on 15 September 2025)	pixels	tiff
Buildings	Microsoft Worldwide Building Footprints (https://planetarycomputer.microsoft.com/dataset/ms-buildings) (accessed on 15 September 2025)	polygon	shp

This geospatial database enables the evaluation of accessibility across the road network in the Vărbilău Catchment and the quantification of road segments affected by flooding. Building data were stored in polygon shapefile format and were used to quantify the number of flooded structures within the study area. Elevation information was provided by a Digital Elevation Model (DEM) in raster format, with a spatial resolution of 30 × 30 m. For the development of accessibility maps, shapefiles (shp) were required, including with point-type data representing the locations of intervention facilities. Using this dataset, a comprehensive road accessibility map was created. All maps were generated by the authors using ArcMap environment, with corrections applied to improve the accuracy and usability of the road network. To ensure the functional road network model, several adjustments were necessary:

- Assignment of travel speeds: Each road segment was attributed a realistic travel speed (according to the legislation and the road situation).
- Verification of intersections: Road connections were carefully checked to confirm that segments converged at a single junction point. Segments that extended beyond intersections without a common node were not recognized in the GIS network, which would otherwise prevent accurate accessibility calculations.

Figure 5a illustrates a road segment with an incorrect network topology. Each line representing a road must terminate at a junction point to ensure proper connectivity, meaning each road should exist as a segment between two junctions. At a cross-shaped ("x") intersection, for example, all road segments must meet at a single point.

Errors of this type were identified using the Topology extension in ArcToolbox (Data Management Tools) and were manually corrected with the Split Tool in the GIS Editor. Hundreds of such corrections were made. Figure 5b shows a properly structured road segment. The Vărbilău watershed polygon was defined using the *Merge* function from the Geoprocessing toolbox, which combined the Slănic and Bertea subbasins with Vărbilău into a single, unified basin. The corrected road network was then integrated into this watershed polygon, with each segment characterized by multiple attributes relevant to the accessibility model. Hydrological and elevation data were incorporated using the *Clip* function, which extracted the river network and DEM layers for the study area.

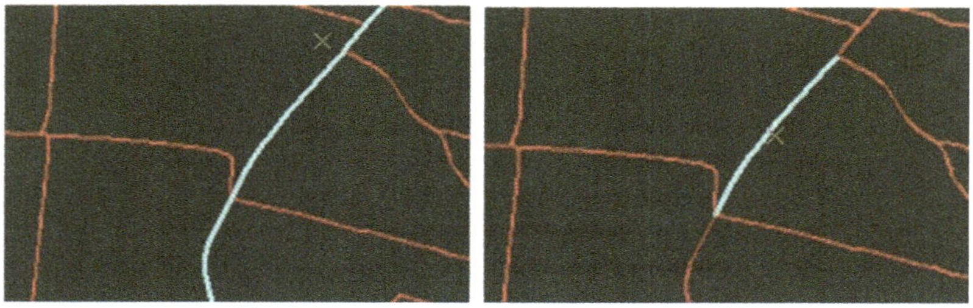

(a) (b)

Figure 5. (**a**) A road section with an error and (**b**) a road segment without errors.

The length of each road segment was automatically calculated based on distances between intersections or from road edges to the nearest junction. Travel time was estimated by adding a new "minutes" field to the road attribute table and computing values with the Field Calculator. A new network dataset was then created in the GIS Catalog, using the minutes field as a cost attribute for network analysis [27].

The Service Area feature of ArcGIS Network Analyst was employed to calculate the regions accessible within specified time intervals from each emergency facility. Four facilities were defined: Câmpina, Vălenii de Munte, and Ploieşti (serving as entry points into the Vărbilău Catchment), and Slănic (the only ISU center located within the catchment). Default time thresholds [28] were set at <10 min, 10–20 min, 20–30 min, 30–40 min, and >40 min. These intervals represent estimated driving times from each facility into surrounding areas. Based on these parameters, service area maps were generated, allowing for the identification of potential coverage gaps and regions at risk of delayed intervention.

2.3. Data Collection and Analysis

To quantify accessibility, a scoring system was applied to travel times: a score of 5 was assigned to areas reachable in less than 10 min, 4 to areas within 10–20 min, 3 to 20–30 min, 2 to 30–40 min, and 1 to areas requiring more than 40 min [29]. Raster layers were generated in GIS for each of the four ISU emergency units, resulting in individual accessibility maps. A composite accessibility map for the entire Vărbilău Catchment was then created using the Raster Calculator tool. This map was obtained by summing the raster layers of the four ISU centers, thereby integrating their respective service areas into a single representation of emergency accessibility. The composite map reflects the combined influence of the road network and the spatial distribution of intervention facilities, offering a comprehensive overview of accessibility conditions across the catchment [30].

To refine this model, a weighting system was incorporated into the raster-based analysis by applying multipliers that accounted for the relative importance and operational capacity of each intervention center [31], i.e.,

$$R = MapS + MapC + 2 \times MapV + 4 \times MapP \quad (1)$$

where

- R = the sum of rasters of the time travel maps for each emergency point, taking into account the road network (general accessibility map);
- MapS = the raster for Slănic ISU point;
- MapC = the raster for Câmpina ISU point;
- MapV = the raster for Vălenii de Munte ISU point;
- MapP = the raster for Ploieşti ISU point.

No multiplier was applied to the Slănic center, as it is a minor facility located within the Vărbilău Catchment. Similarly, the Câmpina center, although a major facility, was not weighted because of its distance from the catchment (approximately 20 km) and its limited accessibility via a winding, hilly route. In contrast, the Vălenii de Munte center, located only 5 km from the catchment boundary and benefiting from easier access, was assigned a multiplier of 2. The Ploieşti center, situated about 12 km away and hosting two major ISU detachments, was given the highest weight, with a multiplier of 4 [32]. These multipliers were integrated in the raster calculation, allowing the composite accessibility model to more accurately represent the relative influence and capacity of each intervention.

The general accessibility map was generated using a raster calculation formula, resulting in values ranging from 8 (indicating low accessibility) to 28 (indicating high accessibility). To classify these values, five intervals were established using the *Natural Breaks* classification method within the GIS environment. This method identifies natural groupings in the data distribution to define class boundaries. The resulting intervals were then imported into the KNIME tool, where they were reclassified into five accessibility categories using the *Rule Engine* node. This node functions similarly to a formula in Excel, allowing conditional logic to reassign values. The reclassified output is presented in Table 2. Reclassification was essential to improve the interpretability of accessibility data.

Table 2. General accessibility value reclassification.

Intervals	Reclassification
8–9	1
9–11	2
11–16	3
16–22	4
22–28	5

The classification scheme is as follows:

- Value 1 (8–9): Least served areas;
- Value 2 (9–11): Poorly served areas;
- Value 3 (11–16): Moderately served areas;
- Value 4 (16–22): Well-served areas;
- Value 5 (22–28): Best served areas.

This step simplifies the visualization and interpretation of accessibility levels across the Vărbilău Catchment, enabling decision-makers to more easily identify areas where improvements to the emergency response infrastructure may be required.

The following section presents the methodology for analyzing two scenarios related to emergency road accessibility. These scenarios simulate potential constraints in the availability of intervention centers, providing insight into how emergency coverage may shift under limited-resource conditions.

- Scenario 1 assumes the unavailability of the Ploieşti intervention center. This situation could arise if teams from Ploieşti are deployed to other emergencies outside the catchment, forcing reliance on the remaining centers.
- Scenario 2 considers a more severe case in which both the Ploieşti and Vălenii de Munte teams are unavailable. This reflects a situation where simultaneous emergencies in neighboring areas require units from other locations—such as Câmpina or Slănic—to assume full responsibility for the Vărbilău Catchment.

For Scenario 1, the general accessibility formula was adjusted to exclude the Ploieşti multiplier and raster layer. The formula applied was as follows:

$$R = MapS + MapC + 2 \times MapV \quad (2)$$

where

- R = composite raster of travel-time maps for available intervention centers (general accessibility map);
- MapS = the raster for Slănic ISU point;
- MapC = the raster for Câmpina ISU point;
- MapV = the raster for Vălenii de Munte ISU point.

In this case, the Vălenii de Munte ISU center was assigned a weight of 2 to reflect its greater relative importance compared to the Slănic and Câmpina centers, owing to its proximity to the Vărbilău Catchment and its strategic role in emergency response. Individual accessibility maps were first generated for each of the three centers, then reclassified into the five-class scale. These reclassified rasters were then subsequently merged to produce a composite accessibility map representing the intervention capacity in the absence of the Ploieşti unit.

For Scenario 2, the general accessibility formula was further reduced to exclude both the Ploieşti and Vălenii de Munte centers. The formula applied was as follows:

$$R = MapS + MapC, \quad (3)$$

where

- R = composite raster of travel-time maps for the available intervention centers (general accessibility map);
- MapS = the raster for Slănic point;
- MapC = the raster for Câmpina point.

In this scenario, only the Slănic and Câmpina centers were considered, reflecting a severe limitation of emergency resources within the catchment. To refine the results, obstacles were introduced in GIS to simulate conditions such as blocked roads, thereby generating updated accessibility maps. Subsequently, hydraulic modeling was performed in Hydrologic Engineering Center River Analysis System (HEC-RAS) to delineate flooded surfaces. These outputs were exported into GIS and overlaid with the accessibility layers, enabling the assessment of how flooding further constrains emergency response capacity under this worst-case scenario.

3. Results

3.1. Accessibility from the Slănic ISU Center

The Slănic ISU center is strategically located within the Vărbilău Catchment, making it highly effective in serving Slănic town and its immediate surroundings, typically within 10 min. This efficiency is supported by the area's favorable topography, high road density, and modernized infrastructure. The extensive and well-maintained network ensures broad coverage across the locality. However, accessibility declines with increasing distance from the center. The least served zones are in the southern, western, and northern parts of the catchment.

Travel times to nearby localities such as Bertea and Ştefeşti are 10–20 min. Although Ştefeşti is geographically close, its hilltop location forces emergency teams to either use a narrow-elevated road or a longer detour; both routes result in similar travel times. More distant localities, including Vărbilău, Scurteşti (north of Ştefeşti), and Aluniş, require

20–30 min. A major concern in the area is the vulnerability of the main road between Slănic and Vărbilău to flooding. Peripheral areas face additional challenges. In western Aluniş, Lutu Roşu, and southern Vărbilău, unpaved or degraded roads increase travel times to 30–40 min. The most remote areas, such as Dumbrăveşti and Sfârleanca, require over 40 min to reach.

The maps of travel time from Slănic and Câmpina ISU are presented in Figure 6. Despite these limitations, the Slănic center holds strategic value. Centrally located near key assets such as the salt mine and the train station, it enables rapid intervention in the basin's most populated areas.

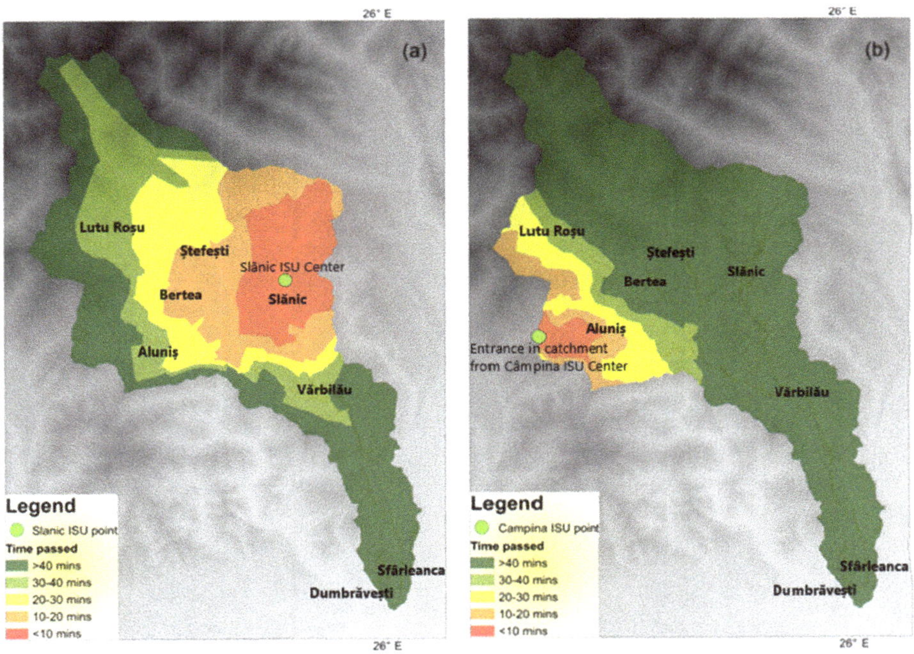

Figure 6. Time travel maps: (**a**) accessibility from Slănic; (**b**) accessibility from Câmpina.

However, its equipment and resources are limited, and large-scale emergencies—particularly flooding—are likely to require intervention from the better-equipped Vălenii de Munte team.

3.2. Accessibility from the Câmpina ISU Center

The Câmpina ISU center lies west of the Vărbilău Catchment and serves as a primary entry point for emergency units. Access is via the Aluniş–Câmpina road, a winding and narrow route that requires 40–50 min to reach the catchment entrance. Once inside the catchment, the best-served locality is western Aluniş (~10 min from the entrance). Given similar response times from Câmpina and Vălenii de Munte (approximately 60 min total), the Câmpina unit is likely to respond here if the Vălenii team is unavailable. Second-tier accessibility zones (10–20 min) include eastern Aluniş, the flood-prone Strâmba neighborhood, and Ostrovu village [3].

Other areas are less accessible. Lutu Roşu, prone to landslides, requires 20–30 min under favorable weather, but poor conditions can significantly delay response. Livadea, between Aluniş and Vărbilău, lies within the 30–40 min range. The least served areas from

Câmpina include Slănic, Vărbilău, and Dumbrăveşti, due to their distance from the western entry point. Bertea is particularly disadvantaged, requiring 30–40 min of travel inside the catchment in addition to the 40–50 min needed to reach the entrance. Accessibility here is further reduced by a sparse, lower-quality road network and terrain constraints.

Despite its robust equipment and operational capabilities, the Câmpina center's major limitation is its location outside the catchment and the time-consuming route required to access it. As outlined in the Section 2.2, the Slănic and Câmpina centers were not assigned multipliers in the emergency accessibility maps, reflecting their secondary role compared to the more influential Ploieşti and Vălenii de Munte facilities.

3.3. *Accessibility from the Vălenii de Munte ISU Center*

The Vălenii de Munte ISU center, located at the southeastern entrance of the catchment, provides relatively rapid intervention to the nearby localities. The best-served area is Coţofeneşti village (southern Vărbilău), located only about 5 km from the entrance. The southeastern locality of Poiana Vărbilău can also be reached in under 10 min, making it the most accessible settlement from this unit. The southern section of Vărbilău village lies within the 10–20 min zone, while central Vărbilău falls into the 20–30 min range. Northern Vărbilău and Dumbrăveşti, despite their denser road networks, require 30–40 min. Overall, a significant portion of the central and northern catchment remains underserved by Vălenii de Munte due to distance and network limitations. Travel times exceed 40 min for more distant areas such as Slănic, Bertea, and Ştefeşti—exceeding 50 min when accounting for the 10 min approach from Vălenii to the catchment entrance.

The elongated north–south alignment of the southern catchment, combined with a single access road and hilly terrain, further constrains emergency coverage. Nevertheless, Vălenii de Munte remains closer than Câmpina, providing a partial advantage despite these challenges.

Figure 7 presents the time travel maps from Vălenii de Munte and Ploieşti.

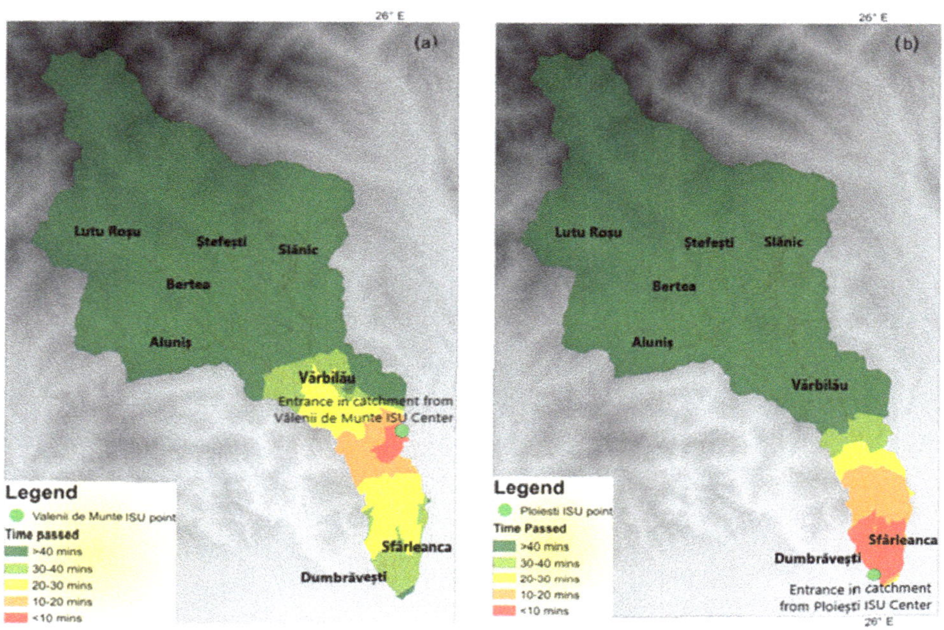

Figure 7. Time travel maps: (**a**) accessibility from Vălenii de Munte; (**b**) accessibility from Ploieşti.

3.4. Accessibility from the Ploieşti ISU Center

The Ploieşti ISU facility, located just south of Dumbrăveşti, serves the southern part of the catchment efficiently. Benefiting from a dense, modern road network and flat terrain, it ensures travel times of less than 10 min to Dumbrăveşti and Sfârleanca. Settlements immediately north of Dumbrăveşti fall within the 10–20 min zone, while Poiana Vărbilău lies within 20–30 min. However, travel times to central and northern settlements, including Vărbilău village and Slănic, exceed 40 min, owing to longer distances and topographic barriers. The approach from Ploieşti to the catchment entrance itself takes ~25 min.

Ploieşti hosts two fully equipped ISU emergency centers, reflecting its status as a major urban center. Given its resources and relative proximity, Ploieşti is expected to play a primary role (together with Vălenii de Munte) in managing hazard events in the catchment. In contrast, Slănic and Câmpina are more likely to play only secondary or localized roles.

According to the emergency accessibility methodology, Ploieşti was assigned a multiplier of 4, reflecting its dual facilities, superior equipment, and relatively short distance to the catchment. Vălenii de Munte received a multiplier of 2, acknowledging its good equipment and closer location, albeit with only one center.

Based on Figures 6 and 7, the southern Vărbilău village, Dumbrăveşti, Slănic, and western Aluniş represent areas that are reachable within 10 min from the nearest intervention center. Figure 8 presents the accessibility index map, created using GIS raster analysis tools such as the Raster Calculator and reclassification methods, which synthesize the time travel results into a comprehensive spatial index.

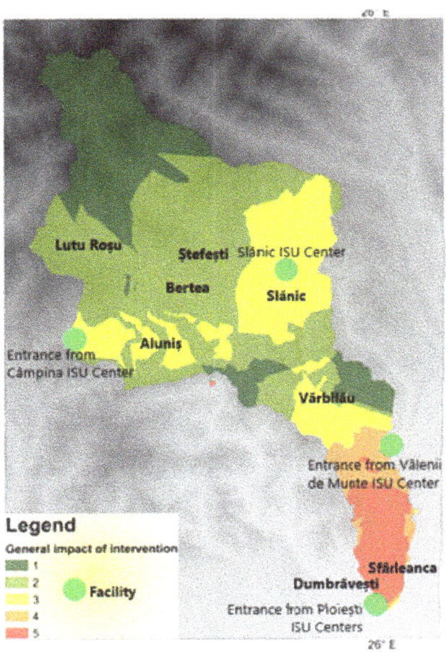

Figure 8. Emergency accessibility map of the Vărbilău Catchment. Marks are given from 1 (lowest accessibility) to 5 (highest accessibility).

The highest accessibility zones (assigned scores of 4 and 5) are concentrated in the southern part of the Vărbilău catchment, particularly around Dumbrăveşti. This area benefits from a dense and high-quality road network, short travel distances, and proximity to well-equipped ISU centers in Ploieşti and Vălenii de Munte, which hold greater operational significance compared to the smaller facilities at Slănic and Câmpina. Moderately accessible areas (scoring 3) are distributed across the northeastern, central, and parts of the western regions of the catchment. These intermediate scores are influenced by longer travel distances from Câmpina, the lower strategic importance of the Slănic team, and reduced road density in key localities such as Vărbilău and Aluniş. The least served areas (score 2) include Ştefeşti, Bertea, and parts of the Vărbilău commune, where hilly terrain, lower road density and quality, and greater distance from emergency facilities pose significant challenges for timely intervention. Finally, the least accessible or inaccessible zones (scoring 1) are found primarily in the northern part of the catchment, where mountainous terrain dominates, as well as in the eastern and western outskirts of the Vărbilău commune, which are also located in rugged, hilly landscapes. These areas present the highest risk for delayed or obstructed emergency response.

The accessibility in the Vărbilău Catchment is summarized in Table 3.

Table 3. Accessibility in the Vărbilău Catchment from different points.

Village/Area	Accessibility Score	Travel Time (min)	Nearest/Primary ISU Center	Notes
Dumbrăvești	5	<10	Ploiești	High-quality roads, flat terrain, dense network
Sfârleanca	5	<10	Ploiești	High accessibility
Southern Vărbilău	5	<10	Ploiești Vălenii de Munte	Well-served, dense road network
Western Aluniș	5	<10	Câmpina	Best served from western entry
Coțofănești	4–5	<10	Vălenii de Munte	Close to SE catchment entrance
Poiana Vărbilău	4–5	<10	Vălenii de Munte	High accessibility in SE catchment
Central Vărbilău	3	20–30	Vălenii de Munte	Moderate accessibility, low road density
Eastern Aluniș Strâmba, Ostrovu	3–4	10–20	Câmpina	Flood-prone areas, moderate roads
Bertea	2	20–40	Slănic Vălenii de Munte	Hilly terrain, low road density
Ștefești	2	20–40	Slănic Vălenii de Munte	Located over hill, limited road access
Lutu Roșu	2	20–30 (good weather)	Câmpina	Landslide-prone, access sensitive to weather
Slănic town	3	10–20	Slănic	Proximity compensates or lower equipment capacity
Northern Vărbilău outskirts	1	40+	Vălenii de Munte	Mountainous, highest risk for delayed response
Eastern/western outskirts of Vărbilău	1	40+	Slănic Vălenii de Munte	Rugged, hilly landscape
Livadea	3	30–40	Câmpina	Between Aluniș and Vărbilău
Aluniș (eastern)	3–4	10–20	Câmpina	Moderate accessibility, western part better served

3.5. *Accessibility in Scenario 1*

In Scenario 1, where the ISU Ploiești center is unavailable, the best-served areas (scoring 4 and 5) are concentrated in the central–southern part of the catchment, especially within Vărbilău commune (including villages such as Coţofeneşti and Poiana Vărbilău). Their good accessibility is largely due to their proximity to the Vălenii de Munte center, which proves more strategically important than the Slănic and Câmpina facilities. Nevertheless, these high-accessibility areas remain limited in extent, constrained by sparse road networks, inconsistent road quality, and restricted access from the remaining operational ISU centers.

By contrast, the least served zones are located mainly in the northern and central sectors, such as Bertea, Ştefeşti, and Lutu Roşu. These areas are defined by poor road infrastructure and geographical isolation in hilly or forested terrain, resulting in intervention times exceeding 40 min and accessibility scores of only 1 or 2. In such cases, emergency response would most likely depend on the Slănic and Câmpina intervention teams. The Slănic area itself scores moderately (3) but, due to its relative proximity to several vulnerable locations, remains the most probable intervention point in this scenario. This outcome underscores that local proximity can offset lower institutional capacity, emphasizing the importance of accessibility in effective emergency planning.

Overall, Scenario 1 results in a reduced yet still functional emergency service coverage compared to the baseline scenario that includes Ploiești. In this context, the Vălenii de Munte ISU team is the most likely to be mobilized, owing to its proximity to the Vărbilău basin and its stronger equipment and operational capacity relative to the Slănic center.

Accessibility scores have been reclassified on a 1–5 scale, and the accessibility map corresponding to Scenario is presented in Figure 9a.

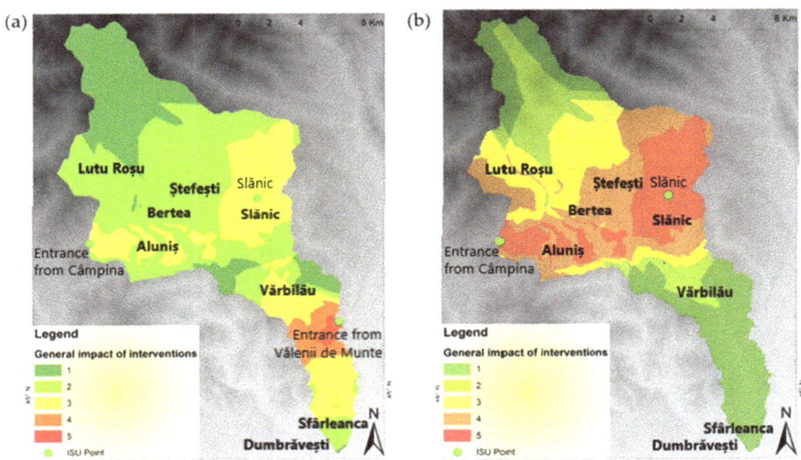

Figure 9. (**a**) Accessibility when ISU Ploieşti unit is unavailable; (**b**) accessibility when both ISU units from Ploieşti and Văleni are unavailable. Marks are given from 1 (lowest accessibility) to 5 (highest accessibility).

3.6. Accessibility in Scenario 2

Scenario 2 is illustrated in Figure 9b, which shows the reclassified accessibility map when only the Slănic and Câmpina ISU centers are operational. The best-served areas (scores 4 and 5) include Aluniş, Ostrovu, and Slănic, benefiting from their proximity to the Slănic center and the Câmpina entrance point. However, this does not imply superior service compared to Scenario 1, when the Ploieşti and Vălenii de Munte centers were still active. In this scenario, the southern part of the catchment becomes the least served (scores 1 and 2), since two critical centers—Ploieşti and Vălenii de Munte, both essential due to their equipment and closeness to densely populated zones—are excluded. For instance, Dumbrăveşti and Vărbilău village, with higher population densities, experience a substantial decline in accessibility. Vărbilău, in particular, scores only 1–2, corresponding to intervention times exceeding 40 min from Slănic [33]. By contrast, Aluniş can be reached within 10–20 min from the Câmpina entrance, while Slănic and its surroundings are accessible in under 10 min. Within this limited context, the Slănic ISU emerges as the most likely facility to be mobilized, mainly due to geographic proximity, despite its restricted resources.

This scenario highlights the importance of maintaining multiple, well-equipped emergency facilities. Reclassifying travel times into accessibility scores (1–5) proves crucial for visualizing relative service levels, enabling planners to quickly identify underserved areas. It also reinforces the critical role of facility location, road network quality, and response time in emergency planning. In general, a greater number of operational centers translates into faster response times and improved redundancy.

4. Discussion

Following the above analyses, flood extent and impact are examined for selected areas of the Vărbilău Catchment to demonstrate the importance of maintaining good accessibility and understanding the time required for emergency interventions.

4.1. Critical Accessibility Cases—Lutu Roșu and Sfârleanca

Two examples further illustrate the compounding effects of hydrologic hazards on accessibility within the Vărbilău catchment: Lutu Roșu and Sfârleanca. In Lutu Roșu, a landslide destroyed the only access road (DJ 101T), completely isolating the village. As shown in Figure 10a, no alternative route connects its approximately 200 households with the neighbor village, Bertea, making emergency access impossible for ISU vehicles during such events.

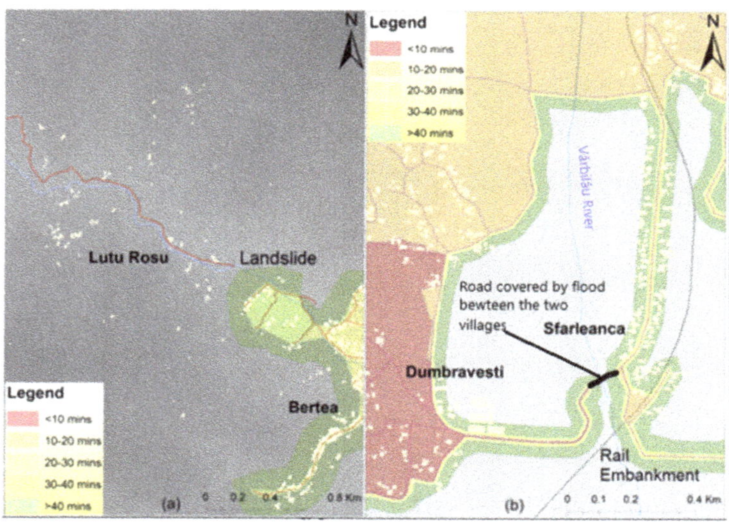

Figure 10. Discontinuities in the ISU accessibility caused by (**a**) landslide at Lutu Roşu and (**b**) flood at Dumbrăveşti.

Field observations confirmed that during heavy rainfall, even when the access road to Lutu Roșu remains technically passable, it often becomes muddy and unsafe, forcing authorities to close it temporarily. Although a modernized road now connects the village externally, the internal road network remains gravel-paved, offering poor accessibility under adverse weather conditions. This chronic isolation, compounded by weak infrastructure and limited economic opportunities, has contributed to gradual depopulation, with many residents relocating to urban areas.

In Sfârleanca, a different hazard—flooding—periodically cuts off road access between Dumbrăveşti and the village, particularly near a railway embankment (Figure 10b). Under normal conditions, ISU crews from Ploieşti can reach Dumbrăveşti in less than 10 min and Sfârleanca within 10–20 min. However, when floodwater submerges the road, crews are forced to reroute northward, extending travel times to approximately 40 min. Flood modeling [25] confirms that water accumulation near the railway embankment expands the floodplain, severely disrupting accessibility.

4.2. Flooding in Vărbilău

Another event that is briefly presented to support is the flooding that took place on July 2018 in the Vărbilău village (served either by Slănic ISU Center of Vălenii de Munte), when the recorded discharges were 98.6 m^3/s for the Slănic River and 199 m^3/s for the Vărbilău River before their confluence. Out of 28,398 m of roads, 2029 m (7.14%) were flooded, while 139 buildings (4.69%) out of 2963 were affected. The total flooded area was 0.75 km^2, with greater expansion downstream of the confluence. Simulations of this event and two other scenarios are presented in detail in [34], indicating the following:

- For a reduced flow (50 m^3/s for the Slănic River and 100 m^3/s for the Vărbilău River), the flood extent decreased slightly, affecting 5.61% of roads and 3.75% of buildings, with a total area of 0.62 km^2.
- For an increased flow (150 m^3/s for the Slănic River and 300 m^3/s for the Vărbilău River): Flooding impacted 2274 m of roads and 151 buildings, covering 0.81 km^2.

4.3. Flooding in Aluniş (June 2021)

In June 2018 and June 2021, flash-flood events happened at the confluence between the Strâmba and Aluniș Rivers, affecting the Aluniș village. Simulation conducted for the last event, conducted for steady flow conditions, with the measured discharges of 50 m^3/s for Strâmba, 100 m^3/s for Aluniş before the confluence, and 150 m^3/s after the confluence, indicated that flooding affected a significant number of buildings due to the villages' location along narrow valleys. The confluence area, relatively flat and exposed to flows from both north and west, shows the largest flood extent. In Strâmba, flooding exhibited flash-flood characteristics due to the steep slopes, isolating several houses as road sections became submerged. The Civic Center was the most affected, e.g., [3]:

- The school (830 m^2), town hall (248 m^2), and cultural hall (526 m^2) completely flooded.
- The church partially flooded, with ~220 m^3 of water entering the building.
- A total of 285 houses and annexes were inundated, with a combined flooded perimeter of nearly 20,000 m.
- Flooded road length reached 3.5 km, leaving many houses completely isolated.

4.4. Comparison of Scenarios 1 and 2 and Linkage to Resilience Literature

The outcomes of Scenarios 1 and 2 (synthesized in Table 4) can be seen within broader resilience theory, particularly through the framework of the 4Rs—robustness, redundancy, resourcefulness, and rapidity—proposed by Bruneau et al. [35].

Robustness refers to the inherent strength of a system to withstand disruption without significant loss of function. In Scenario 1, the continued operation of the Vălenii de Munte center demonstrates robustness, as its higher equipment levels and strategic positioning buffer against the loss of Ploiești. Conversely, Scenario 2 reveals a fragile system, where reliance on only Slănic and Câmpina leads to substantial accessibility gaps in high-population areas such as Vărbilău and Dumbrăveşti.

Table 4. Comparison of Scenarios 1 and 2.

Aspect	Scenario 1	Scenario 2
Main intervention hub	Vălenii de Munte—most likely mobilized, better equipment & capacity. Typical times: 10–20 min for nearby Vărbilău basin; <10 min for the closest villages.	Slănic becomes primary by proximity. Typical times: <10 min in Slănic town; 20–30 min for peripheral Slănic-area sites.
Best-served areas	Vărbilău commune (Coţofeneşti, Poiana Vărbilău)—concentrated central-south. Typical times: <10–20 min.	Aluniş, Ostrovu, Slănic—close to Slănic and Câmpina. Typical times: <10 min—Slănic, 10–20 min—Aluniş.
Least-served areas	Bertea, Ştefeşti, Lutu Roşu—northern/central sectors; poor roads, hilly/forested. Typical times: >40 min or 30–40 min.	Southern catchment (Dumbrăveşti, Vărbilău village). Typical times: >40 min (Vărbilău often >40).
Slănic area performance	Moderate. Being relatively near several vulnerable spots, it remains a likely intervention point for some incidents. Typical times: 20–30 min for many surrounding locations.	Strong locally (score 4–5 in town) but cannot fully compensate for lost centers. Typical times: <10–20 min depending on exact spot.
Accessibility for high-population zones	Relatively preserved in the Vărbilău basin (thanks to Vălenii de Munte), <10–20 min for many populated villages.	Declines sharply. Example: Vărbilău scores 1–2 → >40 min from Slănic; Dumbrăveşti similarly affected.
Overall service level/coverage	Reduced but functional. Pockets of high accessibility remain (central-south). Many peripheral/hilly areas still exceed 40 min.	Much weaker than Scenario 1. Major underserved zones in the south; higher population areas suffer long delays (>40 min).
Key limiting factors	Sparse road network, inconsistent road quality, limited access from remaining centers → drives some locations into 30–40 min ranges.	Loss of two strategic centers increases travel distances and forces reliance on Slănic/Câmpina → many places move into the >40 min band.

Redundancy—the presence of substitutable components that provide backup during failure—is central to both scenarios. Scenario 1 illustrates how redundancy through Vălenii de Munte compensates for the absence of Ploiești, ensuring that emergency response remains viable across much of the catchment. In Scenario 2, the erosion of redundancy (loss of both Ploiești and Vălenii de Munte) leaves the system highly vulnerable, with response times exceeding 40 min in several zones.

Resourcefulness denotes the capacity to mobilize assets and reconfigure operations under constrained conditions.

In both scenarios, the relative importance of smaller units (Slănic, Câmpina) increases, showing how geographic proximity can partly offset lower institutional strength. This adaptive use of available centers exemplifies resourcefulness but also underscores its limitations when facilities lack sufficient capacity.

Rapidity concerns the speed with which service is restored, or response is delivered. The accessibility scoring demonstrates that rapidity is spatially uneven: while Slănic ensures responses under 10 min locally, other areas such as Bertea or Dumbrăveşti fall well outside acceptable thresholds (>40 min), highlighting inequities in resilience across the network. These findings align with the broader literature on resilience, which stresses that both network redundancy and robustness are critical in mitigating hazard impacts [36,37].

Redundancy ensures flexibility and fallback options when nodes fail, while robustness determines whether remaining components can sustain acceptable performance levels. The case study illustrates how strategic facility location, road connectivity, and institutional capacity interact to determine resilience outcomes. As Holling [38] emphasized, resilient systems are those that can absorb shocks and reorganize without collapsing; in this case, Scenario 1 approximates such resilience, while Scenario 2 exposes systemic fragility.

4.5. Sensitivity Analysis

The comparison regarding road accessibility for intervention teams was also conducted to evaluate how travel speed influences accessibility. In GIS, speeds of 50–60 km/h were initially assigned to modernized roads (Figure 11a). These values were then reduced to 30 km/h on the same roads to observe changes in served areas and to assess the overall impact of reduced travel speed on accessibility. For non-modernized and forest roads, travel speeds remained unchanged.

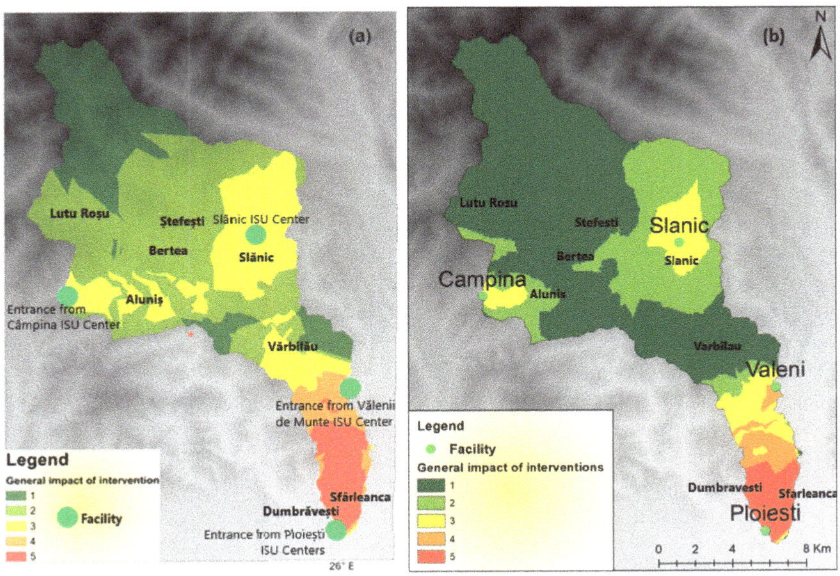

Figure 11. Accessibility comparison in the case of 50–60 km on modernized roads (**a**) and modified speeds (30 km/h) on modernized roads (**b**).

The speed of 50 km/h is the legal limit inside localities, but Romanian legislation permits an increase of up to 10 km/h in emergency situations. The limit of 30 km/h was selected based on the travel speeds experienced during flooding periods in the region.

After recalculating accessibility using the same raster-based method applied in the initial analysis, the new spatial distribution revealed notable differences. Specifically, well-served areas decreased in size (Figure 11b), while less-served areas expanded. From this point, the discussion will focus on the results obtained with the reduced travel speed (30 km/h), since the baseline situation was already presented.

At a travel speed of 30 km/h, the area with accessibility index 5 (very good accessibility) is reduced to approximately half of the original extent (Figure 11b). Under these conditions, only the villages of Dumbrăveşti and Sfârleanca remain very well served by ISU teams. This is due to the high density of modernized roads and the proximity of three ISU centers (Vălenii de Munte, Ploieşti 1, and Ploieşti 2). The area with accessibility index 4 (well-served) also decreases compared to the initial situation. It now largely overlaps with the northern half of the previous best-served area, covering the north of Dumbrăvești and the south of Vărbilău commune.

A small well-served area is also visible near the access road from Vălenii de Munte. Under these conditions, interventions in index 4 areas would most likely be carried out from Vălenii de Munte.

The area with accessibility index 3 (moderate accessibility) is significantly smaller than in the initial analysis. Notably, the center of Vărbilău commune has declined from index 3 to index 1 (very poorly served). Remaining index 3 areas are found only in Slănic, Aluniş, and the southern part of Vărbilău, where interventions would likely be covered by the ISU centers in Vălenii de Munte, Slănic, and Câmpina. This reduction is mainly explained by greater distances from intervention points, the limited number of ISU centers in the north and center, and the poorer quality of the road network.

Areas with index 2 surround those with index 3, particularly around Aluniş and Slănic, as well as a smaller section of Vărbilău commune. However, their extent is also smaller compared to the initial situation, when index 2 covered the largest area.

In contrast, the area with accessibility index 1 (very poorly served) increases substantially in the 30 km/h scenario. These poorly served zones now dominate much of the north and center, including Ştefeşti, Vărbilău, Ostrovu, and Bertea. This is explained by the low travel speed, poor road quality, low road density, and the long distances from ISU centers.

As seen above, an important consideration in an analysis such as ours concerns the choice of weights assigned to road types, travel speeds, and facility importance. While the weighting scheme was informed by field observations, it inevitably involves a degree of subjectivity. Different weighting assumptions can shift the spatial pattern of accessibility scores, with direct implications for where ISU teams are expected to intervene first during emergencies. For instance, assigning greater weight to main roads emphasizes connectivity along the southern corridor, while increasing the influence of secondary and tertiary routes may highlight vulnerabilities in northern communes. This sensitivity suggests that emergency planning must recognize weighting as not only a technical decision but also a strategic one, since it determines priority zones for infrastructure investment, resource allocation, and mobilization protocols. To address this limitation, future research could conduct sensitivity analyses or apply multi-criteria decision-making (MCDM) approaches to test how different weighting schemes alter accessibility outcomes. Such efforts would enhance the robustness of planning tools and provide ISU teams with more flexible guidance under varying hazard conditions.

5. Conclusions

This study assessed road accessibility for ISU intervention teams in the Vărbilău Catchment, Romania, using GIS-based modeling and field observations. The results provide actionable insights for emergency preparedness in flood- and landslide-prone regions. The core findings are as follows:

- Intervention speed is determined primarily by the proximity of ISU centers, but the road quality and matter as much as the number of ISU centers.
- Southern communes such as Dumbrăvești achieved the highest accessibility due to flat terrain and strong links to Ploiești and Vălenii de Munte.
- Northern and central communes remained poorly served because of steep terrain and sparse road infrastructure.
- Scenario testing showed that maintaining multiple operational ISU units—especially Ploiești and Vălenii de Munte—is critical for system resilience.

Pros and cons of the approach are the following:

- Strengths: Integrates GIS modeling with field validation, produces practical accessibility maps, and enables scenario-based evaluation.
- Limitations: Relies on static travel-time assumptions, does not capture real-time hazard dynamics (e.g., floods, blockages), and simplifies operational factors such as team capacity.

The following policy recommendations are issued from the analysis:

- Safeguard redundancy by ensuring continuous operation of major ISU centers and setting up new centers in the worst served areas.
- Upgrade road infrastructure in underserved northern and central communes to reduce systemic vulnerabilities.
- Regularly update accessibility maps to reflect evolving hazard and infrastructure conditions.

Future research directions will include the following:

- Incorporate real-time traffic and hazard data into dynamic modeling.
- Simulate cascading infrastructure failures to test network robustness under compound hazards.
- Apply the methodology to other watersheds with different characteristics (e.g., mountainous basins, floodplains) to evaluate its transferability and refine its applicability.

Author Contributions: Conceptualization, C.P. and A.B.; methodology, C.P. and A.B.; software, C.P.; validation, A.B.; formal analysis, A.B.; investigation, C.P. and A.B.; resources, C.P.; data curation, C.P.; writing—original draft preparation, C.P. and A.B.; writing review and editing C.P. and A.B.; visualization, C.P.; supervision, A.B.; project administration, A.B.; funding acquisition, A.B. All authors have read and agreed to the published version of the manuscript.

Funding: This research received no external funding.

Data Availability Statement: The original contributions presented in this study are included in the article material. Further inquiries can be directed to the corresponding author.

Conflicts of Interest: The authors declare no conflicts of interest.

References

1. Chirițescu, R.-V.; Luca, E.; Iorga, G. Observational study of major air pollutants over urban Romania in 2020 in comparison with 2019. *Rom. Rep. Phys.* **2024**, *76*, 702.
2. Dumitru, A.; Olaru, E.-A.; Dumitru, M.; Iorga, G. Assessment of air pollution by aerosols over a coal open-mine influenced region in southwestern Romania. *Rom. J. Phys.* **2024**, *69*, 801. [CrossRef]
3. Popescu, C.; Bărbulescu, A. Flooding simulation at the confluence between the Strâmba and Aluniș Rivers in the Vărbilău Catchment, in Romania. *Rom. Rep. Phys.* **2025**, *77*, 705. [CrossRef]
4. Zang, Y.; Huang, J.; Wang, H. Dynamic impact assessment of urban floods on the compound spatial network of buildings-roads-emergency service facilities. *Sci. Total Environ.* **2024**, *926*, 172007. [CrossRef]
5. Berdica, K. An introduction to road vulnerability: What has been done, is done and should be done. *Transp. Policy* **2002**, *9*, 117–127. [CrossRef]
6. Borowska-Stefańska, M.; Wiśniewski, S. The Role of Road Transportation in the Flood Evacuation Process. In *Oxford Research Encyclopedia of Natural Hazard Science*; Oxford University Press: Oxford, UK, 2022.
7. Papilloud, T.; Keiler, M. Vulnerability patterns of road network to extreme floods based on accessibility measures. *Transp. Res. Part D Transp. Environ.* **2021**, *100*, 103045. [CrossRef]
8. Gangwal, U.; Siders, A.R.; Horney, J.; Michael, H.; Dong, S. Critical facility accessibility and road criticality assessment considering flood-induced partial failure. *Sustain. Resilient Infrastruct.* **2023**, *8*, 337–355. [CrossRef]
9. Florath, J.; Chanussot, J.; Keller, S. Road Accessibility during Natural Hazards Based on Volunteered Geographic Information Data and Network Analysis. *ISPRS Int. J. Geo-Inf.* **2024**, *13*, 107. [CrossRef]
10. Loreti, S.; Ser-Giacomi, E.; Zischg, A.; Keiler, M.; Barthelemy, M. Local impacts on road networks and access to critical locations during extreme floods. *Sci. Rep.* **2022**, *12*, 1552. [CrossRef]

11. Morelli, A.B.; Cunha, A.L. Measuring urban road network vulnerability to extreme events: An application for urban floods. *Transp. Res. D Transp. Environ.* **2021**, *93*, 102770. [CrossRef]
12. Ford, A.C.; Barr, S.L.; Dawson, R.J.; James, P. Transport Accessibility Analysis Using GIS: Assessing Sustainable Transport in London. *ISPRS Int. J. Geo-Inf.* **2015**, *4*, 124–149. [CrossRef]
13. Ertugay, K.; Duzgun, S.H. Understanding Accessibility: Accessibility Modeling with Geographical Information System (GIS). In *Intelligent Transportation and Planning*; Information Resources Management Association, IGI Global Scientific Publishing: New York, NY, USA, 2018; pp. 576–608. [CrossRef]
14. Stancalie, G.; Diamandi, A.; Corbus, C.; Catana, S. Application of EO data in flood forecasting for the Crisuri basin, Romania. In *Flood Risk Management: Hazards, Vulnerability and Mitigation Measures*; Schanze, J., Zeman, E., Marsalek, J., Eds.; NATO Science Series; Springer: Dordrecht, The Netherlands, 2006; Volume 67, pp. 101–113. [CrossRef]
15. Minea, G. Assessment of the Flash Flood Potential of Bâsca River Catchment (Romania) Based on Physiographic Factors. *Cent. Eur. J. Geosci.* **2013**, *5*, 344–353. [CrossRef]
16. Popa, M.C.; Peptenatu, D.; Drăghici, C.C.; Diaconu, D.C. Flood Hazard Mapping Using the Flood and Flash-Flood Potential Index in the Buzău River Catchment, Romania. *Water* **2019**, *11*, 2116. [CrossRef]
17. Iosub, M.; Minea, I.; Chelariu, O.E.; Ursu, A. Assessment of flash flood susceptibility potential in Moldavian Plain (Romania). *J. Flood Risk Manag.* **2020**, *13*, e12588. [CrossRef]
18. Romanescu, G.; Hapciuc, O.E.; Minea, I.; Iosub, M. Flood vulnerability assessment in the mountain–plateau transition zone: A case study of Marginea village (Romania). *J. Flood Risk Manag.* **2018**, *11*, S502–S513. [CrossRef]
19. Chelariu, O.-E.; Iațu, C.; Minea, I. A GIS-Based Model for Flood Shelter Locations and Pedestrian Evacuation Scenarios in a Rural Mountain Catchment in Romania. *Water* **2022**, *14*, 3074. [CrossRef]
20. Puie, M.; Mihai, B.-A. Integrated Analysis of Rockfalls and Floods in the Jiului Gorge, Romania: Impacts on Road and Rail Traffic. *Appl. Sci.* **2024**, *14*, 10270. [CrossRef]
21. Bălteanu, D.; Chendeş, V.; Sima, M. GIS landslide hazard map of Romania, tool for disaster management and spatial planning. *GIM Int.* **2009**, *23*, 13–15.
22. Olinic, E.; Manea, S.; Boti, I.; Coman, C.; Burlacu, C.; Pantel, G. On the elaboration of landslide hazard and risk maps in Romania. *Ce/Pap.* **2018**, *2*, 235–242. [CrossRef]
23. Erdeli, G.; Dumitrache, L. *Population Geography*; University Publishing House: Bucharest, Romania, 2010. (In Romanian)
24. Ghinea, D. *Enciclopedia geografică a României*; Editura Enciclopedică: Bucureşti, Romania, 2000.
25. Popescu, C.; Bărbulescu, A. A Practical Approach on Reducing the Flood Impact: A Case Study from Romania. *Appl. Sci.* **2024**, *14*, 10378. [CrossRef]
26. Prăvălie, R.; Costache, R. The analysis of the susceptibility of the flash-floods genesis in the area of the hydrographical basin of Basca Chiojdului river. *Forum Geogr.* **2014**, *XIII*, 39–49. [CrossRef]
27. Nicoara, M.; Haidu, I. Creation of the roads network as a network dataset within a geodatabase. *Geogr. Tech.* **2011**, *2*, 81–86.
28. Bulai, M.; Ursu, A. Creating, testing and applying a GIS road travel cost model for Romania. *Geogr. Tech.* **2012**, *1*, 8–18.
29. Sedeghi-Niaraki, A.; Varshosaz, M.; Kim, K.; Jung, J. Real world representation of a road network for route planning in GIS. *Expert Syst. Appl.* **2011**, *38*, 11999–12008. [CrossRef]
30. Caniupan, M.; Torres-Aviles, R.; Gutierrez-Bunster, T.; Lepe, M. Efficient computation of map algebra over raster stored in the k2-acc compact data structure. *GeoInformatica* **2022**, *26*, 95–123. [CrossRef]
31. Malczewski, J. On the Use of Weighted Linear Combination Method in GIS: Common and Best Practice approaches. *Trans. GIS* **2002**, *4*, 5–22. [CrossRef]
32. Slopek, J.; Netzel, P. Comparison of different implementations of a raster map calculator. *Comput. Geosci.* **2021**, *154*, 104824. [CrossRef]
33. Popescu, C.; Bărbulescu, A.; Dumitriu, C.Ș. Modeling Road Accessibility in a Flood-Prone Area in Romania. *Eng. Proc.* **2023**, *39*, 22. [CrossRef]
34. Popescu, C.; Bărbulescu, A. Flash Flood Simulation Between Slănic and Vărbilău Rivers in Vărbilău Village, Prahova County, Romania, Using Hydraulic Modeling and GIS Techniques. In *Modelling and Development of Intelligent Systems. MDIS 2022. Communications in Computer and Information Science (CCIS)*; Simian, D., Stoica, L.F., Eds.; Springer: Cham, Switzerland, 2023; Volume 1761, pp. 309–327.
35. Bruneau, M.; Chang, S.E.; Eguchi, R.T.; Lee, G.C.; O'Rourke, T.D.; Reinhorn, A.M.; Shinozuka, M.; Tierney, K.; Wallace, W.A.; von Winterfeldt, D. A framework to quantitatively assess and enhance the seismic resilience of communities. *Earthq. Spectra* **2003**, *19*, 733–752. [CrossRef]
36. Cutter, S.L.; Barnes, L.; Berry, M.; Burton, C.; Evans, E.; Tate, E.; Webb, J. A place-based model for understanding community resilience to natural disasters. *Glob. Environ. Change* **2008**, *18*, 598–606. [CrossRef]

37. Tierney, K. *The Social Roots of Risk: Producing Disasters, Promoting Resilience*; Stanford University Press: Redwood City, CA, USA, 2014.
38. Holling, C.S. Resilience and stability of ecological systems. *Annu. Rev. Ecol. Evol. Syst.* **1973**, *4*, 1–23. [CrossRef]

water

MDPI

Article

Spatiotemporal Dynamics of Drought–Flood Abrupt Alternations and Their Delayed Effects on Vegetation Growth in Heilongjiang River Basin

Haoyuan Ma [1,2,†], Jianyu Jing [1,2,†], Changlei Dai [1,*], Yijun Xu [3], Peng Qi [2,*] and Hao Song [1,2]

[1] Institute of Water Conservancy and Electric Power, Heilongjiang University, Harbin 150080, China; hss_mahaoyuan@163.com (H.M.); hss_jingjianyu@126.com (J.J.); 15256166727@163.com (H.S.)
[2] State Key Laboratory of Black Soils Conservation and Utilizations, Northeast Institute of Geography and Agroecology, Chinese Academy of Sciences, No. 4888, Shengbei Street, Changchun 130102, China
[3] Coastal Studies Institute, Louisiana State University, Baton Rouge, LA 70803, USA; yjxu@lsu.edu
* Correspondence: hd_daichanglei@126.com (C.D.); qipeng@iga.ac.cn (P.Q.)
† These authors contributed equally to this work.

Abstract: Drought–flood abrupt alternations (DFAAs) have a greater impact on ecosystems and socioeconomic environments than lone droughts or floods. Despite the significant impact of DFAAs, research has paid little attention to their evolutionary characteristics, particularly in relation to vegetation growth in the Heilongjiang River Basin. Therefore, this study focuses on the Heilongjiang River Basin and employs the DFAA Index to identify and analyze abrupt alternation events from 1970 to 2019. It also examines the annual and interannual distributions of vegetation growth changes from 2000 to 2019, based on the Normalized Difference Vegetation Index. Lastly, it utilizes correlation analysis to investigate the responsive relationship between vegetation growth and DFAA events. The results indicate the following: (1) Within the Heilongjiang River Basin, the number of drought-to-flood events increased over time, whereas the number of flood-to-drought events decreased over time. The frequency of mutation was relatively high in the northern region, low in the eastern region, elevated in spring and summer, and reduced in winter. (2) The Normalized Difference Vegetation Index was lowest in January, highest in July, and approximately 0 during the winter. The vegetation coverage reached its peak during the summer. (3) Vegetation changes in response to DFAAs exhibited a significant time lag. Vegetation changes in spring–summer lagged behind DFAA events by 3–4 months, while in summer–autumn, the lag was approximately 3 months. These results are of great significance for the early warning and prevention of DFAAs in the Heilongjiang River Basin.

Keywords: drought–flood abrupt alteration; NDVI; CWatM; Heilongjiang River Basin

1. Introduction

Drought–flood abrupt alternation (DFAA) events are defined as "the phenomenon wherein drought and flood occur alternately over time or coexist within a spatial context", and the alterations in frequency and intensity present a significant threat to ecosystems, social economy, and regional sustainable development [1,2]. Recently, the spatiotemporal patterns of DFAAs have become increasingly complex. This is, in part, due to the intensification of global climate change. Global warming alters the atmospheric circulation pattern, markedly influences the spatiotemporal distribution of precipitation and the frequency of extreme climate events, and complicates regional water resource regulation [3–5]. Human activities, including but not limited to land use change, urbanization, and vegetation degradation, further exacerbate DFAA risk, rendering their evolutionary characteristics more prominent, both locally and regionally [6]. Vegetation growth is an important factor in ecosystem stability and the hydrological cycle. Its spatiotemporal variation serves as a direct manifestation of climate change and human activities and exerts a feedback effect on DFAAs [7]. Nevertheless, the responses of various vegetation to drought–flood abrupt alternation events differ significantly. Certain vegetation types maintain their stability under DFAA conditions via long-term adaptation. However, other types (such as rice and cotton) suffer severe degradation during extreme climate events [8,9]. Therefore, it is crucial to investigate the response of vegetation growth to DFAA events.

In recent decades, this issue has attracted significant attention from experts and scholars. Research on crops like corn and rice has revealed that DFAA events augment the loss of nitrogen and phosphorus in the soil, curtail the absorption of these elements by crops, and consequently diminish crop yields [10–12]. Several studies have demonstrated that DFAA events can significantly affect vegetation growth by altering the availability of sunlight and soil nutrients, thereby disrupting photosynthesis processes and leading to a reduction in vegetation coverage and species diversity. These effects are particularly pronounced in areas with frequent flooding, where the synergistic and antagonistic impacts of drought and flood alternation exacerbate the stress on vegetation [13,14]. Although previous research has highlighted the widespread impact of DFAA events, their temporal distribution and spatial extent remain inadequately understood, particularly in terms of the abrupt transitions between drought and flood conditions and their lag effects on vegetation. Further research is needed to clarify the specific characteristics of DFAA events and their influence on vegetation dynamics, especially regarding their long-term effects on ecosystem recovery and biodiversity.

Recently, drought–flood abrupt alternation has emerged as a prominent topic within the broader field of global climate change research. The current research regarding DFAAs primarily relies on observational data and numerical simulation [15,16]. Index construction is the chief methodology in DFAA research, and indices such as DFAAI, LDFAAI, and SDFAAI have been widely employed in quantitative analysis [14,17,18]. The spatiotemporal evolution analysis predicated by these indices, in conjunction with climate models, can facilitate deeper exploration into the mechanisms that underlie DFAAs. Multi-scale data fusion techniques, combined with both satellite remote sensing and ground observation, have also been extensively utilized in regional research [19,20]. Recently, more attention has been devoted to the spatiotemporal distribution of DFAAs across diverse regions. Studies have revealed that global warming has augmented the frequency and magnitude of drought and flood events, with the Asian monsoon region and the Midwest region of North America labeled as high-risk regions. Studies have uncovered significant correlations between DFAAs and climate factors, including the El Nino–Southern Oscillation (ENSO), the Arctic Oscillation (AO), and the Indian Ocean dipole (IOD). These factors are closely related to the seasonal regulation of precipitation and evaporation [21,22]. By improving hydrological and climate models, the simulation accuracy and predictive capabilities for DFAAs have been enhanced. However, terrain, soil moisture, and environmental alterations cause substantial discrepancies in the local DFAA prediction capabilities of different models [23,24]. Although a plethora of methods have been employed to conduct quantitative analyses on DFAA events, there remains a scarcity of studies on DFAA spatiotemporal evolution in specific regions. Investigating the region-specific spatiotemporal distributions and changing trends of DFAAs helps bridge gaps in regional research and deepen the scientific understanding of their spatiotemporal characteristics.

The Heilongjiang River Basin is situated in the northeastern region of China and the Russian far East. It represents a crucial ecological and agricultural area, wherein its hydrological processes and ecosystems exhibit an acute sensitivity to global climate change. Recently, the frequency and intensity of DFAA events in the Heilongjiang River Basin have undergone significant changes due to climate warming and human activities [25]. Despite the importance of these events, there remains a critical gap in research on their spatiotemporal distribution and their impact on vegetation growth in the basin. This study aims to address this gap by analyzing the spatiotemporal patterns of DFAA events, integrating these patterns with vegetation growth characteristics, and examining the coupling mechanisms between DFAA events and vegetation responses. Thus, this study will contribute to a deeper understanding of the effects of climate change and human activities on the ecosystem and provide valuable insights for ecological protection and sustainable development in the Heilongjiang River Basin.

2. Research Area and Data

2.1. Location of the Study Area

The Heilongjiang River Basin, located at coordinates (41°42′–55°56′ N, 107°31′–141°14′ E), lies in eastern Eurasia and is one of the ten largest rivers in the world. It encompasses an area of approximately 184.3×10^4 km^2 and represents the largest transboundary basin in northeast Asia. Most of its eastern and western portions are situated in Russia, while a small western section lies in Mongolia, and the remainder is located in the Heilongjiang Province, Inner Mongolia, and the Jilin Province in China. In spring, it is often windy; in summer, it is humid; in autumn, the weather is erratic; and in winter, it is cold and dry. The average annual temperature ranges from −14.37 °C to 6.75 °C, demonstrating significant seasonal and regional variations. Much of the precipitation occurs between June and September, with the precipitation in July accounting for nearly 70% of the annual precipitation. The average annual precipitation varies significantly throughout the year and across different regions of the basin. For example, the annual precipitation in the western region ranges from 200 to 400 mm, while that in the eastern coastal region ranges from 500 to 800 mm. Moreover, the precipitation exhibits substantial interannual variability. For instance, in 2016, the precipitation amounted to 632.30 mm. In contrast, it was only 270.15 mm in 1999, representing a difference of approximately 2.3 times. The Heilongjiang River Basin is partitioned into eight zones, based on characteristics of the basin's water system. These areas include the Heilongjiang River's main stream area, the Jieya River Basin, Brea River Basin, the Amgon River Basin, the Shilka River Basin, the Erguna River Basin, the Songhua River Basin, and the Ussuri River Basin. Figure 1 provides a general overview of the research area.

2.2. Data Sources

2.2.1. Runoff Data

The runoff data used in this study were obtained from the daily runoff simulation run by the Community Water Model (CWatM), which is a large-scale hydrological and water resources model with a spatial resolution of 0.5°. The model can simulate global and regional hydrological processes within a daily timestep range of 30 arcmin to 30 arcsec. CWatM encompasses general surface water and groundwater hydrological processes, while also accounting for human activities—such as water use and reservoir regulation—by calculating water demand, water use, and backflow [26]. Reservoirs and lakes are also incorporated into the model framework. In this study, the runoff data spanning from 1970 to 2019 were selected for analysis.

In this study, runoff and evapotranspiration data from 2007 to 2012 were used for calibration, and data from 2012 to 2017 were used for validation. The model's performance was evaluated using the Kling–Gupta Efficiency (KGE) metric, which provided a comprehensive assessment of the simulation accuracy [27,28].

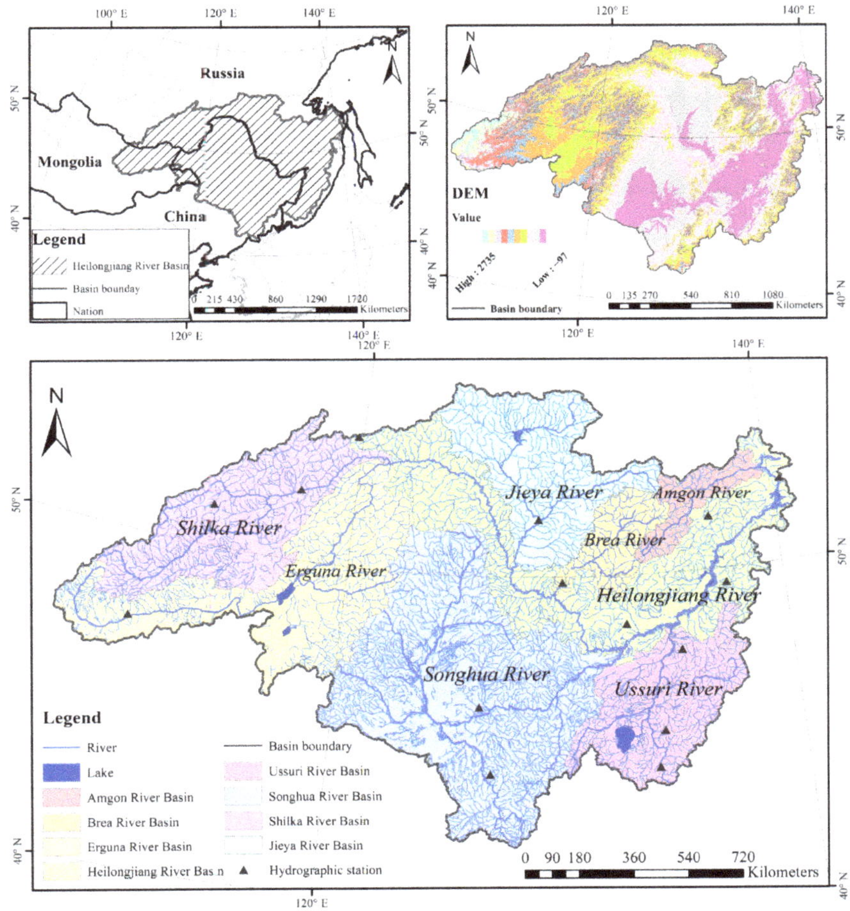

Figure 1. Overview of study area.

2.2.2. Normalized Difference Vegetation Index (NDVI)

The Normalized Difference Vegetation Index (NDVI), initially proposed by Rouse, can accurately reflect the physical mechanisms of vegetation growth, which lie in the disparity between chlorophyll absorption in the red wavelength and the green scattering in the near-infrared wavelength. GIMMS NDVI 3g, SPOT NDVI, and MODIS NDVI (MOD13 and MYD13) represent the most prevalently utilized vegetation index datasets. The MODIS Normalized Difference Vegetation Index, provided by the National Aeronautics and Space Administration (NASA) (https://lpdaac.usgs.gov/products/mod13a3v061/, accessed on 13 October 2024), complements the NDVI product of NOAA's Advanced Very High Resolution Radiometer (AVHRR) and ensures the continuity of historical timeseries applications. The MOD13A3v061 dataset from 2000 to 2019, with a spatial resolution of 1 km, was used in this study. Firstly, the vegetation coverage was estimated using the pixel binary model. Then, a series of extractions, mosaicking, projections, cropping, null value removals, and resampling processes were conducted. The monthly NDVI data were subsequently converted into the TIFF format. Lastly, the monthly NDVI raster data from 2000 to 2019 were extracted based on the Heilongjiang River Basin's vector boundary.

2.2.3. Unified Data Resolution and Analysis Scale Selection

In this study, the original spatial resolution of the runoff data was 0.5°, which served as the main basis for analysis. The NDVI data, with an original resolution of 1 km, was primarily used to characterize vegetation conditions within the study area, rather than for a pixel-by-pixel correlation analysis. Therefore, all correlation analyses were conducted at the 0.5° resolution of the runoff data.

Regarding the resolution difference, although the NDVI data offer higher spatial precision, this study focused on hydrological–ecological processes at the basin scale (i.e., the 0.5° grid scale), making it both reasonable and necessary to adopt the runoff data's 0.5° resolution as the unified analytical scale. Moreover, aggregating the high-resolution NDVI data to the 0.5° scale helped retain more of the original spatial heterogeneity of vegetation. Compared to directly using lower-resolution vegetation data, this approach improved the representativeness and spatial accuracy of vegetation indicators after aggregation.

3. Method

3.1. Community Water Model (CWatM)

Within a grid cell, "runoff concentration" refers to the process through which the runoff generated within the cell converges towards the river path. The runoff originating from each unit is directed to the corners of the corresponding unit. The concentration time (peak time) is determined using the land cover grade, slope, and runoff group (including surface runoff, confluence runoff, or base flow). Subsequently, the triangular weighting function is employed to compute the total runoff of the grid cells [26]. The function is as follows:

$$Q(t) = \sum_{landcover} \sum_{runoff} \sum_{i}^{max} c(i) Q_{runoff}(t - i + 1) \tag{1}$$

where $Q(t)$ is the total runoff of the grid cell with a timestep; runoff is the runoff component (surface, interflow, base flow); Q_{runoff} is the land cover runoff of runoff component; t is the time (1 d); and $c(i)$ is a triangular function, which can be expressed as follows:

$$c(i) = \int_{i-1}^{i} \frac{2}{max} - \left| u - \frac{max}{2} \right| \cdot \frac{4}{max^2} du \tag{2}$$

The hydrographic station was located at the outlet of the Heilongjiang River Basin. The calibration period encompassed 37 years of observational data, with the results provided in Figure 2. With the revised Kling–Gupta efficiency (KGE) serving as the objective function, a KGE value closer to 1 indicates a better fit [29]. The formula is as follows:

$$\text{KGE}' = 1 - \sqrt{(r-1)^2 + (\beta - 1)^2 + (\gamma - 1)^2} \tag{3}$$

where r is the correlation coefficient between simulated and observed values; β is the bias ratio; and γ is the variability ratio.

3.2. Drought–Flood Abrupt Alteration Index (DFAAI)

In this study, the drought–flood abrupt alteration index (DFAAI) was used to quantitatively describe DFAA events. The DFAAI is defined as follows:

$$\text{DFAAI} = (NQ_i - NQ_{i-1}) \cdot (|NQ_i| + |NQ_{i-1}|) \cdot \alpha^{-|NQ_i + NQ_{i-1}|} \ (i = 2, 4, 3, \ldots, n) \tag{4}$$

$$NQ_i = \frac{Q_i - \overline{Q}}{\sigma} \quad (5)$$

where NQ_i and NQ_{i-1} represent the normalized monthly runoff for month i and month $i-1$; Q_i is the average monthly runoff; $\overline{Q}$ and σ are the mean and standard error of Q_i; α is the monthly weight coefficient (a value of 3.2 is more appropriate); i is the month number; and n is the total number of months in the study period [18,30]. The item $(NQ_i - NQ_{i-1})$ represents the intensity of the DFAAs; the item $(|NQ_i| + |NQ_{i-1}|)$ denotes the magnitude of the droughts and floods; and $\alpha^{-|NQ_i + NQ_{i-1}|}$ is the weight coefficient (which may decrease the weight of droughts or floods in two consecutive months and increase the weight of the DFAAs). DFAAI values greater than 1 (>1) or less than −1 (<−1) are defined as DTF or FTD, respectively. The greater the absolute value of the DFAAI, the stronger the DFAA.

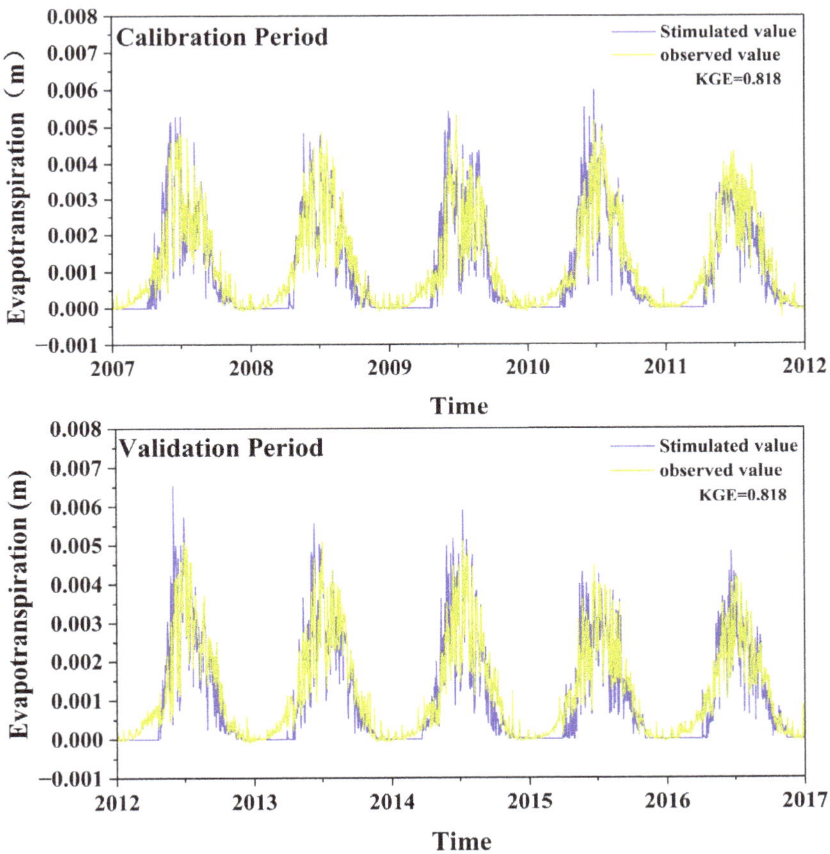

(a)

Figure 2. *Cont.*

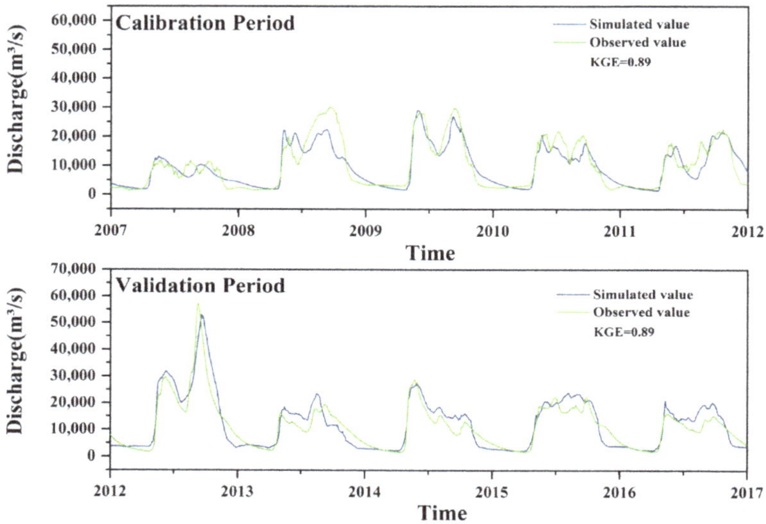

(b)

Figure 2. CWatM model verification results for (**a**) evapotranspiration and (**b**) discharge.

3.3. Wavelet Analysis

For time series *x*(*t*), the Continuous Wavelet Transform (CWT) is defined as follows:

$$W_x(s,\tau) = \int_{-\infty}^{\infty} x(t)\psi^*(\frac{t-\tau}{s})dt \tag{6}$$

where $Wx(s,\tau)$ is the CWT coefficient on the time scale (τ,s); s is a scale parameter (inversely proportional to frequency); τ is the time translation parameter; $\psi(t)$ is the parent wavelet function; and $\psi^*(t)$ is the complex conjugate.

For time series *x*(*t*) and *y*(*t*), the Cross Wavelet Transform (XWT) is defined as follows:

$$W_{xy}(s,\tau) = W_x(s,\tau) \cdot W_y^*(s,\tau) \tag{7}$$

where $Wx(s,\tau)$ and $Wy(s,\tau)$ are the wavelet transforms of *x*(*t*) and *y*(*t*); $Wy*(s,\tau)$ is the complex conjugate of the *y*(*t*) wavelet transform; and $Wxy(s,\tau)$ is the cross wavelet coefficient of *x*(*t*) and *y*(*t*) on the time scale (τ,s). Cross Wavelet Transform (XWT) is concerned with the joint energy distribution of two sequences in time and scale, but it cannot provide a normalized correlation measure [31,32].

Wavelet Coherence (WTC) is a standardized measure of correlation that provides the local linear correlation between two sequences in time and scale, compensating for XWT's inability to distinguish between strong energies and correlations. It is an effective method for studying the scale correlation between two geophysical variables and is used to analyze the coherence of two time series in the time-frequency domain, thereby examining their covariance. The wavelet coherence coefficient is equivalent to the local correlation coefficient in the time-frequency domain and is defined as follows:

$$R_{xy}^2(s,\tau) = \frac{\left|S(s^{-1}W_{xy}(s,\tau))\right|^2}{S(s^{-1}\left|W_x(s,\tau)\right|^2) \cdot S(s^{-1}\left|W_y(s,\tau)\right|^2)} \tag{8}$$

where s is the scale; τ is the time; $Wx(s,\tau)$ and $Wy(s,\tau)$ are the wavelet transform of time series x and y; $Wxy(s,\tau)$ is the Cross Wavelet Transform; s is the smoothing operator used to smooth the local wavelet coefficients in time and scale; and $|W_x(s,\tau)|^2$ and $|W_y(s,\tau)|^2$ is the local wavelet energy of two time series [33,34].

3.4. Pearson's Correlation Analysis

For the variables $X = \{x_1, x_2, \ldots, x_n\}$ and $Y = \{y_1, y_2, \ldots, y_n\}$, Pearson's correlation coefficient r is calculated as follows:

$$r = \frac{\sum_{i=1}^{n}(x_i - \overline{x})(y_i - \overline{y})}{\sqrt{\sum_{i=1}^{n}(x_i - \overline{x})^2 \sum_{i=1}^{n}(y_i - \overline{y})^2}} \tag{9}$$

where x_i and y_i are the ith observed values of the two variables; $\overline{x} = \frac{1}{n}\sum_{i=1}^{n} x_i$ is the mean of x; $\overline{y} = \frac{1}{n}\sum_{i=1}^{n} y_i$ is the mean of y; numerator $\sum_{i=1}^{n}(x_i - \overline{x})(y_i - \overline{y})$ is covariance of two variables; denominator $\sqrt{\sum_{i=1}^{n}(x_i - \overline{x})^2 \sum_{i=1}^{n}(y_i - \overline{y})^2}$ is the product of the standard deviations of two variables; Pearson's correlation coefficient r ranges from $[-1,1]$; $r = 1$ indicates a perfect positive correlation (the two variables have a perfect linear positive relationship); $r = -1$ indicates a perfect negative correlation (a perfectly linear inverse relationship between the two variables); and $r = 0$ means no linear correlation (there is no linear relationship between the two variables, but other nonlinear relationships may exist) [5].

4. Results

4.1. Spatiotemporal Analysis of DFAAs

4.1.1. Spatial Distribution of DFAAs

To quantify the spatial distribution of DFAA events in the Heilongjiang River Basin, we tallied the amounts of drought-to-flood (DTF) and flood-to-drought (FTD) events in each sub-basin from 1970 to 2019. The proportion of these events was computed to illustrate the distribution frequency of DFAAs in the region. In Figure 3, HeilongjiangR illustrates the overall pattern of the region to highlight the general trend; arrows pointing to the individual sub-basins were not included in the figure to avoid confusion with the detailed analysis of the sub-basins. Figure 3 demonstrates that over the past 50 years, the occurrences of DFAA events in the Heilongjiang River main stream area (HLJRMS), Jieya River Basin (JieyaR), Brea River Basin (BreaR), Amgon River Basin (AmgonR), Shilka River Basin (ShilkaR), Erguna River Basin (ErgunaR), Songhua River Basin (SHR), and Ussuri River Basin (UssuriR) measured 14.15%, 10.72%, 7.02%, 5.79%, 14.31%, 14.74%, 18.92%, and 14.36%, respectively. Of these, the greatest frequency of DTF events was observed in the Songhua River Basin (SHR), while the lowest was in the AmgonR. The SHR also had the most FTD events, while the Brea River Basin (BreaR) had the least.

Differences were observed in the frequencies of the DTF and FTD events in the different sub-basins. Generally, the DTF events exhibited an upward trend, increasing at a rate of 3.51% per decade. In the AmgonR, however, the frequency of DTF events decreased by 1.17% per decade, making it the only sub-basin with a declining trend. The remaining seven sub-basins demonstrated an upward trend, with HLJRMS, JieyaR, BreaR, ShilkaR, ErgunaR, SHR, and UssuriR increasing at rates of 7.33%/decade, 2.14%/decade, 1%/decade, 2.67%/decade, 1.8%/decade, 8.48%/decade, and 1.29%/decade, respectively. FTD events demonstrated a downward trend, with a decreasing rate of 4.52% per decade. The frequency of FTD events in ShilkaR (1.9%/decade) and UssuriR (0.91%/decade) increased slightly. However, the other six sub-basins exhibited a downward tendency. The decreasing rates of HLJRMS, JieyaR, BreaR, AmgonR, ErgunaR, and SHR were 8.46%/decade, 9.4%/decade, 13.33%/decade, 13.33%/decade, 1.9%/decade, and 2.78%/decade, respectively.

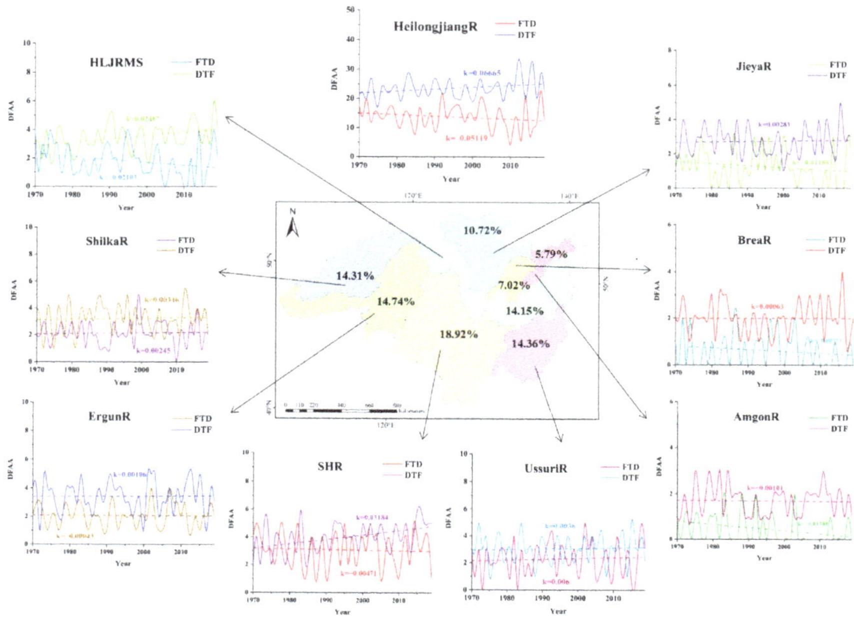

Figure 3. Distribution of DFAAs in study area's sub-basins.

The DTF events exhibited an overall upward trend from 1970 to 2019, whereas the FTD events demonstrated a downward trend. This reflects the spatiotemporal heterogeneity of DFAAs. Spatially, differences existed in both the frequency and trend of DFAA events. The frequency of alternations was relatively higher in the northern regions (i.e., the ErgunaR and the ShilkaR) and relatively lower in the eastern regions (i.e., JieyaR and the BreaR). This discrepancy is likely related to the climatic conditions, topographic features, and hydrological environments of each sub-basin.

4.1.2. Time Distribution of DFAA

To understand the trends of DFAAEs' overextended time scales and identify the diverse characteristics of seasonal droughts and floods, we analyzed the occurrences of DFAA events in the Heilongjiang River Basin over the course of 5 decades (1970–1979, 1980–1989, 1990–1999, 2000–2009, and 2010–2019) and across four different seasons (spring, summer, autumn, and winter).

According to the seasonal distribution shown in Figure 4, clear seasonal differences were observed in the spatial distribution characteristics of DFAAs and their associated events (FTD and DTF) in the Heilongjiang River Basin. Spring and summer were the peak periods for DFAA events, which were characterized by broad spatial distributions and extensive influence ranges. In contrast, the frequencies of such events were relatively low in autumn and winter and were accompanied by a reduction in spatial coverage. The spatial coverage of FTD events in spring, summer, autumn, and winter were 52.88%, 48.77%, 40.94%, and 0.05%, respectively. The spatial coverage of DTF events were 39.29%, 26.28%, 20.58%, and 0.16%. In spring, the regions with high DFAAI values were concentrated in the middle and southern parts of the basin, particularly in the SHR and the HLJRMS. This indicates that DFAAs occurred more frequently there. FTD events were predominantly distributed in the southern region of the SHR and the northern region of the ErgunaR, whereas DTF events were more concentrated in the central region of the SHR and the

ShilkaR. During summer, northern ErgunaR and JieyaR had high DFAAI values. FTD events occurred more frequently in the northern and eastern basins covering the ErgunaR and UssuriR. Meanwhile, the DTF events were concentrated in the southern part of the SHR and the UssuriR. In autumn, the high-value area of DFAAI was primarily located in the southern SHR and the UssuriR. However, the high-frequency area of the FTD events shrank to the middle region of the river basin. The DTF events remained concentrated in certain sections of the SHR and the UssuriR, yet their frequencies decreased significantly, compared to those in spring and summer. In winter, the spatial distribution range of DFAAI was the narrowest, and the high-value area was confined to the southern part of the SHR. The FTD events were scattered across a few areas within the basin, whereas the DTF events rarely occurred.

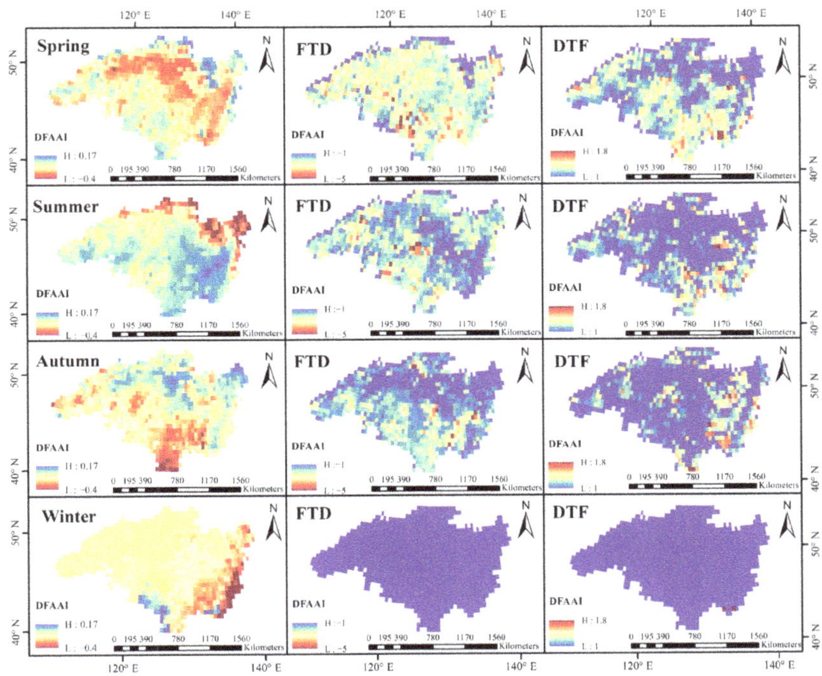

Figure 4. Seasonal spatial distribution of DFAA events.

Generally, the spatial coverage of DFAA events was broadest in spring and summer. In autumn, the events were mainly concentrated in the southern region. In winter, their frequency reached the lowest point accompanied by a significant reduction in spatial range. Moreover, FTD events were more prevalent in the northern and central basins, whereas DTF events were usually observed in the southern and eastern basins. These seasonal distribution differences might be caused by the seasonal variations in precipitation, snowmelt, the topographic features of the watershed, and human activities.

As depicted in Figure 5, during the period from 1970 to 1979, DFAAs were predominantly concentrated in the northern and central regions of the Heilongjiang River Basin, particularly in the ErgunaR, ShilkaR, and SHR. In these regions, the DFAA frequency was relatively high. As time progressed, these events gradually expanded towards the eastern part of the basin, with regions like the UssuriR and AmgonR also exhibiting a relatively higher mutation frequency. The general trend indicated that the spatial distribution range of DFAA events gradually expanded, while the mutation frequency in the southern region significantly increased. The temporal evolution of the high-value area showed that, in the 1970s, the high-frequency area was concentrated in the northern region and gradually expanded to cover the entire basin.

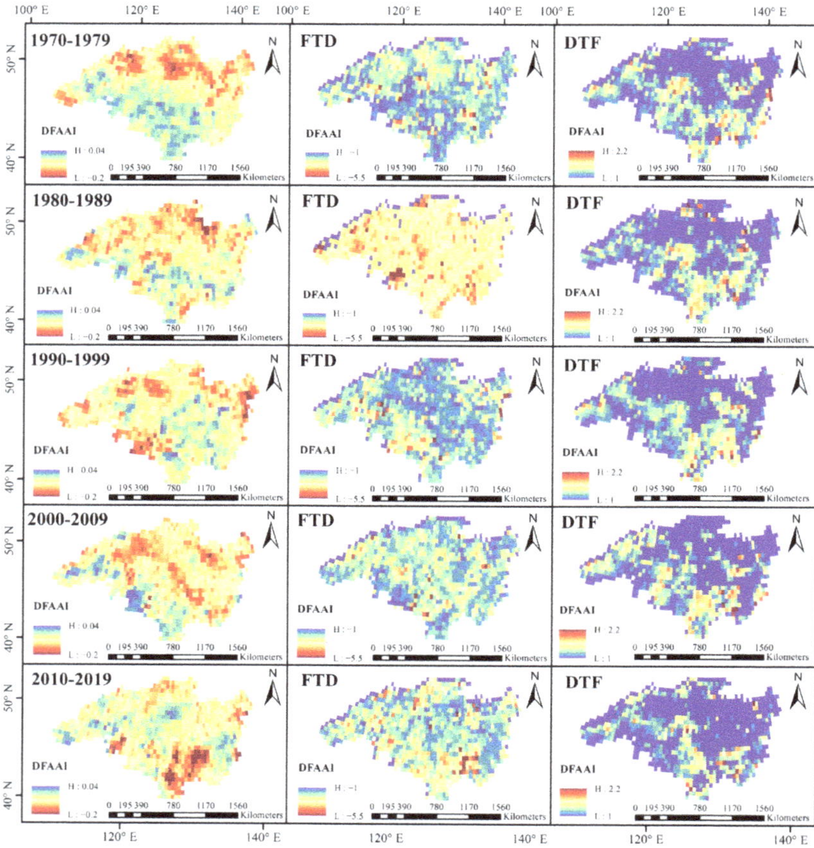

Figure 5. Spatial distribution of DFAA events per decade.

The spatial coverage of FTD events increased from 90.67% (1970–1979) to 93.69% (2010–2019). The highest frequencies in the periods of 1970–1979, 1980–1989, 1990–1999, 2000–2009, and 2010–2019 were 20, 17, 22, 20, and 23, respectively. During the period from 1970 to 1979, the FTD events were predominantly concentrated in the central and northern regions of the SHR and the ErgunaR, forming a relatively concentrated high-frequency area. From 1980 to 1999, the spatial range of the FTD events expanded significantly, with the high-frequency area gradually encompassing the ShilkaR and a part of the HLJRMS. From 2000 to 2019, the FTD events exhibited an overall downward trend, with high-frequency areas diminishing but remaining concentrated in the northern and central basins.

The spatial coverage of DTF events declined from 55.7% (1970–1979) to 40.62% (2010–2019). The highest frequencies registered during 1970–1979, 1980–1989, 1990–1999, 2000–2009, and 2010–2019 were 27, 29, 27, 27, and 33, respectively. During the period from 1970 to 1979, DTF events occurred on a relatively limited scale and were concentrated in the southern and central regions of the basin. Over time, the area with high values gradually expanded and extended towards the north, encompassing regions like the ErgunaR and the JieyaR. From 2000 to 2019, the spatial distribution of the DTF events became more expansive, and the high-frequency areas increased substantially (specifically within the SHR and the UssuriR).

The spatial distribution of DFAAs expanded gradually. Once concentrated in the northern region, they eventually covered the entire basin. The DTF events showed a more pronounced increase in the southern and eastern regions, whereas the FTD events experienced a gradual decline in the high-frequency northern areas. The DFAA events within the Heilongjiang River Basin exhibited significant temporal variations, with these changes likely due to climate warming in the basin, alterations in the spatiotemporal distribution of precipitation, and human activities.

4.2. *Analysis of NDVI Evolution Law*

4.2.1. Annual Variation

A significant positive correlation existed between the NDVI, ecosystem parameters, biomass, and vegetation growth within the Heilongjiang River Basin. Specifically, a high NDVI value corresponded to greater vegetation coverage and better growth conditions. The multi-year average NDVI data from January to December (2000 to 2019) revealed both the annual NDVI variations and the temporal dynamic characteristics of vegetation growth in the basin. As shown in Figure 6, the vegetation coverage within the basin exhibited an initial increase and then a decrease over the course of a year. The coverage was greatest in July and lowest in January. This pattern was primarily influenced by climate shifts and moderate precipitation. During winter, the NDVI values were relatively low, approaching 0 in most areas. This was especially true in the northern regions. These results reflected sparse vegetation and/or low activity levels, which align with the characteristics of cold weather, snow cover, and vegetation dormancy. In spring, the NDVI values gradually rose, and vegetation recovered in the southern and central regions. In summer, the NDVI values reached their annual peak (close to 1 in most areas). Extensive vegetation covered the SHR, the UssuriR, and the southern plain areas. Meanwhile, NDVI increased in the ErgunaR and the HLJRMS in the north. In autumn, the NDVI values gradually declined, vegetation activity waned, and some areas entered a state of dormancy or decay. NDVI's spatial distribution exhibited significant differences between the northern and southern regions. Generally, the NDVI values in the southern areas (the SHR and the UssuriR) were higher throughout the year, likely due to the warm, humid climate conditions. In contrast, in the northern regions (the ErgunaR), the vegetation growth and the growing season were constrained by the high altitudes and cold climate. Along the Heilongjiang River's main stream area and its main tributaries, the NDVI values were relatively higher in spring and summer, and the vegetation thrived under the water supply. In summer, the areas with high NDVI values were concentrated in the southern and plain regions. These areas generally shifted to low values in winter, indicating that seasonal variations significantly influenced vegetation activity. Overall, the distribution of NDVI within the Heilongjiang River Basin exhibited significant seasonal and spatial differences. Vegetation growth was active, with extensive coverage in spring and summer. Moreover, the NDVI values in the southern regions were significantly higher than those in the northern regions, and in winter, vegetation growth was largely stagnated. This was likely due to the climatic, topographic, and hydrological conditions of the basin.

4.2.2. Interannual Variation

This study used monthly NDVI data to calculate a multi-year average NDVI dataset to examine the spatial patterns of vegetation growth within the Heilongjiang River Basin. As indicated by the legend, a darker color represents a higher NDVI value, signifying greater vegetation coverage and better growth conditions. Conversely, lighter colors signify lower vegetation vitality, which could correspond to bare land, construction land, or lakes. As illustrated in Figure 7, the multi-year average of NDVI from 2000 to 2019 reveals generally high NDVI values across the entire Heilongjiang River Basin, particularly in the central and southern parts of the basin. In these areas, the NDVI values are close to 1, suggesting high vegetation cover density and lush growth. In contrast, the NDVI values in certain cold, high-altitude areas in the northern and western parts of the basin were relatively low. In some regions, they were close to 0, indicating sparse vegetation coverage, possibly due to the environmentally restrictive terrain, lakes, and vegetation.

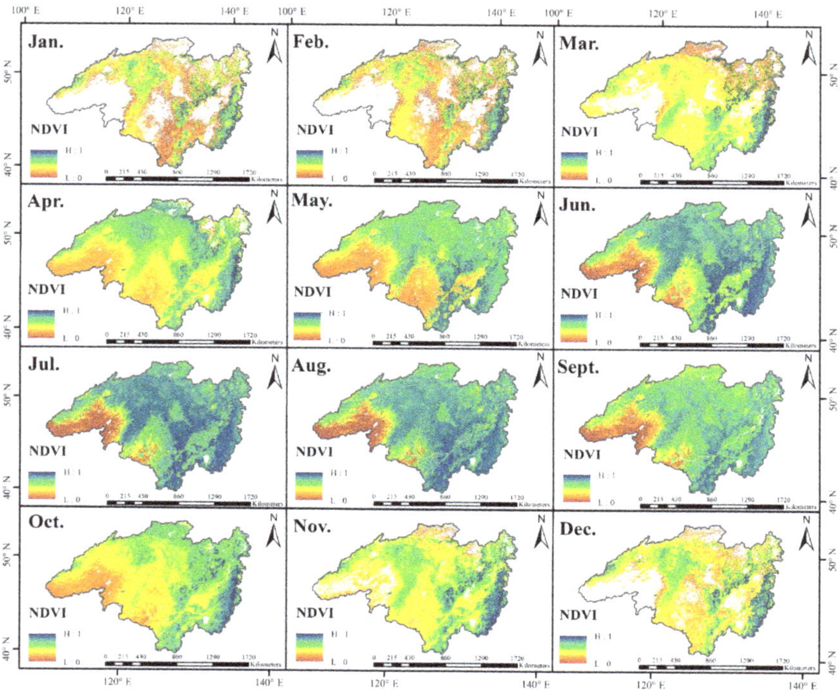

Figure 6. Annual distribution of NDVI.

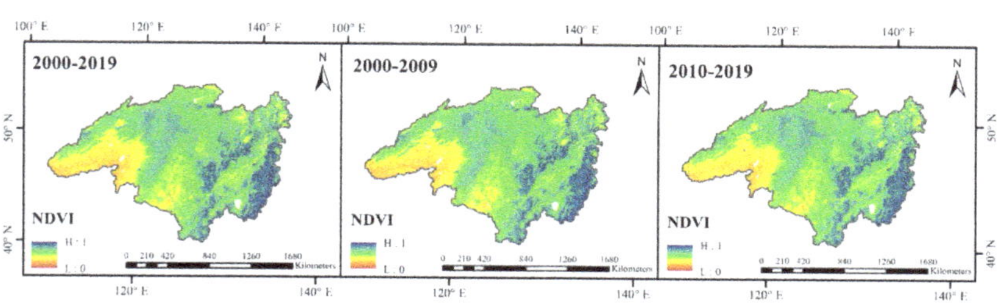

Figure 7. Interannual distribution of NDVI.

During the period from 2000 to 2009, the overall NDVI values within the basin were lower compared to those of the subsequent decade. Specifically, the area with low NDVI values in the northern region was relatively larger, suggesting that vegetation growth was somewhat restricted. This situation might be associated with the climatic conditions, regional economic development activities, or land use changes that occurred in the late

20th and early 21st centuries. During the period from 2010 to 2019, the NDVI values increased significantly. The high-value areas in the middle of the basin and in the plain areas expanded further, specifically into the Songnen Plain and Sanjiang Plain regions. Meanwhile, the area with low NDVI values decreased, indicating that vegetation growth was enhanced. This improvement could stem from the ecological restoration policies implemented by the state (i.e., returning farmland to forest, wetland protection, and the amelioration of climate conditions). Comparing the NDVI distribution from 2000 to 2009 with that from 2010 to 2019, the average NDVI increase in the basin during the 2010–2019 period was 0.54%. Moreover, the area of vegetation growth improvement accounted for 17.29% of the total basin area, whereas the area of degradation accounted for only 3.05%. The area of regions with NDVI values > 0.8 increased, while the area of low-value regions with NDVI values < 0.3 decreased significantly.

In summary, the vegetation growth within the Heilongjiang River Basin exhibited an overall upward trend over the past 20 years. During the period from 2010 to 2019, the NDVI values experienced a significant increase, accompanied by an overall shrinkage in low-value areas. This alteration reflects both the amelioration of climate conditions (i.e., the extension of the vegetation growing season brought about by climate warming) and the positive impacts of national ecological protection measures (i.e., policies like the conversion of farmland back to forest and wetland protection) on vegetation restoration. Nevertheless, the high altitude, cold climate, and complex terrain in the northern region resulted in relatively slight improvements in vegetation growth.

4.3. Response of NDVI to DFAAE

To explore the specific DFAAI characteristics influencing the NDVI, we analyzed the time-frequency relationship between the DFAAI and NDVI using Cross Wavelet Transform and Wavelet Coherence analysis. The univariate wavelet spectra demonstrate the periodic characteristics of the DFAAI and NDVI. The Wavelet Coherence analysis revealed the correlation between the two time series within the frequency domain. Cross wavelet analysis isolated the common wave pattern shared by the two.

As seen in Figure 8, the DFAAI and NDVI in the Heilongjiang River Basin exhibited a significant correlation in spring and summer, particularly within the 8–16-year cycle. In the DFAAI figure, the regions with high energy values were concentrated in the 8–16-year range, specifically at the positions ranging from 40 to 80 on the time axis. This demonstrates strong cyclical fluctuations and contrasts with the NDVI figure, where the high energy values were concentrated within the 8–12-year cycle range. This suggests that the vegetation cover index experiences significant volatility during this cycle. Wavelet Coherence analysis revealed that the time-frequency coherence between the DFAAI and NDVI was primarily concentrated within the mesoscale period range of 8–12 years, specifically at the positions ranging from 40 to 80. Moreover, the coherence and common fluctuations of the two were the most pronounced during this range. During this period, the arrows predominantly pointed toward the lower right, indicating that the DFAAI slightly preceded NDVI, implying that DFAAs might have a lag effect on vegetation growth. The significant coherent regions in summer and autumn remained concentrated within the mesoscale period range of 8–12 years, further validating the delayed response relationship between drought–flood events and vegetation growth. Nevertheless, in local areas, the arrows pointed towards the upper right, which could indicate the recovery characteristics of vegetation in response to drought–flood events. The coherence at different periodic scales exhibited a certain phase difference, suggesting that the interaction between the two is complex and nonlinear. Additionally, the DFAAI also exhibited a significant correlation with the NDVI within the 30–40 period. This could be due to extreme climate events like summer and autumn floods. Overall, this correlation implies that drought and flood events induced by climate change will likely significantly impact vegetation growth within the Heilongjiang River Basin, particularly in the medium- and long-term cycles. Consequently, the sensitivity and response mechanism of vegetation to climate change warrant further attention and research.

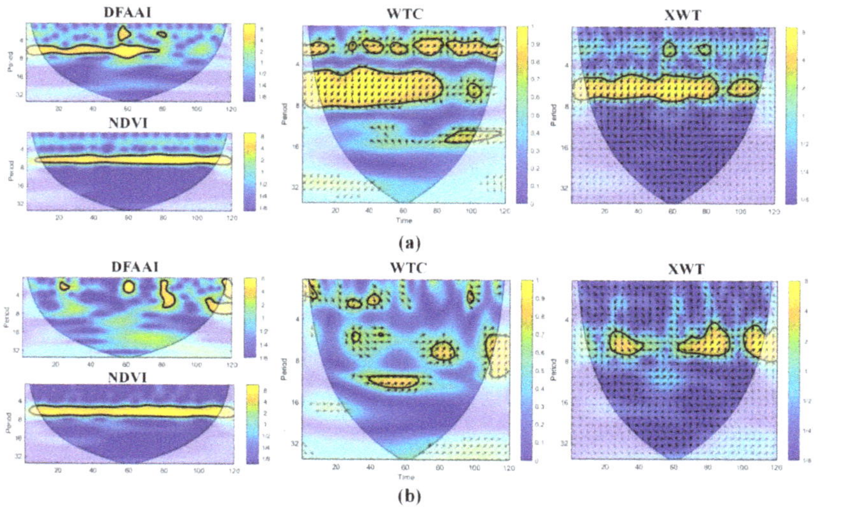

Figure 8. WTC and XWT of DFAAI and NDVI in (**a**) spring–summer seasons; (**b**) summer–autumn seasons.

In the spring–summer seasons (Figure 9), the red and blue regions in the figure were intertwined. They displayed different color intensities, and the overall color tone was relatively light, suggesting that the correlation in most regions had not fully manifested. After a 1–2-month lag, the dark colored area gradually expanded, the distribution of red and blue became more distinct, and the area of significant correlation broadened considerably. The dark blue area expanded and gradually became dominant. After a 2-month lag, the spatial proportion of the blue area was 6.49% greater than that of the red area, signifying that the adverse impacts of drought or flood had accumulated to a peak, potentially leading to a large-scale decline and the destruction of vegetation. Particularly, in flood-prone regions, the lag effect was more pronounced, indicating that the flood damage to vegetation had further accumulated. Meanwhile, the red area gradually expanded, and the dark red area became more concentrated, showing that the promoting effect of appropriate precipitation on vegetation was emerging. After a 3-month lag, the expansion and intensification of the deep red area indicated a maximum positive effect, implying that the recovery response of vegetation to water conditions in some areas required time to activate. The dark blue area remained significant, and the negative effects of droughts and floods persisted in some areas. After a 4-month lag, the correlation between the drought–flood mutation index and the vegetation cover index weakened to the point of being invalid. This could indicate that the lag effect had already passed, or that the vegetation change during this period was primarily driven by other factors, such as temperature and light. Overall, the significant correlation with a 3–4-month lag during the spring–summer period was the strongest and most extensive, suggesting that the lag effect of drought–flood fluctuations on vegetation had reached its peak.

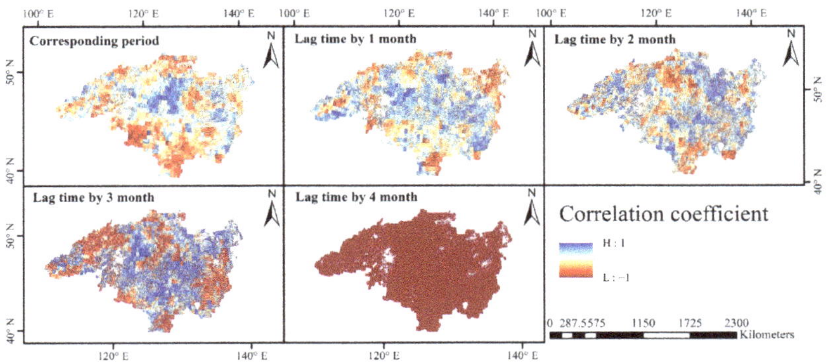

Figure 9. Correlation analysis between DFAAI and NDVI in spring–summer.

In the summer–autumn season (Figure 10), the correlation during the period was rather weak, and the colors of most regions were relatively light, suggesting that the overall correlation distribution was not pronounced. After a 1–2-month lag, the red area gradually expanded, signifying that the lagged effect of summer precipitation exerted a clear, positive effect on vegetation growth. This could be due to the precipitation accumulation improving the soil moisture or the positive response of vegetation to precipitation as it enters the vigorous growth stage. The blue areas on the diagram became darker, particularly in areas affected by extreme precipitation (i.e., flooding). This indicates more substantial damage to vegetation or cumulative damage over an extended period by extreme drought and flood conditions. At this juncture, the overall correlation distribution was significantly strengthened, and the lag effect reached its peak. Following a lag of 3 months, the blue region further expanded, occupying 1.21% more area than the red region. The negative correlation became dominant, while the positive correlation weakened, suggesting that the cumulative effect of droughts and flooding on vegetation destruction was approaching its peak, particularly in ecologically sensitive areas. Nevertheless, the persistent effects of precipitation continued to have a positive influence on vegetation in certain regions. After a 4-month lag. the overall color gradually lightened, and the distribution of red and blue became more uniform. Theoretically, the restoration mechanism of vegetation could gradually take effect, diminishing the negative impacts of droughts and floods while reducing the positive effects of precipitation. Consequently, the lag effect may dissipate, and the correlation could return to a lower level.

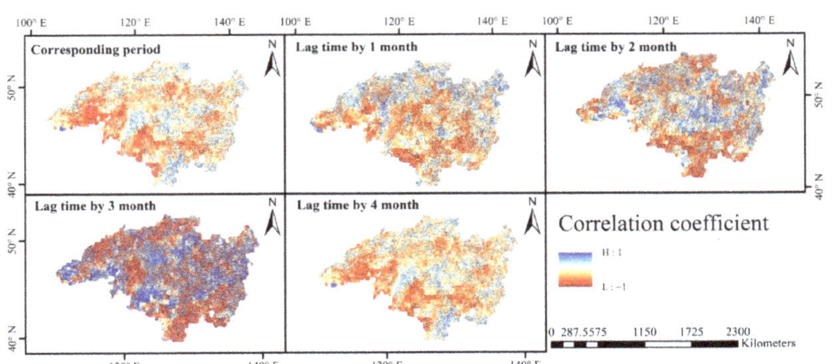

Figure 10. Correlation analysis between DFAAI and NDVI in summer–autumn.

Within a lag of 4 months, the correlation increased gradually, and the dark-colored area expanded continuously, indicating a significant lag effect from droughts and floods on vegetation. In the spring–summer period, both the positive and negative correlations reached their peaks during a lag of 3–4 months, suggesting that this specific time window was when droughts and floods significantly disrupted vegetation. In contrast, in the summer–autumn period, the correlation reached its peak during the 3-month lag. Moreover, the lag effect was more pronounced in the summer–autumn period than in the spring–summer period, and the spatial range, as well as the intensity of both the positive and negative correlations, were more conspicuous. The distinct color distribution reflects the differences in the response mechanisms to droughts and floods in different regions within the basin. For instance, dark red areas might correspond to wet environments like forests, whereas dark blue areas could correspond to more vulnerable grasslands or cultivated lands. The delayed response of DFAAs to vegetation presents obvious temporal and spatial heterogeneity. This suggests that ecological management should focus on the cumulative effects over time and develop tailored recovery strategies based on regional characteristics.

5. Discussion

5.1. Direct Effects of DFAAE on Vegetation Growth

Moderate flood events can briefly augment the regional soil water and nutrient supply, particularly in wetland ecosystems or for plants tolerant to waterlogging. When floodwaters recede, floods can introduce a substantial amount of sediment and nutrients into the soil, thereby facilitating the rapid recovery and expansion of vegetation [1,35]. Nevertheless, persistent flooding can result in a reduction in soil oxygen, root decay, and the significant degradation of water-intolerant vegetation [36]. Drought events decrease vegetation cover by reducing soil moisture and restricting plant transpiration and photosynthesis. Prolonged drought can completely kill vegetation and even alter affected vegetation types [37]. Our study demonstrates that there might be a positive or negative correlation between DFAA events and vegetation growth. This implies that the interaction between the two could have synergistic or antagonistic effects. Other studies have corroborated this finding, positing that in certain ecosystems, moderate droughts and floods might promote vegetation diversity and maintain ecosystem balance. For example, during dry periods, the number of non-drought-tolerant plants decreases, thereby providing more space for drought-tolerant plants to thrive. Meanwhile, during flood periods, water is supplied for other types of vegetation [1,38–42]. Thus, certain vegetation types can demonstrate their adaptability in long-term DFAAs, allowing vegetation growth to fluctuate periodically yet remain generally stable [43]. However, the combination of droughts and floods can also pose challenges for vegetation in terms of adaptation. Certain types of vegetation might not have time to recover after a drought, and subsequent flooding could further damage the soil and roots. There could also be significant competition among different types of vegetation in response to DFAAs. For example, drought-tolerant plants can be damaged during flood events, while water-tolerant plants may be degraded during drought events. Such competition can result in fluctuations in vegetation growth and serious regional degradation [44].

The impacts of DFAAs on vegetation growth varied significantly across different regions and ecosystems. The positive effect was more pronounced in humid regions, while the negative effect prevailed in more arid regions [45]. Meanwhile, climate warming likely increased the frequency and intensity of DFAAs, thereby introducing greater uncertainty into the dynamic changes of vegetation growth. Increasing research on the ability of vegetation to adapt to DFAAs is crucial, particularly in areas that are prone to frequent extreme events. In areas that experience severe vegetation degradation, adaptive management measures should be adopted in accordance with the characteristics of DFAAs. For instance, drought-tolerant or water-tolerant plants should be planted to enhance the resilience of the ecosystems.

5.2. Feedback Mechanism of Vegetation Growth Change to DFAAs

Based on the WTC and XWT figures presented in Section 4.3 of this study, most of the arrows within the significant region of the 8–16-year period were positioned either in the upper right or lower right. This suggests that the change in DFAA plays a significant role in influencing vegetation growth over a specific period of time. Nevertheless, in the lower period of 4–8 years, the arrows were positioned in the upper left or lower left, indicating that vegetation growth might exert a certain feedback effect on DFAAs. During droughts, reduced NDVI may result in soil drying and a decrease in evaporation, which, in turn, indirectly impacts the performance of DFAAs. Changes in vegetation growth can impact regional hydrological conditions by affecting soil moisture and local climate regulation, which in turn can provide feedback to DFAA. Similar situations have also been observed in previous studies. Specifically, an increase in vegetation cover can enhance the efficiency of the local water cycle via evapotranspiration and precipitation recycling. The leaf interception and root water absorption functions of vegetation help stabilize soil moisture and reduce the intensity of droughts and floods [46,47]. Furthermore, the hydrological regulatory effects of vegetation (i.e., rainwater retention, soil stabilization, and enhanced surface infiltration) are capable of significantly reducing surface runoff and soil erosion. This decreases the intensity of flooding [48]. A reduction in vegetation cover weakens evapotranspiration and diminishes the efficiency of local precipitation recycling, elevating the risk of drought. The reduction in surface vegetation can result in the deterioration of soil structure and a decrease in permeability, facilitating the formation of surface runoff and flooding. During the dry season, the water-holding capacity of the soil also weakens, further aggravating the risk of flooding [49].

Vegetation restoration measures, including the reversion of farmland back to forest and wetland restoration, can significantly improve water retention and regional climate regulation at the ecosystem level and reduce the frequency and intensity of DFAAs. For example, China's Three-North Shelterbelt Project, by means of large-scale vegetation restoration, strengthened the stability of the regional water cycle and mitigated the amplifying effect of climate change on DFAAs [50]. An increase in vegetation diversity can also enhance ecosystems' abilities to endure extreme climate events. Under DFFAs, highly diverse ecosystems are capable of achieving functional maintenance through the species complementarity effect [51].

Although the conversion of forests into farmland might boost food production in the short term, it will undermine regional hydrological regulation. For instance, farmland is less capable of withstanding flooding and less resilient in the face of drought compared to natural vegetation [52]. Furthermore, feedback conflicts might arise among different types of vegetation, due to water competition. As an example, deep-rooted trees could impact the recovery of herbaceous vegetation by over-consuming groundwater resources. In the context of farmland conversion projects in northern India, certain revegetation projects have decreased the risk of flooding while augmenting the risk of regional drought [53].

Through climate warming, the feedback mechanism through which vegetation growth changes impact DFAAs has intensified. High temperatures accelerate transpiration and exacerbate drought effects, while extreme precipitation events heighten soil erosion and flood risks. Meanwhile, accounting for changes in atmospheric circulation and surface albedo, the feedback of vegetation growth on droughts and floods could extend from the local level to regional or even global scales, amplifying its overall impact.

5.3. Uncertainty of CWatM Simulation

The uncertainty in CWatM simulations is a significant issue in the research and application of hydrological modeling. It primarily stems from the complexity and ambiguity associated with numerous factors. Firstly, the input of vague or faulty data directly influences the model's results [54]. For instance, discrepancies in spatial resolution and accuracy among different data sources, including observation, reanalysis, and remote sensing data, could give rise to significantly different simulation results. Errors in the input data (i.e., precipitation and temperature) will directly impact the simulation of evapotranspiration, runoff, and water resource supply and demand. Moreover, data projections related to land use change, population growth, and economic activity frequently contain errors that can result in inaccurate estimates of irrigation demand and water use [55]. Secondly, uncertainty surrounding the model's structure can also influence the simulation accuracy. During the simulation, CWatM simplifies several complex hydrological and management processes, such as the interactions between groundwater and surface water and the rapid response to extreme events. Such simplification could lead to certain regions or specific events being inaccurately represented [56]. Simultaneously, the parameters within the model, namely soil moisture and evapotranspiration coefficient, rely on empirical formulas or regional calibration and could result in substantial simulation errors in uncalibrated regions [57].

Apart from the uncertainties associated with the data and model structure, the uncertainty surrounding scenario assumptions also presents challenges to the reliability of the results [58]. Predictions of future climate conditions vary considerably among climate models like CMIP6 and emission scenarios like the RCP or SSP pathways. Socioeconomic scenarios including population growth, policy changes, and technological advances are equally challenging to accurately predict. This increases the uncertainty of future water supply and demand assessments [59–61]. Meanwhile, the uncertainties related to spatial and temporal scales also warrant special attention. Due to the complexity of topography, land use, and water resources distribution across different regions, low-resolution simulations will likely be unable to capture local characteristics. Furthermore, simulations conducted over long time scales, such as those involving interannual variations, may fail to accurately depict the short-term dynamics of extreme events [62].

In light of the aforementioned insecurities, the following measures can be adopted to mitigate the errors: (1) Integrate diverse meteorological data and land use data, utilize remote sensing data to correct crucial input parameters like precipitation and evapotranspiration, and conduct a comprehensive evaluation of the impact of data errors on model results [63]. Local parameter calibration should be conducted for different regions. For example, the evapotranspiration and discharge data of the Heilongjiang River Basin were utilized to calibrate the model in this study, with the aim of enhancing the accuracy of the simulation. (2) Multi-model ensemble analysis should be carried out jointly with other global hydrological models to diminish the uncertainty resulting from a single-model structure. (3) Simulate multiple climate and socioeconomic scenarios and quantify the uncertainty of projections by comparing the ranges of outcomes among different scenarios. (4) Optimize the model grid resolution and utilize high time resolution data to conduct finer-grained simulations of complex regions. These methods can effectively mitigate the errors induced by inconsistencies in CWatM simulations.

Notwithstanding certain factors, the regional and global applications of CWatM demonstrate that its integrated and versatile architecture can support multi-scenario and multi-scale analyses. With improvements in the input data quality, the optimization of model structure, and the advancements in uncertainty quantification methods, CWatM holds significant potential for widespread application in water resources management and climate change.

6. Conclusions

In this study, the spatiotemporal distribution of drought–flood abrupt alternation events within the Heilongjiang River Basin was analyzed. The dynamic characteristics of vegetation growth were investigated through the utilization of NDVI, and the response mechanism of vegetation growth to drought–flood abrupt alternation events was identified. The following conclusions were drawn:

(1) From 1970 to 2019, the DTF events were most frequent in the SHR and least frequent in the AmgonR, with an upward trend (3.51%/decade). The FTD events decreased (4.52%/decade) over time, were more frequent in northern regions like the ErgunaR and the ShilkaR, and less frequent in eastern areas like the JieyaR and the BreaR. Temporally, events peaked in spring and summer, were concentrated in the south in autumn, and were least frequent in winter. High-frequency areas expanded from the north in the 1970s–1980s into the entire basin. From 2010 to 2019, the DTF events were concentrated in the south and east, while the FTD events showed a decline in the northern high-frequency areas.

(2) Annual NDVI values were lowest in January, highest in July, and close to 0 in winter. Vegetation coverage was greatest in summer. Southern areas like SHR and the UssuriR had relatively higher NDVI values all year. Northern regions, like ErgunaR, were more climate restricted. From 2000 to 2019, vegetation growth markedly increased. The NDVI values were lower from 2000 to 2009, especially in the northern and central high altitudes, but rose significantly from 2010 to 2019, with a reduction in low-value areas and an expansion of high-value areas.

(3) Wavelet Coherence analysis revealed that drought and flood events have a 1–4-month lag effect on vegetation. The positive and negative correlations were most significant in the spring–summer period with a 3–4-month lag. In the summer–autumn period, the peak lagged by 3 months.

In conclusion, significant spatiotemporal heterogeneity exists in both drought–flood abrupt alternation events and vegetation growth within the Heilongjiang River Basin. The delayed response of vegetation to DFAAs implies the necessity of strengthening ecological restoration strategies to address the impact of climate change on ecosystems.

Author Contributions: H.M. and J.J. contributed equally to this manuscript. Conceptualization, H.M., J.J. and P.Q.; methodology, H.M., J.J. and P.Q.; software, H.M., J.J. and H.S.; validation, H.M., J.J., P.Q. and H.S.; formal analysis, Y.X., P.Q. and C.D.; investigation, H.M. and J.J.; resources, C.D.; data curation, H.M., J.J. and P.Q.; writing—original draft preparation, H.M. and J.J.; writing—review and editing, P.Q. and C.D.; visualization, H.M., J.J. and P.Q.; funding acquisition, C.D. All authors have read and agreed to the published version of the manuscript.

Funding: This research was supported by the National Natural Science Foundation of China (42371037) and the Yunnan Provincial Key Laboratory of International Rivers and Transboundary Ecological Security Open Fund (NO.2022KF03).

Data Availability Statement: Some or all the data that support the findings of this study are available from the corresponding authors upon reasonable request.

Acknowledgments: The authors would like to express their sincere gratitude to the Key Laboratory of International Rivers and Trans-boundary Ecological Security in Yunnan Province and the Heilongjiang Provincial Department of Science and Technology for their valuable support. We also appreciate the assistance provided by our classmates and teachers for their insightful discussions and technical support. Special thanks go to the institutions and individuals who contributed data and resources essential for this study. Their support has been instrumental in the successful completion of this research.

Conflicts of Interest: The authors have no competing interests to declare that are relevant to the content of this article.

References

1. Bi, W.; Weng, B.; Yan, D.; Zhang, D.; Liu, C.; Shi, X.; Jing, L.; Yan, S.; Wang, H. Response of summer maize growth to drought-flood abrupt alternation. *Front. Earth Sci.* **2023**, *11*, 1086769. [CrossRef]
2. Liu, T.T.; Zhu, X.F.; Tang, M.X.; Guo, C.H.; Lu, D.Y. Multi-model ensemble bias-corrected precipitation dataset and its application in identification of drought-flood abrupt alternation in China. *Atmos. Res.* **2024**, *307*, 107481. [CrossRef]
3. Shi, W.Z.; Huang, S.Z.; Liu, D.F.; Huang, Q.; Han, Z.M.; Leng, G.Y.; Wang, H.; Liang, H.; Li, P.; Wei, X.T. Drought-flood abrupt alternation dynamics and their potential driving forces in a changing environment. *J. Hydrol.* **2021**, *597*, 126179. [CrossRef]
4. Bi, W.; Li, M.; Weng, B.; Yan, D.; Dong, Z.; Feng, J.; Wang, H. Drought-flood abrupt alteration events over China. *Sci. Total Environ.* **2023**, *875*, 162529. [CrossRef]
5. Zhou, W.; Liu, D.; Zhang, J.; Jiang, S.; Xing, S.; Wang, J.; Cheng, Y.; Chen, N. Identification and frequency analysis of drought–flood abrupt alternation events using a daily-scale standardized weighted average of the precipitation index. *Front. Environ. Sci.* **2023**, *11*, 1142259. [CrossRef]
6. Chakraborty, T.; Lee, X. A simplified urban-extent algorithm to characterize surface urban heat islands on a global scale and examine vegetation control on their spatiotemporal variability. *Int. J. Appl. Earth Obs. Geoinf.* **2019**, *74*, 269–280. [CrossRef]
7. Shi, X.L.; Yang, Y.Q.; Yuan, Z.; Ding, H.; Zhang, Y.; Chen, J.J.; Shi, M.Q. Integrating SWAP and SIF anomaly to assess the responses of vegetation to the drought-flood abrupt alternation in the middle and lower reaches of the Yangtze River basin, China. *J. Hydrol. Reg. Stud.* **2024**, *52*, 101726. [CrossRef]
8. Qian, L.; Meng, H.Y.; Chen, X.H.; Tang, R. Evaluating agricultural drought and flood abrupt alternation: A case study of cotton in the middle-and-lower Yangtze River, China. *Agric. Water Manag.* **2023**, *283*, 108313. [CrossRef]
9. Huang, J.; Hu, T.; Yasir, M.; Gao, Y.; Chen, C.; Zhu, R.; Wang, X.; Yuan, H.; Yang, J.J.E.; Botany, E. Root growth dynamics and yield responses of rice (*Oryza sativa* L.) under drought—Flood abrupt alternating conditions. *Environ. Exp. Bot.* **2019**, *157*, 11–25. [CrossRef]
10. Bi, W.X.; Weng, B.S.; Yan, D.H.; Wang, M.K.; Wang, H.; Wang, J.J.; Yan, H.L. Effects of drought-flood abrupt alternation on phosphorus in summer maize farmland systems. *Geoderma* **2020**, *363*, 114147. [CrossRef]
11. Bi, W.; Weng, B.; Yan, D.; Wang, M.; Wang, H.; Jing, L.; Yan, S. Soil phosphorus loss increases under drought-flood abrupt alternation in summer maize planting area. *Agric. Water Manag.* **2022**, *262*, 107426. [CrossRef]
12. Zhu, J.; Li, A.; Zhang, J.; Sun, C.; Tang, G.; Chen, L.; Hu, J.; Zhou, N.; Wang, S.; Zhou, Y.J.E.; et al. Effects of nitrogen application after abrupt drought-flood alternation on rice root nitrogen uptake and rhizosphere soil microbial diversity. *Environ. Exp. Bot.* **2022**, *201*, 105007. [CrossRef]
13. Zhu, R.; Wu, F.; Zhou, S.; Hu, T.; Huang, J.; Gao, Y. Cumulative effects of drought–flood abrupt alternation on the photosynthetic characteristics of rice. *Environ. Exp. Bot.* **2020**, *169*, 103901. [CrossRef]
14. Shi, W.; Huang, S.; Zhang, K.; Liu, B.; Liu, D.; Huang, Q.; Fang, W.; Han, Z.; Chao, L. Quantifying the superimposed effects of drought-flood abrupt alternation stress on vegetation dynamics of the Wei River Basin in China. *J. Hydrol.* **2022**, *612*, 128105. [CrossRef]
15. Liu, Y.; Shan, F.; Yue, H.; Wang, X.; Fan, Y. Global analysis of the correlation and propagation among meteorological, agricultural, surface water, and groundwater droughts. *J. Environ. Manag.* **2023**, *333*, 117460. [CrossRef]
16. Weng, X.; Zhu, J.; Wang, D.; Chen, H.; Wang, S.; Qing, Y.J.G. Natural Hazards; Risk. Exploring the relationship between drought-flood abrupt alternation and soil erosion over Guangdong, China through a convection-permitting model. *Geomat. Nat. Hazards Risk* **2024**, *15*, 2383779. [CrossRef]
17. Li, X.H.; Yuan, C.Y.; Sun, T.; Fan, H.X. Identifying the spatiotemporal patterns of drought-flood alternation based on IMERG product in the humid subtropical Poyang Lake basin, China. *J. Hydrol.-Reg. Stud.* **2024**, *54*, 101912. [CrossRef]
18. Li, X.H.; Zhang, Q.; Zhang, D.; Ye, X.C. Investigation of the drought-flood abrupt alternation of streamflow in Poyang Lake catchment during the last 50 years. *Hydrol. Res.* **2017**, *48*, 1402–1417. [CrossRef]
19. Ren, J.X.; Wang, W.G.; Wei, J.; Li, H.B.; Li, X.L.; Liu, G.S.; Chen, Y.L.; Ye, S.L. Evolution and prediction of drought-flood abrupt alternation events in Huang-Huai-Hai River Basin, China. *Sci. Total Environ.* **2023**, *869*, 161707. [CrossRef]
20. Wang, L.; Guo, S.L.; Wang, J.; Chen, Y.B.; Qiu, H.; Zhang, J.; Wei, X.J. A novel multi-scale standardized index analyzing monthly to sub-seasonal drought-flood abrupt alternation events in the Yangtze River basin. *J. Hydrol.* **2024**, *633*, 130999. [CrossRef]
21. Yang, P.; Zhang, S.Q.; Xia, J.; Zhan, C.S.; Cai, W.; Wang, W.Y.; Luo, X.G.; Chen, N.C.; Li, J. Analysis of drought and flood alternation and its driving factors in the Yangtze River Basin under climate change. *Atmos. Res.* **2022**, *270*, 106087. [CrossRef]
22. Song, X.Y.; Lei, X.P.; Ma, R.; Hou, J.Z.; Liu, W.B. Spatiotemporal variation and multivariate controls of short-cycle drought-flood abrupt alteration: A case in the Qinling-Daba Mountains of China. *Int. J. Climatol.* **2023**, *43*, 4756–4769. [CrossRef]

23. Chen, H.; Wang, S.; Zhu, J.; Zhang, B. Projected Changes in Abrupt Shifts Between Dry and Wet Extremes Over China Through an Ensemble of Regional Climate Model Simulations. *J. Geophys. Res.-Atmos.* **2020**, *125*, e2020JD033894. [CrossRef]
24. Zhang, Y.Q.; You, Q.L.; Ullah, S.; Chen, C.C.; Shen, L.C.; Liu, Z. Substantial increase in abrupt shifts between drought and flood events in China based on observations and model simulations. *Sci. Total Environ.* **2023**, *876*, 162822. [CrossRef]
25. Bai, X.; Zhao, C.; Tang, Y.; Zhang, Z.; Yang, B.; Wang, Z. Identification, physical mechanisms and impacts of drought–flood abrupt alternation: A review. *Front. Earth Sci.* **2023**, *11*, 1203603. [CrossRef]
26. Burek, P.; Satoh, Y.; Kahil, T.; Tang, T.; Greve, P.; Smilovic, M.; Guillaumot, L.; Zhao, F.; Wada, Y. Development of the Community Water Model (CWatM v1.04)—A high-resolution hydrological model for global and regional assessment of integrated water resources management. *Geosci. Model Dev.* **2020**, *13*, 3267–3298. [CrossRef]
27. Burek, P.; Smilovic, M.; Guillaumot, L.; de Bruijn, J.; Greve, P.; Satoh, Y.; Islaam, A.; Virgen-Urcelay, A.; Tang, T.; Kahil, T.; et al. *Community Water Model CWatM Manual*; IIASA: Laxenburg, Austria, 2020.
28. Hinton, R.; Smilovic, M.; Fridman, D.; Willaarts, B.; Banda, L.; Macleod, K.; Troldborg, M.; Kalin, R. A stakeholder driven, holistic water resources model for Malawi: Applying the CWatM hydrological model. In Proceedings of the European Geosciences Union General Assembly 2024 (EGU24), Vienna, Austria, 14–19 April 2024; p. 5783.
29. Kling, H.; Fuchs, M.; Paulin, M. Runoff conditions in the upper Danube basin under an ensemble of climate change scenarios. *J. Hydrol.* **2012**, *424–425*, 264–277. [CrossRef]
30. Wu, Z.; Li, J.; He, J.; Jiang, Z. Large-scale atmospheric singularities and summer long-cycle droughts-floods abrupt alternation in the middle and lower reaches of the Yangtze River. *Chin. Sci. Bull.* **2006**, *51*, 2027–2034. [CrossRef]
31. Labat, D. Recent advances in wavelet analyses: Part 1. A review of concepts. *J. Hydrol.* **2005**, *314*, 275–288. [CrossRef]
32. Grinsted, A.; Moore, J.C.; Jevrejeva, S. Application of the cross wavelet transform and wavelet coherence to geophysical time series. *Nonlinear Process. Geophys.* **2004**, *11*, 561–566. [CrossRef]
33. Torrence, C.; Compo, G.P. A Practical Guide to Wavelet Analysis. *Bull. Am. Meteorol. Soc.* **1998**, *79*, 61–78. [CrossRef]
34. Hu, W.; Si, B. Improved partial wavelet coherency for understanding scale-specific and localized bivariate relationships in geosciences. *Hydrol. Earth Syst. Sci.* **2021**, *25*, 321–331. [CrossRef]
35. Shi, L.L.; Wang, Y.Y.; Jia, Y.F.; Lu, C.; Lei, G.C.; Wen, L. Vegetation Cover Dynamics and Resilience to Climatic and Hydrological Disturbances in Seasonal Floodplain: The Effects of Hydrological Connectivity. *Front. Plant Sci.* **2017**, *8*, 2196. [CrossRef] [PubMed]
36. Huang, J.P.; Yu, H.P.; Guan, X.D.; Wang, G.Y.; Guo, R.X. Accelerated dryland expansion under climate change. *Nat. Clim. Change* **2016**, *6*, 166–171. [CrossRef]
37. Anyamba, A.; Tucker, C.J. Analysis of Sahelian vegetation dynamics using NOAA-AVHRR NDVI data from 1981–2003. *J. Arid. Environ.* **2005**, *63*, 596–614. [CrossRef]
38. Bi, W.; Wang, M.; Weng, B.; Yan, D.; Yuheng, Y.; Wang, J. Effects of Drought-Flood Abrupt Alternation on the Growth of Summer Maize. *Atmosphere* **2019**, *11*, 21. [CrossRef]
39. Zuo, X.; Li, X.; Yue, P.; Guo, A.; Yue, X.; Xu, C.; Knapp, A.K.; Smith, M.D.; Luo, W.; Allington, G.R.H.; et al. Drought-driven shifts in relationships between plant biodiversity and productivity in temperate steppes. *Funct. Ecol.* **2022**, *36*, 2917–2928. [CrossRef]
40. Van den Hoof, C.; Lambert, F. Mitigation of drought negative effect on ecosystem productivity by vegetation mixing. *J. Geophys. Res. Biogeosci.* **2016**, *121*, 2667–2683. [CrossRef]
41. Borisade, T.V.; Odiwe, A.I.; Akinwumiju, A.S.; Uwalaka, N.O.; Orimoogunje, O.O. Biodiversity. Impact of flooding frequency on the diversity and structures of riparian forests in Southwestern, Nigeria. *Earth Hist. Biodivers.* **2025**, *3*, 100018.
42. Fischer, F.M.; Wright, A.J.; Eisenhauer, N.; Ebeling, A.; Roscher, C.; Wagg, C.; Weigelt, A.; Weisser, W.W.; Pillar, V.D. Plant species richness and functional traits affect community stability after a flood event. *Philos. Trans. R. Soc. B Biol. Sci.* **2016**, *371*, 20150276. [CrossRef]
43. Kumar, S.; Sachdeva, S.; Bhat, K.; Vats, D.S. *Plant Responses to Drought Stress: Physiological, Biochemical and Molecular Basis. Biotic and Abiotic Stress Tolerance in Plants*; Springer: Singapore, 2018; pp. 1–25.
44. Blom, C.W.P.M.; Voesenek, L.A.C J. Flooding: The survival strategies of plants. *Trends Ecol. Evol.* **1996**, *11*, 290–295. [CrossRef] [PubMed]
45. Yang, Y.; Weng, B.; Bi, W.; Xu, T.; Yan, D.; Ma, J. Climate Change Impacts on Drought-Flood Abrupt Alternation and Water Quality in the Hetao Area, China. *Water* **2019**, *11*, 652. [CrossRef]
46. Jackisch, C.; Knoblauch, S.; Blume, T.; Zehe, E.; Hassler, S.K. Estimates of tree root water uptake from soil moisture profile dynamics. *Biogeosciences* **2020**, *17*, 5787–5808. [CrossRef]
47. Spracklen, D.V.; Arnold, S.R.; Taylor, C.M. Observations of increased tropical rainfall preceded by air passage over forests. *Nature* **2012**, *489*, 282–285. [CrossRef]
48. Feng, X.M.; Fu, B.J.; Piao, S.; Wang, S.H.; Ciais, P.; Zeng, Z.Z.; Lü, Y.H.; Zeng, Y.; Li, Y.; Jiang, X.H.; et al. Revegetation in China's Loess Plateau is approaching sustainable water resource limits. *Nat. Clim. Change* **2016**, *6*, 1019–1022. [CrossRef]

49. Zeng, N.; Neelin, J.D.; Lau, K.M.; Tucker, C.J. Enhancement of Interdecadal Climate Variability in the Sahel by Vegetation Interaction. *Science* **1999**, *286*, 1537–1540. [CrossRef]
50. Peng, D.; Wu, C.; Zhang, B.; Huete, A.; Zhang, X.; Sun, R.; Lei, L.; Huang, W.; Liu, L.; Liu, X.; et al. The Influences of Drought and Land-Cover Conversion on Inter-Annual Variation of NPP in the Three-North Shelterbelt Program Zone of China Based on MODIS Data. *PLoS ONE* **2016**, *11*, e0158173. [CrossRef]
51. Malhi, Y.; Roberts, J.T.; Betts, R.A.; Killeen, T.J.; Li, W.; Nobre, C.A. Climate Change, Deforestation, and the Fate of the Amazon. *Science* **2008**, *319*, 169–172. [CrossRef] [PubMed]
52. Ruíz-García, V.H.; Borja de la Rosa, M.A.; Gómez-Díaz, J.D.; Asensio-Grima, C.; Matías-Ramos, M.; Monterroso-Rivas, A.I. Forest Fires, Land Use Changes and Their Impact on Hydrological Balance in Temperate Forests of Central Mexico. *Water* **2022**, *14*, 383. [CrossRef]
53. Chaturvedi, R.K.; Gopalakrishnan, R.; Jayaraman, M.; Bala, G.; Joshi, N.; Sukumar, R.; Ravindranath, N.H. Impact of climate change on Indian forests: A dynamic vegetation modeling approach. *Mitig. Adapt. Strateg. Glob. Change* **2011**, *16*, 119–142. [CrossRef]
54. Müller Schmied, H.; Eisner, S.; Franz, D.; Wattenbach, M.; Portmann, F.T.; Flörke, M.; Döll, P. Sensitivity of simulated global-scale freshwater fluxes and storages to input data, hydrological model structure, human water use and calibration. *Hydrol. Earth Syst. Sci.* **2014**, *18*, 3511–3538. [CrossRef]
55. Shah, D.; Shah, H.L.; Dave, H.M.; Mishra, V. Contrasting influence of human activities on agricultural and hydrological droughts in India. *Sci. Total Environ.* **2021**, *774*, 144959. [CrossRef]
56. Guillaumot, L.; Smilovic, M.; Burek, P.; de Bruijn, J.; Greve, P.; Kahil, T.; Wada, Y. Coupling a large-scale hydrological model (CWatM v1.1) with a high-resolution groundwater flow model (MODFLOW 6) to assess the impact of irrigation at regional scale. *Geosci. Model Dev.* **2022**, *15*, 7099–7120. [CrossRef]
57. Kushwaha, A.P.; Tiwari, A.D.; Dangar, S.; Shah, H.; Mahto, S.S.; Mishra, V. Multimodel assessment of water budget in Indian sub-continental river basins. *J. Hydrol.* **2021**, *603*, 126977. [CrossRef]
58. Hanus, S.; Schuster, L.; Burek, P.; Maussion, F.; Wada, Y.; Viviroli, D. Coupling a large-scale glacier and hydrological model (OGGM v1.5.3 and CWatM V1.08)—Towards an improved representation of mountain water resources in global assessments. *Geosci. Model Dev.* **2024**, *17*, 5123–5144. [CrossRef]
59. Eyring, V.; Bony, S.; Meehl, G.A.; Senior, C.A.; Stevens, B.; Stouffer, R.J.; Taylor, K.E. Overview of the Coupled Model Intercomparison Project Phase 6 (CMIP6) experimental design and organization. *Geosci. Model Dev.* **2016**, *9*, 1937–1958. [CrossRef]
60. Dione, P.M.; Faye, C.; Mohamed, A.; Alarifi, S.S.; Mohammed, M.A. Assessment of the impact of climate change on current and future flows of the ungauged Aga-Foua-Djilas watershed: A comparative study of hydrological models CWatM under ISIMIP and HMF-WA. *Appl. Water Sci.* **2024**, *14*, 163. [CrossRef]
61. Cheng, W.; Feng, Q.; Xi, H.; Yin, X.; Cheng, L.; Sindikubwabo, C.; Zhang, B.; Chen, Y.; Zhao, X. Modeling and assessing the impacts of climate change on groundwater recharge in endorheic basins of Northwest China. *Sci. Total Environ.* **2024**, *918*, 170829. [CrossRef]
62. de Bruijn, J.A.; Smilovic, M.; Burek, P.; Guillaumot, L.; Wada, Y.; Aerts, J. GEB v0.1: A large-scale agent-based socio-hydrological model-simulating10 million individual farming households in a fully distributed hydrological model. *Geosci. Model Dev.* **2023**, *16*, 2437–2454. [CrossRef]
63. He, Y.F.; Xiong, J.H.; Guo, S.L.; Zhong, S.R.; Yu, C.T.; Ma, S.G. Using Multi-Source Data to Assess the Hydrologic Alteration and Extremes under a Changing Environment in the Yalong River Basin. *Water* **2023**, *15*, 1357. [CrossRef]

water

MDPI

Article

Understanding Climate Change Impacts on Streamflow by Using Machine Learning: Case Study of Godavari Basin

Ravi Ande [1,*], Chandrashekar Pandugula [2], Darshan Mehta [3], Ravikumar Vankayalapati [4], Prashant Birbal [5], Shashikant Verma [6], Hazi Mohammad Azamathulla [5] and Nisarg Nanavati [3]

1 Department of Fabric and Apparel Science, Lady Irwin College, University of Delhi, New Delhi 110001, Delhi, India
2 Lowe's Inc., 1000 Lowe's Blvd., Mooresville, NC 28117, USA; chandrashekar.pandugula.de@gmail.com
3 Department of Civil Engineering, Dr. S. & S. S. Ghandhy Government Engineering College, Surat 395001, Gujarat, India; darshanmehta2490@gmail.com (D.M.); nanavatinisharg2011@gmail.com (N.N.)
4 Equinix Inc., Dallas, TX 75207, USA; ravikumar.vankayalapti.research@gmail.com
5 Department of Civil and Environmental Engineering, University of the West Indies, St. Augustine 330110, Trinidad and Tobago; birbal.prashant58@gmail.com (P.B.); hazi.azamathulla@uwi.edu (H.M.A.)
6 Department of Civil Engineering, National Institute of Technology, Raipur 492010, Chhattisgarh, India; sverma.phd2018.ce@nitrr.ac.in
* Correspondence: ravi.ande@lic.du.ac.in

Abstract: The study aims to assess future streamflow forecasts in the Godavari basin of India under climate change scenarios. The primary objective of the Coupled Model Intercomparison Project Phase 6 (CMIP6) was to evaluate future streamflow forecasts across different catchments in the Godavari basin, India, with an emphasis on understanding the impacts of climate change. This study employed both conceptual and machine learning models to assess how changing precipitation patterns and temperature variations influence streamflow dynamics. Seven satellite precipitation products CMORPH, Princeton Global Forcing (PGF), Tropical Rainfall Measuring Mission (TRMM), Climate Prediction Centre (CPC), Infrared Precipitation with Stations (CHIRPS), and Precipitation Estimation from Remotely Sensed Information Using Artificial Neural Networks (PERSIANN-CDR) were evaluated in a gridded precipitation evaluation over the Godavari River basin. Results of Multi-Source Weighted-Ensemble Precipitation (MSWEP) had a Nash–Sutcliffe efficiency (NSE), coefficient of determination (R^2), and root mean square error (RMSE) of 0.806, 0.831, and 56.734 mm/mon, whereas the Tropical Rainfall Measuring Mission had 0.768, 0.846, and 57.413 mm, respectively. MSWEP had the highest accuracy, the lowest false alarm ratio, and the highest Peirce's skill score (0.844, 0.571, and 0.462). Correlation and pairwise correlation attribution approaches were used to assess the input parameters, which included a two-day lag of streamflow, maximum and minimum temperatures, and several precipitation datasets (IMD, EC-Earth3, EC-Earth3-Veg, MIROC6, MRI-ESM2-0, and GFDL-ESM4). CMIP6 datasets that had been adjusted for bias were used in the modeling process. R, NSE, RMSE, and R^2 assessed the model's effectiveness. RF and M5P performed well when using CMIP6 datasets as input. RF demonstrated adequate performance in testing ($0.4 < NSE < 0.50$ and $0.5 < R^2 < 0.6$) and extremely good performance in training ($0.75 < NSE < 1$ and $0.7 < R < 1$). Likewise, M5P demonstrated good performance in both training and testing ($0.4 < NSE < 0.50$ and $0.5 < R^2 < 0.6$). While RF was the best performer for both datasets, Indian Meteorological Department outperformed all CMIP6 datasets in streamflow modeling. Using the Indian Meteorological Department gridded precipitation, RF's NSE, R, R^2, and RMSE values during training were 0.95, 0.979, 0.937, and 30.805 m^3/s. The test results were 0.681, 0.91, 0.828, and 41.237 m^3/s. Additionally, the Multi-Layer Perceptron (MLP) model demonstrated consistent performance across both the

 https://doi.org/10.3390/w17081171

training and assessment phases, reinforcing the reliability of machine learning approaches in climate-informed hydrological forecasting. This study underscores the significance of incorporating climate change projections into hydrological modeling to enhance water resource management and adaptation strategies in the Godavari basin and similar regions facing climate-induced hydrological shifts.

Keywords: artificial neural networks; climate change; Indian rivers; Godavari basin; machine learning; streamflow; tropical rainfall measuring mission; multi-source weighted-ensemble precipitation

1. Introduction

Hydrological models, integral to understanding and simulating natural hydrological processes, serve as decision-making tools, particularly in data-scarce contexts with numerous options [1]. They maximize available data but do not replace field observations. These models require inputs like climatic variables (e.g., precipitation, temperature), watershed characteristics (e.g., drainage network, topography), and more, with complexity depending on the model's design [1,2]. Historically, hydrological modeling has evolved from Mulvany's 1850 Rational Method to contemporary grid-based physically distributed models [3]. Sherman's (1932) Unit Hydrograph (UH) model marked a turning point, predicting flood hydrographs by assuming uniform rainfall at fixed intervals [4]. Later, models like Instantaneous Unit Hydrographs (IUH) and Finite Period Unit Hydrographs (TUH) emerged [5]. Some IUH models were physically based, while others, like the Clark Unit Hydrograph, relied on historical data [6]. The study of the hydrological cycle and its transformations, as well as the prospective consequences for water resources, holds great importance in the areas of sustainable development and the management of water resources [7].

It is crucial to comprehend river hydrology and how it may evolve under future climatic conditions. Nevertheless, it is imperative to recognize that although several studies conducted on a global scale offer valuable insights, they frequently employ spatial resolutions that are insufficient to capture localized climate events [8]. In recent years, the field of streamflow forecasting has witnessed a surge in the popularity of data-driven models, owing to their swift development, minimal data prerequisites, and ease of real-time application [9]. These models encompass diverse approaches, including Linear Time Series Models such as autoregressive moving average (ARMA), autoregressive integrated moving average (ARIMA), and seasonal autoregressive integrated moving average (SARIMA), which presume a probabilistic distribution for streamflow [10]. Nonlinear Time Series Models, particularly chaos-based models, have been employed to capture the intricate nonlinear dynamics inherent in streamflow data [11]. ANNs have demonstrated effectiveness in handling large, dynamic, and nonlinear time series data, especially when the underlying physical processes are not fully understood.

Furthermore, the adoption of machine learning models, particularly using CMIP6 data, represents an innovative approach poised to enhance predictive capabilities in hydrology. In summary, this study's holistic approach aims to advance our understanding of hydrological processes and their responses to climate change, acknowledging the regional variations and specific relevance to the Godavari River basin. Against the backdrop of the 21st century's environmental challenges, understanding the impacts of climate change on hydrological extremes is paramount [12]. The accelerated emission of greenhouse gases has significantly influenced the hydrological cycle, notably altering extreme precipitation patterns and temperature regimes [13]. These climatic shifts have profound repercussions on hydrological extremes, with manifestations varying spatially and temporally across different regions [14]. Environmental alteration is a global phenomenon characterized by a multitude of interconnected factors, including land degradation, loss of biodiversity, shifts in climate patterns, and alterations in hydrology [15]. The consequences of these changes are far-reaching and extend to various aspects of our natural environment, including the hydrological cycle. Studies indicate that while climate change has been linked to phenomena such as reduced water-resource quantity and intensified tropical cyclones, the manifestation of these hydrological extremes is influenced by complex atmospheric, land, and water interactions [16]. The impact of environmental change on hydrology is a critical concern in today's world. Anthropogenic activities have replaced approximately 41% of global natural vegetation with land cover such as croplands and pastures [17]. This transformation, coupled with the escalating emissions of greenhouse gases, has led to alarming levels of global warming, resulting in a steady increase in global temperatures [18].

Physical, conceptual, and data-driven models are the three primary categories of streamflow models that are frequently employed. Physical models, sometimes referred to as process-driven models, include lumped, semi-distributed, and distributed hydrological models. Given the profound socio-economic and environmental implications of these extremes, there is an urgent need for nuanced, region-specific research to devise adaptive and informed strategies for the future [18]. Climate models project changes in precipitation patterns, with some regions expected to receive more precipitation while others receive less. These changes in precipitation are likely to be mirrored in future variations in runoff. The escalating trend in temperature has the potential to lead to increased flooding in various regions worldwide, including South Asia, which is particularly vulnerable to the adverse impacts of climate change.

Streamflow forecasting models are essential in hydrology for predicting the behavior of hydrologic systems. These models are broadly categorized into three types: physical-based models, conceptual models, and black-box models [19]. Physical-Based Models: These models, also known as white-box or process-based models, provide detailed descriptions of hydrological characteristics by solving differential equations based on physical laws. They account for the physical processes governing mass, energy, and momentum conservation [20]. However, they require extensive physical data and complex mathematical representations. Streamflow Modelling and Prediction: The hydrological cycle and the management of water resources depend heavily on streamflow. We can learn more about flood risk and water availability by modelling and forecasting streamflow, which will help with sustainable water resource management. Creative Uses of Machine Learning: The use of machine learning models to estimate streamflow, particularly with CMIP6 data as input, offers a novel strategy that could greatly improve the precision of hydrological forecasts. Studies in India and other countries have used these machine learning models in hydrological research worldwide. They have shown promise in enhancing runoff forecasting, sediment estimation, infiltration modeling, and streamflow predictions [21,22]. Overall, the use of machine learning methods in hydrological modeling offers useful information and resources for the efficient management of water resources and disaster prevention. Examples of literature from various parts of the world that have used AI models in hydrological investigations include the following. Examples of literature from various parts of the world that have applied these machine learning approaches for many uses in hydrological modeling, such as managing water resources, forecasting floods and droughts, and creating water management strategies, demonstrate that precise streamflow prediction is essential. Researchers and decision-makers can evaluate how human activity, climate change, and land use changes affect water availability using streamflow models [22]. These models, which mimic hydrological processes within a catchment region, include IHACRE, HBV,

NRECA, and MIKE-SHE. They frequently call for a large amount of input data [23,24]. These models can be difficult to calibrate and validate, and their simplistic architecture and inaccurate data may add uncertainty. However, data-driven models have become more widely used in streamflow forecasting because of their quick iteration cycles, lower data needs, and ability to be implemented in real time [25–27]. Without specifically outlining the underlying hydrological processes, these models can make accurate predictions about streamflow [28,29]. Hydrological modeling has made use of several machine learning approaches, including ANNs, SVMs, Random Forest (RF), Multilayer Perceptron (MLP), and M5P. ANNs, sometimes known as "black-box models", have demonstrated promise in identifying intricate and nonlinear correlations in streamflow data. They need very few input parameters [30–33].

1.1. Research Gap of the Study Area

The focus on the Godavari River basin, where little research has been carried out on the effects of climate change, highlights how special this place is in dealing with its complexities. Additionally, the use of machine learning models specifically with CMIP6 data represents a novel strategy positioned to improve hydrology's forecasting capacities. There is not much research on how climate change affects streamflow; therefore, the study's primary focus on streamflow modeling offers insights into managing and controlling water resources [34]. The paper also addresses the urgent problem of excessive runoff rates, which exacerbate erosion and flooding, and highlights the usefulness of precise runoff forecasting in disaster mitigation. In conclusion, the holistic approach of this study recognizes regional variability and the particular importance of the Godavari River basin while working to improve our understanding of hydrological processes and their responses to climate change [35,36].

1.2. Objectives of the Research Work

Apply conceptual rainfall-runoff models to several sub-basins of the Godavari River basin using the meteorological and bias-corrected CMIP6 datasets. Examine gridded precipitation products in the Godavari River basin of India in comparison to meteorological data. Using CMIP6 data for scenarios SSP126, SSP245, SSP370, and SSP585, examine Extreme Precipitation indices across India using spatial and temporal analysis, using Ranking and Trend assessments [37–39]. Forecast streamflow in the Godavari River basin's numerous sub-basins by using machine learning models using Indian Meteorological Development and bias-corrected CMIP6 datasets. Using CMIP6 data, combine conceptual models and machine learning techniques to forecast future streamflow across various sub-basins in the Godavari basin [40].

2. Materials and Methods

In this section, the selection of data sources, the application of specialized tools such as Climate Data Operators 26.4 (CDO) (https://www.unidata.ucar.edu/software/netcdf/workshops/2012/third_party/CDO.html, accessed on 7 April 2025), Geographic Information System (GIS 10.4) software (https://www.esri.com/, accessed on 7 April 2025), and Python 3.9 programming (https://docs.python.org/release/3.9.22/whatsnew/3.9.html, accessed on 7 April 2025), and the logical flow of activities tailored to the unique objectives of this study. Using CMIP6 data, the combination of conceptual models and machine learning approaches improves the precision of future streamflow forecasts across several sub-basins within the Godavari River basin. The Godavari River basin's gridded precipitation data demonstrate greater accuracy and dependability in comparison to Indian Meteorological Department (IMD) data [41,42].

2.1. Location

Smaller parts of Karnataka, Madhya Pradesh, Odisha, Telangana, Chhattisgarh, and Andhra Pradesh are included in the Godavari River basin, which spans 312,812 km^2 [43]. It makes up around 9.5% of the country's total land area and is located between latitudes 16°19′ N and 22°34′ N and longitudes 73°24′ E to 83°4′ E. The Godavari River runs 1465 km from its source to its drainage into the Bay of Bengal [44]. The basin is bounded to the east and west by the Eastern and Western Ghats, to the north by the Satmala hills, and to the south by the Mahadeo hills and Ajanta range [45]. The monsoon season (June to September) provides almost 85% of the basin's annual rainfall, contributing to its tropical climate. In the Godavari, rainfall in the research area, divided into 12 sub-basins, rises from Trimbakeshwar in the Nashik area of Maharashtra [46]. Rainfall in the basin varies significantly by place and time each year, ranging from 600 to 3000 mm (Figure 1)

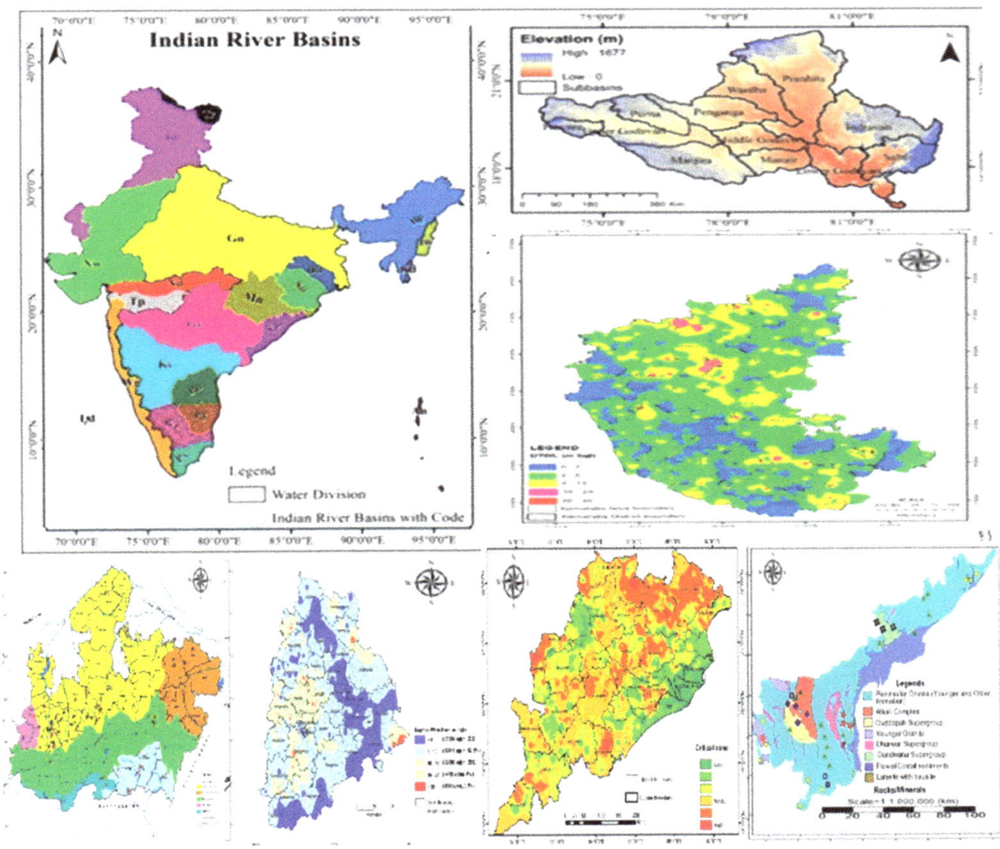

Figure 1. Godavari River basin's geographic location of the study area.

2.2. Data Collection and Sources

The data collection process forms the foundation of the study. Essential climate data were gathered from multiple sources to comprehensively assess climate change impacts in the Godavari River basin. Primary data sources include the CMIP6 climate model data, which simulates Earth's climate under various greenhouse gas concentration scenarios [47]. These scenarios, specifically SSP126, SSP245, SSP370, and SSP585, allow investigation of climate change effects. Additionally, historical precipitation records from the Indian Meteorological Department and gridded satellite precipitation datasets were acquired for validation and complementary analysis, as streamflow datasets [48].

2.2.1. CMIP6 Model Data

The eight extreme precipitation indices created by the ETCCDI for the historical period of 1950–2024 were evaluated using the 17 CMIP6 models used in this study. Due to CMIP6 Model Data, only 17 of the available models were processed for study. The 17 CMIP6 models were employed in this study to assess the eight extreme precipitation indices developed by the ETCCDI for the historical period of 1950–2024. Due to insufficient data, only 17 of the available models were processed for analysis over the selected research region for any of the baseline scenarios. Out of 17 models only five were selected for the subsequent stage, and the SSP126, SSP245, SSP370, and SSP585 models for the near (2015–2040), medium (2041–2070), and far future (2071–2100) were evaluated for future extreme precipitation indices for all baseline scenarios [49]. The archives kept by the ESGF contain these CMIP6 models for inspection. Bilinear interpolation was used to spatially remap all general circulation models and data to a latitude and longitude grid that was like 0.25°*0.25° (Table 1) [50,51].

Table 1. An overview of the 17 CMIP6 models used in the research.

Model	Atmospheric Resolution	Institution
ACCESS-ESM1-5	1.9° × 1.2°	The Commonwealth Scientific and Industrial Research Organization (CSIRO) and the Australian Bureau of Meteorology
ACCESS-CM2	1.87° × 1.25°	The Commonwealth Scientific and Industrial Research Organization (CSIRO) and the Australian Bureau of Meteorology (BOM)
CanESM5	2.81° × 2.79°	Canada's Centre for Climate Modelling and Analysis, The Centre for Climate Change in Europe and the Mediterranean
EC-Earth3	1.3° × 0.9°	Consortium EC-EARTH
EC-Earth3-Veg	0.7° × 0.7°	Consortium EC-EARTH
EC-Earth3-Veg-LR	0.7° × 0.7°	Consortium EC-EARTH
GFDL-ESM4	1.1° × 1.1°	Geophysical Fluid Dynamics Laboratory
INM-CM5-0	1.3° × 1°	The Institute of Numerical Mathematics
IPSL-CM6A-LR	2° × 1.5°	Institute of Pierre-Simon Laplace
MIROC6	1.41° × 1.41°	R-CCS, AORI, NIES, and JAMSTEC
MPI-ESM1-2-HR	0.93° × 0.93°	The Max Planck Institute of Meteorology (MPI-M)
MPI-ESM1-2-LR	0.93° × 0.93°	The Max Planck Institute of Meteorology (MPI-M)
MRI-ESM2-0	0.9° × 1.3°	Meteorological Research Institute
NorESM2-LM	0.9° × 1.3°	Meteorological Research Institute
NorESM2-MM	1.3° × 1°	The Taiwan Earth System Model, Version 1

2.2.2. Conceptual Models

Sacramento

Using daily rainfall and PET data, the Sacramento model is a hydrological model that calculates streamflow daily. The model replicates the catchment region's hydrological balance by incorporating soil moisture [52]. The Sacramento model differs from the AWBM model in that it is more complicated, with a total of five stores and twenty-two distinct attributes [53,54]. The Sacramento model uses the genetic algorithm (GA) to optimize its maximum, minimum, and default parameters. The Sacramento soil's hydrologically active zone is made up of two distinct layers: Figure 2 shows the Sacramento model's schematic form. Surface tension in tension water reservoirs helps to retain water within the soil profile. The only mechanism responsible for the outflow of water from this stratum is evapotranspiration [55]. Water can flow both horizontally and vertically within a free water store before being released as interflow (upper zone) or base flow (lower zone). The Sacramento model separates the catchment into impermeable and pervious groups according to its permeability [56]. Lakes, rivers, and other bodies of water, as well as non-porous surfaces like pavement that are directly connected to the river system, make up the impermeable area. Making use of the Rainfall-Runoff Library's features to model runoff in a particular catchment region by using evapotranspiration and daily rainfall, Manual's Sacramento model is guided by Podger's (2003) extensive instructions (Table 2) [29].

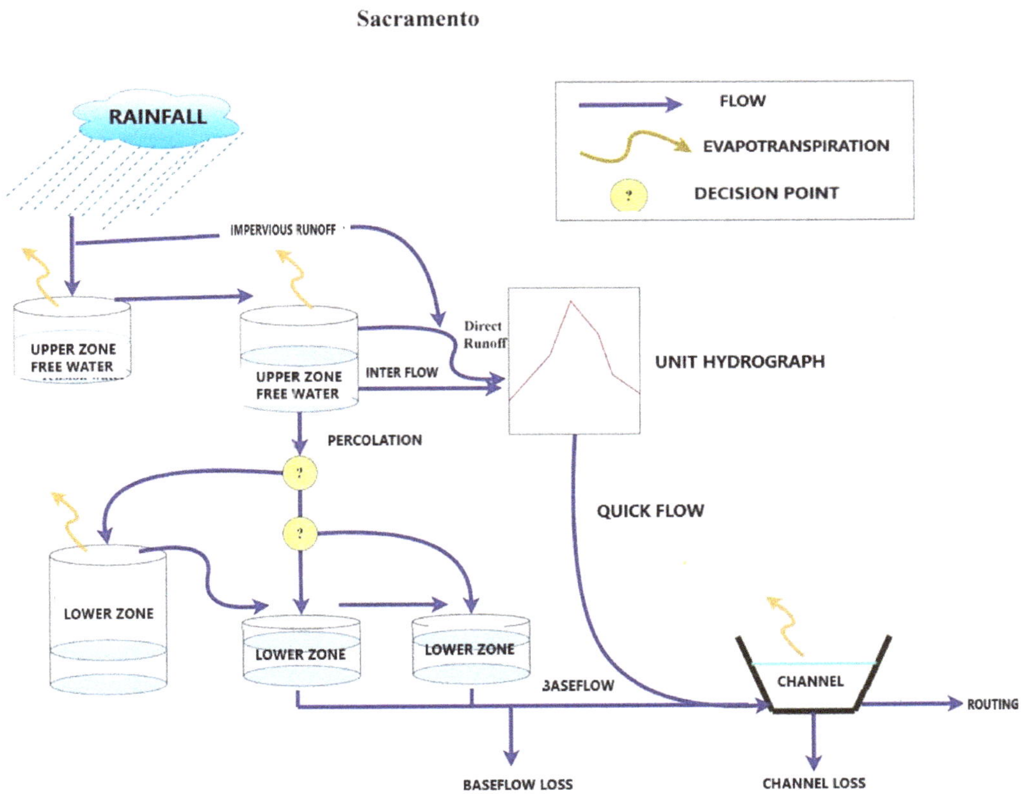

Figure 2. The Sacramento model's schematic representation.

Table 2. The Sacramento model parameters by default.

Parameter	Description	Default	Min	Max
LZPK	The proportion of water in LZFPM that drains daily as base flow.	Zone	Free	Water
LZSK	The proportion of water in LZFSM that drains daily as base flow.	60	40	600
UZK	The percentage of water in UZFWM that drains as daily interflow.	0.06	0	0.5

Table 2. *Cont.*

Parameter	Description	Default	Min	Max
UZTWM	Maximum Water Tension in the Upper Zone.	1	0	3
UZFWM	The storage that serves as the source of water for interflow and the impetus for moving water to greater depths is known as the Upper Zone Free Water Maximum.	40	0	80
LZTWM	Water Maximum for Lower Zone Tension.	0	0	0.8
LZFSM	Maximum Free Water Supplement in the Lower Zone.	0.01	0.001	0.015
LZFPM	Primary Maximum for Lower Zone Free Water.	0.05	0.03	0.2
PFREE	Recharging the lower zone's free water reservoirs requires a minimum percentage of percolation from the upper zone to the lower zone.	0.3	0.2	0.5
REXP	An exponent that calculates how quickly the percolation rate changes as the lower zone water storage changes	50	25	125
ZPERC	The maximal percolation rate is determined by the proportionate increase in Pbase	40	10	75
SIDE	The non-channel base flow ratio.	130	75	300

The Australian Water Balance Model

In hydrological management, the Australian Water Balance Model (AWBM) is a theoretical framework used to evaluate rainfall losses and ascertain the interdependencies among daily rainfall, evapotranspiration, and runoff at the watershed scale. Five separate stores make up the model: a base flow store, a surface runoff routing shop, and three surface stores that mimic partial runoff regions [14]. It has been observed that the Pranhita sub-basin simulates daily streamflow by utilizing the characteristics of the Rainfall-Runoff Library to model runoff in a specific catchment area, utilizing daily rainfall and evapotranspiration. Each storage unit is given a unique storage capacity, and the water balance of each partial surface area of the AWBM is calculated independently, as shown in the schematic model in Figure 3 and Table 3 [21].

Table 3. The Australian Water Balance Model's parameter values by default.

Parameter	Description	Default	Min	Max
KSurf	Recession constant of surface flow	150	0	500
KBase	Recession constant for base flow	70	0	200
C3	Surface store capacity 3 (in mm)	7	0	50
C2	Surface store capacity 2 (in mm)	0.35	0	1
C1	Surface store capacity 1 (in mm)	0.95	0	1
BFI	Index of base flow	0.134	0	1
A2	Surface storage 2's partial area	0.35	0	1
A1	Surface store 1's partial area	0.35	0	1
KSurf	Recession constant of surface flow	0.433	0	1

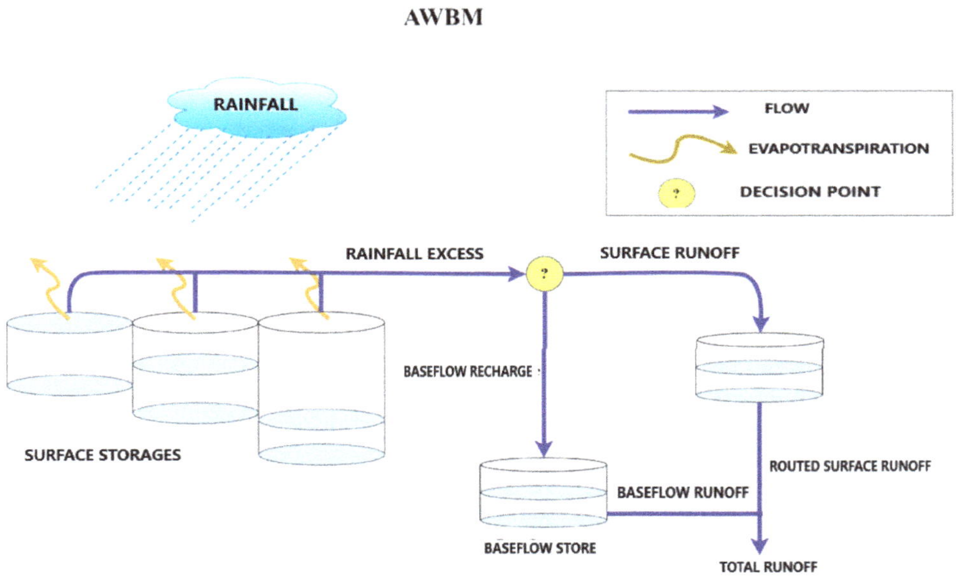

Figure 3. Schematic representation of the Australian Water Balance Model (AWBM).

TANK

This model is easy to use and effective. As seen in Figure 4, the highest tank is where rainwater is collected. The water level gradually decreases due to evaporation; water is taken from the tank beneath it to compensate until all the tanks are empty [12]. The side outlets then release the calculated runoff, with the top tank output representing surface runoff, the second tank output representing intermediate runoff, the third tank output representing sub-base runoff, and the fourth tank output representing base flow [13]. Even though the TANK model is simple, its behavior is intricate, and the contents of each tank have a big influence on how well the model works. Notably, depending on the tanks' storage capacities, the same rainstorm might produce very varied amounts of runoff [15]. The tank model is a widely used method for simulating diurnal streamflow patterns. It is based on daily inputs of evapotranspiration and rainfall and does not require accounting for the initial rainfall loss as the model's nonlinear structure already accounts for its impact [16]. The TANK model's parameters are listed in Table 4 and Figure 4, along with their lowest, maximum, and default values.

Table 4. Parameter values by default for the TANK model.

Parameter	Minimum	Default Value	Maximum
First outlet height of the first tank (H11) (in mm)	0	0	500
Second outlet height of first tank (H12) (in mm)	0	0	300
First outlet height of the second, third, and fourth tanks (H21, H31, and H41) (in mm)	0	0	100
Coefficient of runoff from various tank outlets (a11, a12, a21, a31, and a41)	0	0.2	1
Coefficient of Evaporation (α)	0	0.1	1

Table 4. *Cont.*

Parameter	Minimum	Default Value	Maximum
Coefficient of infiltration in tanks 1, 2, and 3 (b1, b2, and b3)	0	0.2	1
Tank's water level (C1, C2, C3, and C4) (in mm)	0	20	100

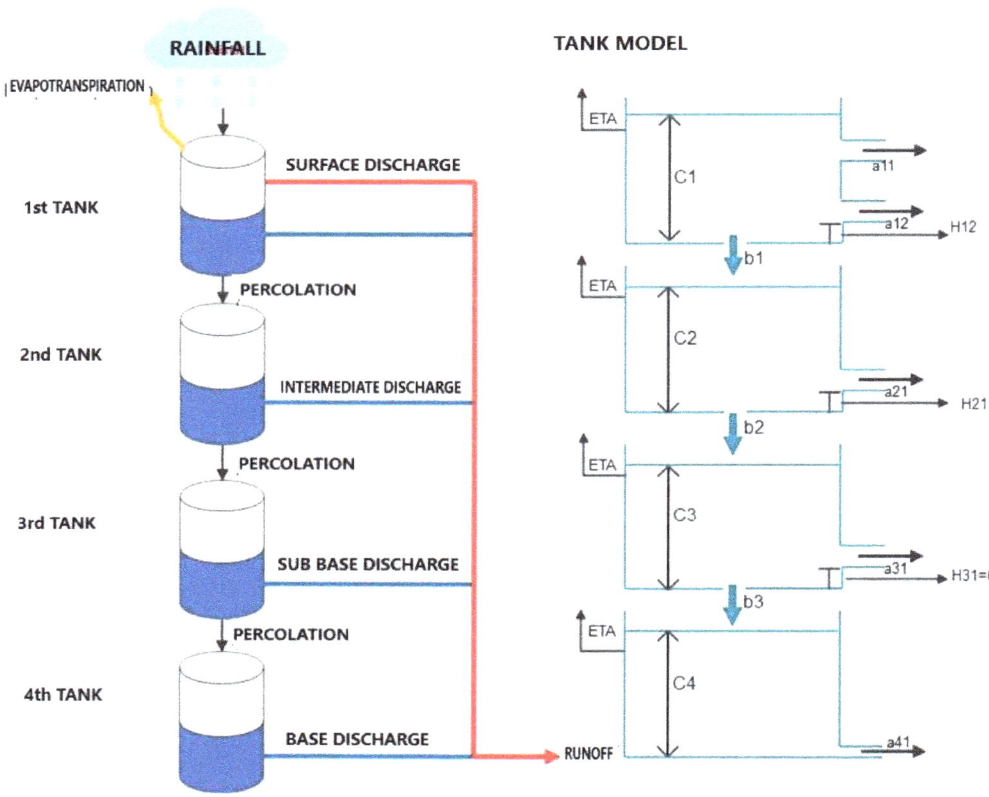

Figure 4. Schematic representation of the TANK model.

SIMHYD

The SIMHYD model predicts the daily discharge at a selected gauging station using daily rainfall and PET data [31]. A more straightforward version of the HYDROLOG model, created in 1972, is SIMHYD. A conceptual model that mimics the physics of rainwater runoff is called HYDROLOG. Furthermore, another model known as MODHYDROLOG [24] exists. HYDROLOG and MODHYDROLOG have seventeen and nineteen parameters, respectively, while SIMHYD only has seven (Table 5 and Figure 5). A system that uses the daily rainfall to replenish the interception storage, which is released daily, is part of the SIMHYD concept. The extra rainfall is subsequently subjected to an infiltration capacity assessment function. Infiltration excess runoff is the term used to describe any additional rainfall that is beyond the infiltration capacity.

Table 5. SIMHYD's default model parameter values.

Parameter	Minimum	Default Value	Maximum
Baseflow coefficient	0.0	0.9	1.0
Unaffected Threshold	0.0	1.5	5.0
Infiltration Coefficient	0.0	0.2	1.0
Infiltration Form	1	320	500
Interflow Coefficient	0.0	0.3	1.0
Prior Fraction	0	1	5
Capacity of the Rainfall Interception Store	0	200	400
Recharge Coefficient	0	3	10
Soil Moisture Store Capacity	0.0	0.1	1.0

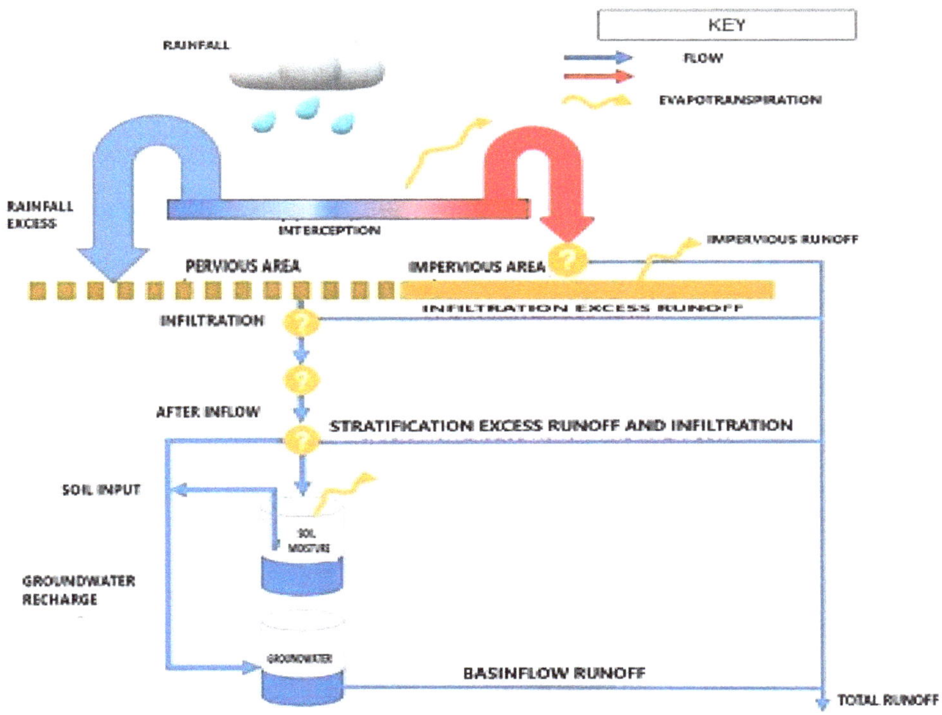

Figure 5. Schematic representation of the SIMHYD model.

3. Results and Discussion

3.1. Findings and Talks

In addition to categorical measures like accuracy, POD, FAR, POFD, and PSS, which are calculated daily using the rainfall threshold of 2.5 mm/day that the Indian Meteorological Department uses to identify a wet day, continuous metrics include NSE, R^2, MBE, MAE, and RMSE. These metrics were computed using monthly rainfall accumulation for each dataset from 1998 to 2020. The entire basin was compared using 427 Indian Meteorological Department gridded points with a spatial precision of 0.25°*0.25°. The average yearly rainfall for several datasets is shown in Figure 6 [25]. The yearly average precipitation varied less in other models; however, TRMM, PERSIANN CDR, and MSWEP nearly matched data from the Indian Meteorological Department [26]. In terms of spatial extent, MSWEP and CHIRPS nearly matched the Indian Meteorological Department, whereas Figure 6 shows the average yearly precipitation over all eight datasets. Except for PGF, CPC, and CMORPH, all products faithfully captured the well-known precipitation features, such as high precipitation areas on the north and east boundaries of the study area. According to the Indian Meteorological Department, the east side of the basin receives the most rainfall on average each year (1887.04 mm), while the western side receives the lowest amount (553.46 mm). From west to east, the average annual rainfall tends to rise [28].

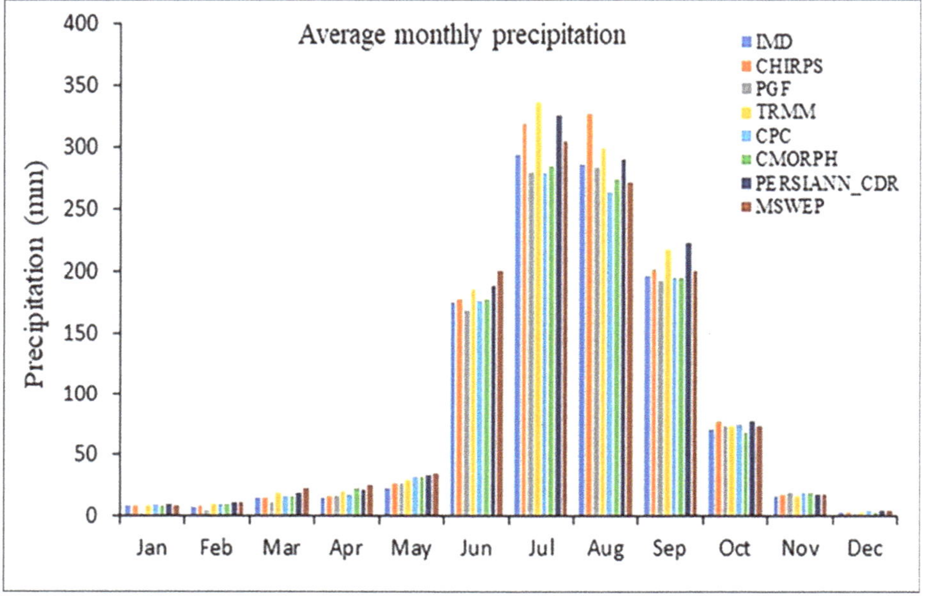

Figure 6. CHIRPS, PGF, TRMM, CPC, CMORPH, PERSIANN_CDR, MSWEP, and the Indian Meteorological Department's monthly average precipitation.

The average monthly precipitation for the entire basin from 1998 to 2024 is shown in Table 6 and Figure 6. The monsoon months of June, July, August, and September bring the majority of the basin's rainfall (around 85%). Monsoon precipitation over the basin was estimated with 83% to 86% accuracy by almost all precipitation products [4]. The findings indicate that TRMM and PERSIANN_CDR were overstated for every month throughout the monsoon season [15]. The data from the Indian Meteorological Department and PERSIANN_CDR differed significantly. PERSIANN_CDR and CHIRPS overestimated precipitation by up to about 350 mm [18]. Moderate precipitation (200–300 mm), CMORPH and CPC were subtitled [32].

The Indian Meteorological Department's seven-satellite precipitation estimates served as the basis for the CDFs. The Indian Meteorological Department gridded dataset was used as a reference for the CDF comparison [24]. MSWEP and the Indian Meteorological Department matched pretty well when compared to the other datasets. However, Figure 7a–d exhibits a slight overestimation of positive bias, particularly for low (between 10 and 65 mm) and moderate (between 110 and 180 mm) precipitation amounts. TRMM displayed a favorable bias for most of the events between 100 and 420 mm; it also matched Indian Meteorological Department data for the remaining section [32].

Table 6. Various gridded precipitation datasets used in this study area.

DATA	Temporal Resolution	Availability	Link
IMD	0.25° and D	1951–2024	The website https://www.imdpune.gov.in/, accessed on 7 April 2025
CHIRPS	0.05° and D and M	1981–present	CHIRPS 2.0: https://data.chc.ucsb.edu/products/, accessed on 7 April 2025
PGF	0.25° and D	1948–2020	https://hydrology.soton.ac.uk/, accessed on 7 April 2025
TRMM	0.25° and 3H, D and M	1998–2023	The summary can be seen at https://disc.gsfc.nasa.gov/datasets/TRMM_3B42_Daily_7, accessed on 7 April 2025
CPC	0.5° and D	1979–present	Balprecip.html https://psl.noaa.gov/data/gridded/data.cpc.glo, accessed on 7 April 2025
CMORPH	0.25° and 30 min, 1H and D	1998–2024	https://www.ncei.noaa.gov/products/climate-data-records, accessed on 7 April 2025
PERSIA NN_CDR	0.25° and D and M	1983–Present	https://chrsdata.eng.uci.edu/, accessed on 7 April 2025
MSWEP	0.1° and 3H, D and M	1979–2024	Gloh2o.org/mswep/'s website

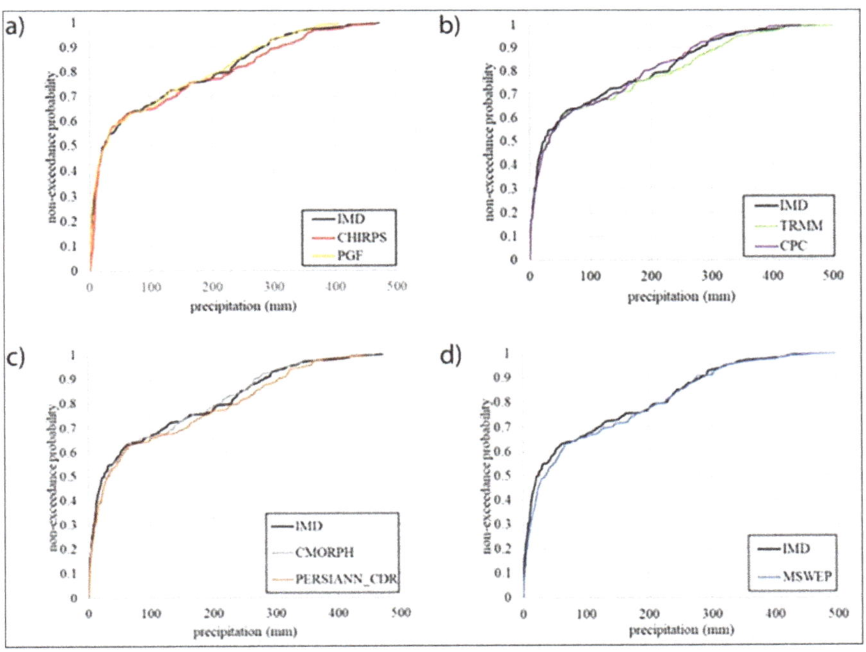

Figure 7. A comparison of the non-exceedance probability of the Indian Meteorological Department with seven-satellite precipitation estimates. (**a**) The precipitation estimates from the IMD and CHIRPS data, (**b**) Matched with IMD data for the remaining portion of TRMM, (**c**) PERSIANN_CDR had a high discrepancy compared to IMD data, (**d**) MSWEP was closely matched with IMD as compared to the remaining datasets.

The gridded dataset from the Indian Meteorological Department served as a source for the CDF comparison. When juxtaposed with the other datasets, Figure 7a–d displays the CDFs based on seven satellite precipitation estimates from the Indian Meteorological Department's gridded precipitation. MSWEP and the Indian Meteorological Department matched quite well. Even so, Figure 7d shows a small positive bias (overestimates), especially for moderate precipitation (110 to 180 mm) and light precipitation (10 to 65 mm) [19]. TRMM had a favorable bias for most of the events that occurred between 100 and 420 mm; for the remaining, it also coincided with Indian Meteorological Department data. PERSIANN CDR data showed a significant disparity from Indian Meteorological Department data. Up to roughly 350 mm of precipitation, PERSIANN CDR and CHIRPS were overestimated. For moderate precipitation (200–300 mm), CMORPH and CPC were understated [22].

3.1.1. Ongoing Metrics Assessment

The monthly precipitation for each grid cell of the seven datasets was computed and compared to the monthly precipitation data from the Indian Meteorological Department, and statistical characteristics such as NSE, R^2, MBE, MAE, and RMSE were evaluated. Continuous measurements for the study area between 1998 and 2024 were used [15]. The NSE coefficients between each precipitation product and the Indian Meteorological Department reference dataset suggest that there might be a conflict of interest between the precipitation products and the Indian Meteorological Department product. MSWEP has higher NSE values over the study area, especially in the western section (Upper Godavari, Pravara, and some portions of Purna), while the other products have low or negative NSE values [23]. The eastern component of the study area includes extremely elevated locations and a few areas of Pranhita, on the northern side. Higher accuracy is indicated by an NSE value greater than 0.7, which was present in 389 out of 427 stations. Comparable to the second-most accurate precipitation dataset after MSWEP, TRMM has 367 stations with an NSE value greater than 0.7 [24]. The lack of rainfall and the large spatial variance at all-time scales are the primary reasons for this notable performance drop. The regional variability of the R^2 along the basin is depicted in Table 7, where all precipitation datasets apart from PGF show robust connections. In particular, MSWEP and TRMM showed the highest correlation among the 427 stations, with R^2 values over 0.8 for 336 and 357 stations, respectively [26].

Table 7. Monthly precipitation was calculated for each of the seven dataset's grid cells.

DATA	NSE	R^2	MBE	MAE	RMSE
CHIRPS	0.768	0.812	7.505	34.324	61.519
PGF	0.72	0.768	−2.433	35.502	66.775
TRMM	0.768	0.846	9.126	31.528	57.413
CPC	0.772	0.801	−0.774	33.247	61.58
CMORPH	0.767	0.815	0.157	32.663	60.16
PERSIANN_CDR	0.667	0.815	9.332	35.934	65.177
MSWEP	0.806	0.831	5.531	31.794	56.734

The mountainous landscapes of the eastern, western, and northern regions did not correlate well. Figure 8 displays the spatial variance of MBE for the study region. All datasets in the eastern portion had negative MBE, suggesting a comparable underestimation of precipitation, especially in extremely and moderately dense woodlands [32]. There was little to no positive bias in the central and northern regions, suggesting that the judgment was accurate. Due to the infrequent rainfall, TRMM and PERSIANN_CDR provide accurate results up to 76° E from the east; nevertheless, the remaining region displays subpar data with negative NSE. The Upper Godavari region receives 800 mm of rainfall annually, which may have been influenced by the leeward side of the Western Ghats [33].

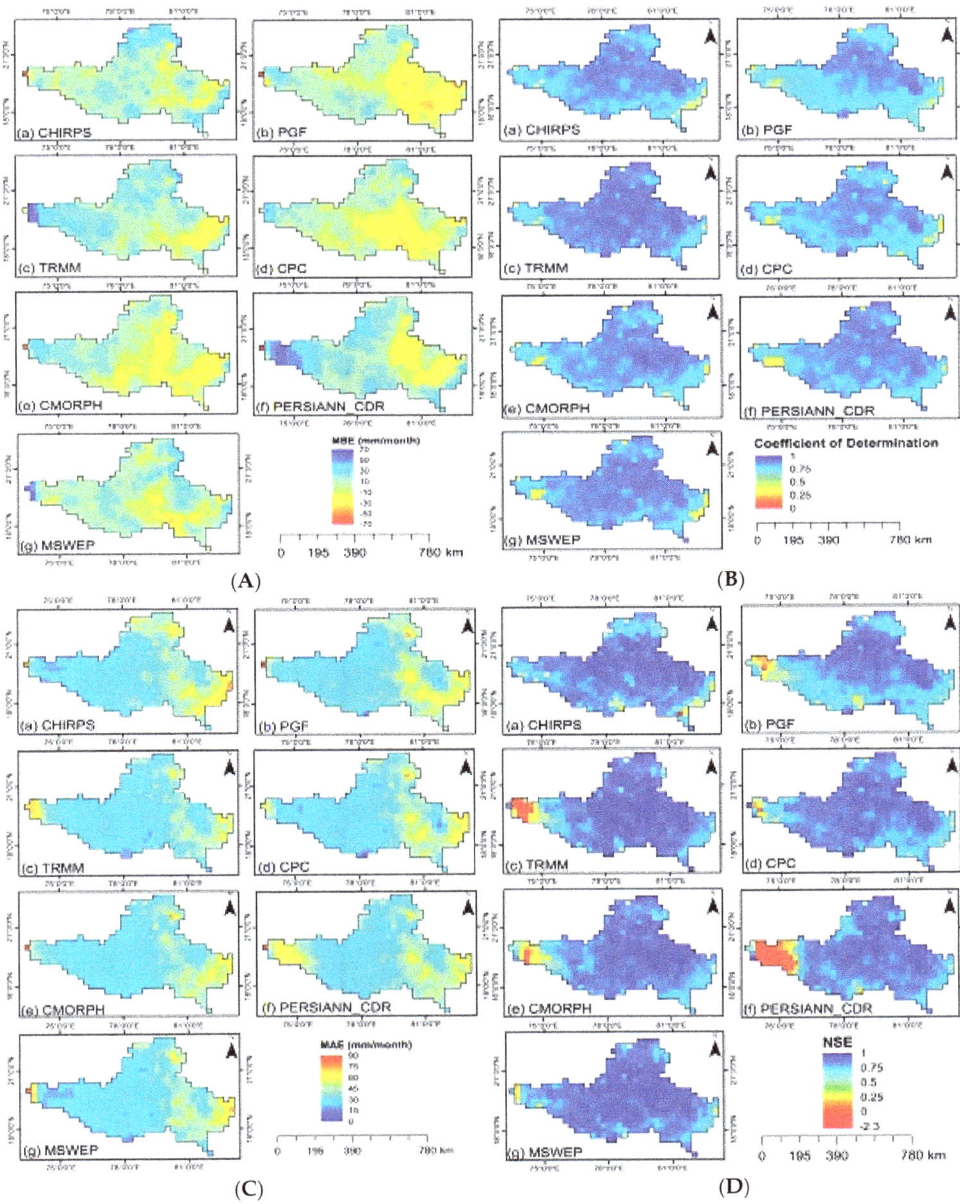

Figure 8. The coefficient of determination of datasets from 1998 to 2024: (**A**) POD of the seven-satellite precipitation estimation, (**B**) NSE of the seven datasets estimation, (**C**) FAR of the seven-satellite precipitation estimation, (**D**) the seven datasets estimation of precipitation.

For the TANK, SIMHYD, AWBM, and Sacramento hydrological models, the standard deviation values of the predicted values are 324.74, 288.44, 296.36, and 332.98, respectively, as shown in Figure 8b. Confidence intervals (CC) for the observed and expected values are 0.87, 0.84, 0.86, and 0.86, respectively [13]. The Taylor diagram also shows the RMSE values, which are 202.61, 226.35, 216.85, and 211.99 m^3/s. Since the Sacramento model is the closest to the observed value, it seems to perform the best based on the standard deviation values. Additionally, the Sacramento model is demonstrated to be the most effective of the daily streamflow simulation models according to the RMSE and CC values during the validation period [15].

However, when compared to the other models, the AWBM and TANK models display smaller standard deviation values, which may indicate that they have a propensity to underestimate the variability of the actual data. Nonetheless, the Sacramento model's standard deviation values of 308.59 were quite similar [19]. In terms of absolute errors, the Sacramento model corresponds with the real data the best, as evidenced by it displaying the lowest RMSE score. The RMSE values for each model fall between 168.04 and 181.44 m^3/s. Figure 8a–d compares observed data for daily streamflow modeling with several hydrological models (TANK, SIMHYD, AWBM, and Sacramento) throughout both the calibration process and validation using Taylor's diagram [22]. A graphical tool called Taylor's diagram is used to compare models to the observed data in Table 8 and Figure 8. It offers a rapid visual evaluation of a model's ability to represent the volatility of and pattern in the data. As seen in Figure 8a, the calibration CC for each model ranged from 0.83 to 0.86, indicating that the model outputs and the observed data have a respectable level of agreement [24].

Table 8. All conceptual models' parameters were calibrated.

Sacramento	AWBM	TANK	SIMHYD
ADIM 0.031	A1 0.014	H11 119.60	base flow Coefficient 0.373
LZFP 49.608	A2 0.433	a11 0.169	Impervious Threshold 4.431
LZFS M 49.608	BFI 0.298	a12 0.204	Infiltration Coefficient 371.76
LZPK 0.118	C1 1.569	a21 0.812	Infiltration Shape 0.196
LZSK 0.729	C2 130.1 96	a31 0.847	Interflow Coefficient 0.000
LZTW M 117.647	C3 252.9 41	a41 0.478	previous Fraction 1.000
PCTI M 0.000	KBase 0.561	alpha 1.000	rainfall Interception Capacity 4.569
PFREE 0.184	KSurf 0.627	b1 0.031	Recharge Coefficient 0.741
REXP 1.529		b2 0.337	Soil Moisture Store Capacity 169.29 0
RSERV 0.300		b3 0.027	
SARVA 0.010		C1 51.765	
SIDE 0.000		C2 18.824	
SSOUT 0.001		C3 52.549	
UZFWM 79.373		C4 26.667	

3.1.2. Evaluation of Categorical Metrics

To assess each precipitation product's capacity to predict a rainy day, five category metrics were calculated. As per the rules set by the Indian Meteorological Department, a day is deemed rainy if it records 2.5 mm or more of precipitation. Evaluations of five categorical criteria for seven products of precipitation between 1998 and 2016 were carried out [12]. With values of 0.844, 0.571, and 0.462, MSWEP had the lowest FAR, the highest accuracy, and the lowest PSS. With accuracy, POD, and PSS of 0.821, 0.705, and 0.549, respectively, PERSIANN_CDR was the second-most dependable precipitation product to identify wet days. With a maximum MBE of 9.332 mm/month, PERSIANN_CDR was likely overestimating the entire research region, which is how rainy days were found [14].

In the eastern Godavari basin (Sabari, parts of lower Godavari), the accuracy of all precipitation products is poorer. When it comes to counting hits within the research region, MSWEP was operating effectively, ignoring false alarms, which ranged from 0.59 to 0.89 [24]. In terms of hit detection, PERSIANN_CDR also outperformed other datasets. In areas with higher yearly precipitation, both of these datasets made accurate predictions in some areas of Sabari and the lower Godavari. Moreover, CHIRPS demonstrated strong performance in identifying rainy days, with 0.823 accuracy, 0.613 POD, and 0.497 PSS. When it comes to predicting wet days, MSWEP performs somewhat better than the other precipitation programs [26].

The average for July will increase significantly in the near future, from 277.41 m^3/s to 481.75 m^3/s. Furthermore, it will increase to 619.29 m^3/s in the medium future and 847.14 m^3/s in the distant future. While the intermediate and far future averages will be 724.62 m^3/s and 965.38 m^3/s, respectively, this will soon increase to 757.67 m^3/s. The historical average for August is 374.57 m^3/s. It will shortly rise to 402.71 m^3/s, with the historical average for September at 252.30 m^3/s [11]. The averages are 528.45 m^3/s in the intermediate future and 433.63 m^3/s in the remote future. Last but not least, the October average is 60.68 m^3/s historically, rising to 85.57 m^3/s in the middle and 118.46 m^3/s in the long term. However, in the near future, it drops somewhat to 58.89 m^3/s [14].

This study rigorously assessed seven satellite precipitation products against Indian Meteorological Department gridded data in the Godavari River basin from 1998 to 2024. The findings revealed that spatial correlations were generally strong, but performance varied with climate regimes. Notably, MSWEP and TRMM stood out, delivering exceptional accuracy, high NSE and R^2 values, and low MAE and RMSE in monthly evaluations. These datasets performed remarkably well during the monsoon and post-monsoon seasons, making them valuable for precipitation detection. However, the spatial assessment showed limitations in elevated, forested, and low-rainfall regions. Overall, MSWEP and TRMM emerged as top performers, especially in data-scarce areas, offering significant potential for hydrological studies and climate scenario downscaling in the future. Further research should consider bias correction and finer-resolution datasets for enhanced accuracy in various water resource applications.

To estimate the monthly and daily precipitation over the Godavari River basin using the Indian Meteorological Department gridded data with 427 grid stations, this paper provides a comprehensive assessment of seven satellite precipitation products from 1998 to 2016: CHIRPS, PGF, TRMM, CPC, CMORPH, PERSIANN_CDR, and MSWEP. The evaluation period was established at 27 years due to the availability and quality. All seven products were interpolated (distance-weighted average remapping) using a comparable geographic resolution of 0.25° and contrasted with gridded data from the Indian Meteorological Department.

Several parameters, including calibrated connections between sensor reflectivity and projected rain rate and sampling error derived from satellite overpass time, were used to evaluate rainfall product category metrics. The algorithms and procedures used in the product created their relationship, and they were connected in some manner. The monthly average streamflow forecasts for the SSP585 scenario were calculated. It will shortly rise to 5.23 m^3/s, which is higher than the historical average for January at 3.94 m^3/s. In the intermediate and distant futures, it will rise to 1.32 m^3/s and 30.34 m^3/s, respectively. The historical average for June will soon rise from 22.67 m^3/s to 39.00 m^3/s. In the intermediate and distant future, it will continue to rise to 98.20 m^3/s and 15.08 m^3/s.

The models were created using data for the years between 1987 and 2024, with 30% going towards validation and 70% going towards calibration. The evaluation metrics used showed that all four conceptual models performed satisfactorily in modeling streamflow. In contrast to the other models, the Sacramento model produced noticeably superior results. The most notable variations in rainfall patterns across various periods are displayed in the SSP585 scenario. The region with the largest relative change in rainfall in the far future is SSP585 (55.02%), followed by SSP370 (46.09%) [18], SSP245 (25.78%), and SSP126 (22.64%). The SSP585 scenario predicts the far future to have the biggest absolute change in annual mean temperature, 3.29 °C. Bias-corrected EC-Earth3 datasets have been used to forecast future streamflow using the Sacramento model. Depending on the situation and time frame, the monthly streamflow changed. Q95 is predicted to rise by 40.09% to 127.06% in the mid-future and 73.90% to 215.13% in the distant future during the wettest July [14].

In terms of evaluation measures, RF and M5P performed well when using CMIP6 datasets as input. RF demonstrated adequate performance in testing ($0.4 < NSE < 0.50$ and $0.5 < R^2 < 0.6$) and extremely good performance in training ($0.75 < NSE < 1$ and $0.7 < R < 1$). Likewise, M5P demonstrated good performance in both training and testing ($0.4 < NSE < 0.50$ and $0.5 < R^2 < 0.6$). The input precipitation dataset for CMIP6 was determined to be MIROC6 for the RF model and MRI-ESM2-0 for the M5P model. Daily estimates: The streamflow estimates for the Kanhargaon, Nowrangpur, and Wairagarh regions show a trend of rising water flow across several SSP scenarios, suggesting both potential flooding difficulties and opportunities in terms of water supply [16]. An average streamflow of 74.16 m^3/s is anticipated at Kanhargaon under the more severe SSP585 scenario, with peaks possibly reaching 10,915.63 m^3/s. Extreme water flows, indicated by this high value, could result in flooding and pose serious problems for flood control. Nowrangpur, which has historically seen larger flows, is also about to undergo significant changes. SSP585 predicts extremely high levels of 11,548.50 m^3/s, which would pose a serious risk of flooding during peak hours. Wairagarh is susceptible to significant changes, even though it began with very small flows in the SSP126 scenario at 48.92 m^3/s. Highs of 6691.33 m^3/s are predicted by the SSP585 scenario, indicating that even regions with normally moderate flows may experience significant flooding difficulties.

4. Conclusions

The seven satellite precipitation products CHIRPS, PGF, TRMM, CPC, CMORPH, PERSIANN_CDR, and MSWEP are thoroughly evaluated in this study. The monthly and daily precipitation over the Godavari River basin is estimated using data from 427 grid stations collected by the Indian Meteorological Department between 1998 and 2024. The availability and quality of the data led to the establishment of the evaluation period at 27 years. Using a similar geographic resolution of 0.25°, all seven products were interpolated (distance-weighted average remapping) and compared to gridded data from the Indian Meteorological Department. The assessment was conducted using both continuous and categorical variables at multiple temporal (daily, monthly, and seasonal) and geographical (basin and grid) resolutions. The following are the key findings:

- **Geographic Correlation**: The satellite precipitation products showed excellent geographic correlation, although the degree of accuracy varied depending on the climate regime and evolving weather patterns influenced by climate change.
- **Monsoon Dominance**: The monsoon season (June to September) remains the dominant precipitation period, accounting for approximately 85% of the basin's total rainfall.

This trend aligns with projections of intensified monsoon activities due to climate change, potentially increasing flood risks and altering water resource availability.

- **Seasonal Performance Variations**: Except for MSWEP and TRMM, none of the datasets accurately captured precipitation during the pre-monsoon (March–May) and winter (January–February) seasons. This gap is critical under climate change scenarios, where shifts in seasonal precipitation patterns are expected to intensify.
- **Performance Under Monsoon and Post-Monsoon Conditions**: MSWEP and TRMM outperformed the other datasets during the monsoon and post-monsoon seasons, exhibiting lower Mean Bias Error (MBE), Mean Absolute Error (MAE), and Root Mean Square Error (RMSE), along with higher Nash–Sutcliffe Efficiency (NSE) and coefficient of determination (R^2). The MSWEP dataset, in particular, achieved an NSE of 0.806, R^2 of 0.831, an MAE of 31.79 mm/month, and an RMSE of 56.73 mm/month, indicating robust performance in tracking precipitation trends amidst climate variability.
- **Accuracy in Precipitation Event Detection**: MSWEP demonstrated the highest accuracy, with a Peirce's Skill Score (PSS) of 0.571, a False Alarm Ratio (FAR) of 0.462, and an overall accuracy of 0.844. Additionally, CHIRPS and PERSIANN_CDR successfully detected rainfall occurrences, proving their utility as reliable backup data sources for identifying extreme precipitation events under changing climate conditions.

These findings highlight the importance of selecting the most reliable precipitation datasets to enhance hydrological modeling and climate adaptation strategies in the Godavari basin. As climate change continues to alter precipitation patterns, such evaluations are crucial for improving flood forecasting, water resource management, and resilience planning in vulnerable regions.

Author Contributions: Conceptualization, R.A. and C.P.; methodology, D.M. and S.V.; software, R.V.; validation, H.M.A. and N.N.; formal analysis, R.A. and C.P.; investigation, D.M. and S.V.; resources, R.V.; data curation, H.M.A. and P.B.; writing—original draft preparation, N.N.; writing—review and editing, R.A.; visualization, D.M.; supervision, R.V. All authors have read and agreed to the published version of the manuscript.

Funding: This research received no external funding.

Data Availability Statement: The raw data supporting the conclusions of this article will be made available by the authors on request.

Conflicts of Interest: Author Chandrashekar Pandugula was employed by the company Lowe's Inc., author Ravikumar Vankayalapati was employed by the company Equinix Inc. The remaining authors declare that the research was conducted in the absence of any commercial or financial relationships that could be construed as a potential conflict of interest.

References

1. Agridiotis, V.; Forster, C.F.; Carliell-Marquet, C. Addition of Al and Fe salts during treatment of paper mill effluents to improve activated sludge settlement characteristics. *Bioresour. Technol.* **2007**, *98*, 2926–2934. [CrossRef] [PubMed]
2. Akar, Ö.; Saralioğlu, E.; Güngör, O.; Bayata, H.F. Semantic segmentation of very-high spatial resolution satellite images: A comparative analysis of 3D-CNN and traditional machine learning algorithms for automatic vineyard detection. *Int. J. Eng. Geosci.* **2024**, *9*, 12–24. [CrossRef]
3. Al-Shannag, M.; Al-Qodah, Z.; Bani-Melhem, K.; Qtaishat, M.R.; Alkasrawi, M. Heavy metal ions removal from metal plating wastewater using electrocoagulation: Kinetic study and process performance. *Chem. Eng. J.* **2015**, *260*, 749–756. [CrossRef]
4. Anselmo, L. The long-term evolution of the Italian satellites in the GEO region and their possible interaction with the orbital debris environment. In Proceedings of the 54th International Astronautical Congress of the International Astronautical Federation, the International Academy of Astronautics, and the International Institute of Space Law, Bremen, Germany, 29 September–3 October 2003; IAC-03-IAA.5.2.05. [CrossRef]
5. Aydın, V.A. Comparison of CNN-based methods for yoga pose classification. *Turk. J. Eng.* **2024**, *8*, 65–75. [CrossRef]

6. Akgül, V.; Görmüş, K.S.; Kutoğlu, Ş.H.; Jin, S. Performance analysis and kinematic test of the BeiDou Navigation Satellite System (BDS) over coastal waters of Türkiye. *Adv. Eng. Sci.* **2024**, *4*, 1–14.
7. Baudhanwala, D.; Mehta, D.; Kumar, V. Machine learning approaches for improving precipitation forecasting in the Ambica River basin of Navsari District, Gujarat. *Water Pract. Technol.* **2024**, *19*, 1315–1329. [CrossRef]
8. Briffa, J.; Sinagra, E.; Blundell, R. Heavy metal pollution in the environment and their toxicological effects on humans. *Heliyon* **2020**, *6*, e04691. [CrossRef]
9. Bakar, A.F.A.; Halim, A.A. Treatment of automotive wastewater by coagulation-flocculation using poly-aluminum chloride (PAC), ferric chloride (FeCl3) and aluminum sulfate (alum). *AIP Conf. Proc.* **2013**, *1571*, 524–529. [CrossRef]
10. Camaj, A.; Haziri, A.; Nuro, A.; Ibrahimi, A.C. Levels of BTEX and Chlorobenzenes in water samples of White Drin River, Kosovo. *Adv. Eng. Sci.* **2024**, *4*, 45–53. Available online: https://publish.mersin.edu.tr/index.php/ades/article/view/1326 (accessed on 12 February 2024).
11. Capizzi, G.; Lo Sciuto, G.; Napoli, C.; Tramontana, E.; Wozniak, M. Automatic classification of fruit defects based on co-occurrence matrix and neural networks. In Proceedings of the 2015 Federated Conference on Computer Science and Information, Lodz, Poland, 13–16 September 2015; pp. 861–867. [CrossRef]
12. Çetin Taş, İ.; Bozdoğan, A.M.; Arıca, S. Application of Convolutional Neural Networks for Watermelon Detection in UAV Aerial Images: A Case Study. *Turk. J. Eng.* **2025**, *9*, 1–11. [CrossRef]
13. Chang, S.; Ahmad, R.; Kwon, D.E.; Kim, J. Hybrid ceramic membrane reactor combined with fluidized adsorbents and scouring agents for hazardous metal-plating wastewater treatment. *J. Hazard. Mater.* **2020**, *388*, 121777. [CrossRef] [PubMed]
14. Dursun, S.; Qasim, M.N. Measurements and modelling of PM2.5 level in summertime period in Novada Main Shopping Centre Konya, Turkey. *Eng. Appl.* **2022**, *1*, 19–32. Available online: https://publish.mersin.edu.tr/index.php/enap/article/view/314 (accessed on 15 January 2023).
15. Demirgül, T.; Yılmaz, C.B.; Zıpır, B.N.; Kart, F.S.; Pehriz, M.F.; Demir, V.; Sevimli, M.F. Investigation of Turkey's climate periods in terms of precipitation and temperature changes. *Eng. Appl.* **2022**, *1*, 80–90. Available online: https://publish.mersin.edu.tr/index.php/enap/article/view/320 (accessed on 18 June 2022).
16. Daberdini, A.; Basholli, F.; Metaj, N. Simulation with the Mirone application for the construction of marine mechanical waves generated by possible seismic events in the territory of the Adriatic and Ionian seas. *Adv. Eng. Sci.* **2023**, *3*, 46–54. Available online: https://publish.mersin.edu.tr/index.php/ades/article/view/856 (accessed on 13 February 2023).
17. Dursun, S.; Sarcan, A. Determination of water quality in Hadim District of Konya (Turkey) and the investigation of disinfection efficiency. *Adv. Eng. Sci.* **2022**, *2*, 101–108. Available online: https://publish.mersin.edu.tr/index.php/ades/article/view/89 (accessed on 16 January 2022).
18. Ekiz, A.; Arica, S.; Bozdogan, A.M. Classification and Segmentation of Watermelon in Images Obtained by Unmanned Aerial Vehicle. In Proceedings of the ELECO 2019—11th International Conference on Electrical and Electronics Engineering, Bursa, Turkey, 28–30 November 2019; pp. 619–622. [CrossRef]
19. El Gaayda, J.; Rachid, Y.; Titchou, F.E.; Barra, I.; Hsini, A.; Yap, P.S.; Akbour, R.A. Optimizing removal of chromium (VI) ions from water by coagulation process using central composite design: Effectiveness of grape seed as a green coagulant. *Sep. Purif. Technol.* **2023**, *307*, 122805. [CrossRef]
20. Ezquerro, O.; Ortiz, G.; Pons, B.; Tena, M.T. Determination of benzene, toluene, ethylbenzene and xylenes in soils by multiple headspace solid-phase microextraction. *J. Chromatogr. A* **2004**, *1035*, 17–22. [CrossRef]
21. Fen, İ.Ü.; Enst, B.; Der, K. Kadir SABANCI ve Ark. Available online: http://www.tuik.gov.tr/bitkiselapp/ (accessed on 30 June 2022).
22. Fu, S.Y.; Wang, Z.R.; Shi, H.L.; Ma, L.H. The application of decommissioned GEO satellites to CAPS. In Proceedings of the IOP Conference Series: Materials Science and Engineering, Kitakyushu City, Japan, 10–13 April 2018; Volume 372, p. 012033. [CrossRef]
23. Gangani, P.; Mangukiya, N.K.; Mehta, D.J.; Muttil, N.; Rathnayake, U. Evaluating the efficacy of different DEMs for application in flood frequency and risk mapping of the Indian Coastal River Basin. *Climate* **2023**, *11*, 114. [CrossRef]
24. Gholami, A. Exploring drone classifications and applications: A review. *Int. J. Eng. Geosci.* **2024**, *9*, 418–442. [CrossRef]
25. Gordani, O.; Simoni, A. Leveraging SVD for efficient image compression and robust digital watermarking. *Adv. Eng. Sci.* **2024**, *4*, 103–112. Available online: https://publish.mersin.edu.tr/index.php/ades/article/view/1496 (accessed on 3 September 2024).
26. Gohil, M.; Mehta, D.; Shaikh, M. An integration of geospatial and fuzzy-logic techniques for multi-hazard mapping. *Results Eng.* **2024**, *21*, 101758. [CrossRef]
27. Hosseini, S.S.; Bringas, E.; Tan, N.R.; Ortiz, I.; Ghahramani, M.; Shahmirzadi, M.A.A. Recent progress in development of high performance polymeric membranes and materials for metal plating wastewater treatment: A review. *J. Water Process Eng.* **2016**, *9*, 78–110. [CrossRef]
28. Huang, J.; Yuan, F.; Zeng, G.; Li, X.; Gu, Y.; Shi, L.; Liu, W.; Shi, Y. Influence of pH on heavy metal speciation and removal from wastewater using micellar-enhanced ultrafiltration. *Chemosphere* **2017**, *173*, 199–206. [CrossRef] [PubMed]

29. Ilhan, F.; Ulucan-Altuntas, K.; Avsar, Y.; Kurt, U.; Saral, A. Electrocoagulation process for the treatment of metal-plating wastewater: Kinetic modeling and energy consumption. *Front. Environ. Sci. Eng.* **2019**, *13*, 1–8. [CrossRef]
30. Kocalar, A.C. Sinkholes caused by agricultural excess water using and administrative traces of the process. *Adv. Eng. Sci.* **2023**, *3*, 15–20. Available online: https://publish.mersin.edu.tr/index.php/ades/article/view/756 (accessed on 18 January 2023).
31. Kalay, E.; Sarıoğlu, H.; Özkul, I. Design parameters of sand filtration systems in wastewater treatment process. *Adv. Eng. Sci.* **2021**, *1*, 34–42. Available online: https://publish.mersin.edu.tr/index.php/ades/article/view/23 (accessed on 24 August 2021).
32. Mehta, D.; Prajapati, K.; Islam, M.N. Watershed delineation and land use land cover (LULC) study of Purna River in India. In *India II: Climate Change Impacts, Mitigation and Adaptation in Developing Countries*; Springer International Publishing: Cham, Switzerland, 2021; pp. 169–181.
33. Mukasa, P.; Wakoli, C.; Faqeerzada, M.A.; Amanah, H.Z.; Kim, H.; Joshi, R.; Suh, H.K.; Kim, J.; Lee, H.; Kim, M.S.; et al. Nondestructive discrimination of seedless from seeded watermelon seeds by using multivariate and deep learning image analysis. *Comput. Electron. Agric.* **2022**, *194*, 106799. [CrossRef]
34. Mehta, D.; Hadvani, J.; Kanthariya, D.; Sonawala, P. Effect of land use land cover change on runoff characteristics using curve number: A GIS and remote sensing approach. *Int. J. Hydrol. Sci. Technol.* **2023**, *16*, 1–16. [CrossRef]
35. Reddy, N.M.; Saravanan, S.; Almohamad, H.; Al Dughairi, A.A.; Abdo, H.G. Effects of Climate Change on Streamflow in the Godavari Basin Simulated Using a Conceptual Model including CMIP6 Dataset. *Water* **2023**, *15*, 1701. [CrossRef]
36. Nguyen, M.K.; Tran, V.S.; Pham, T.T.; Pham, H.G.; Hoang, B.L.; Nguyen, T.H.; Nguyen, T.H.; Tran, T.H.; Ngo, H.H. Fenton/ozone-based oxidation and coagulation processes for removing metals (Cu, Ni)-EDTA from plating wastewater. *J. Water Process Eng.* **2021**, *39*, 101836. [CrossRef]
37. Oymak, A.; Tür, M.R.; Bouchiba, N. PV connected pumped-hydro storage system. *Eng. Appl.* **2022**, *1*, 46–54. Available online: https://publish.mersin.edu.tr/index.php/enap/article/view/316 (accessed on 18 June 2022).
38. Öz, İ.; Yılmaz, Ü.C. Determination of coverage oscillation for inclined communication satellite. *Sak. Univ. J. Sci.* **2020**, *24*, 973–983. [CrossRef]
39. Oz, I. Salınımlı yörünge haberleşme uydularında 2 eksen düzeltmeli kapsama alanı stabilizasyonu. *J. Fac. Eng. Archit. Gazi Univ.* **2022**, *38*, 219–229. [CrossRef]
40. Palansooriya, K.N.; Yang, Y.; Tsang, Y.F.; Sarkar, B.; Hou, D.; Cao, X.; Meers, E.; Rinklebe, J.; Kim, K.H.; Ok, Y.S. Occurrence of contaminants in drinking water sources and the potential of biochar for water quality improvement: A review. *Crit. Rev. Environ. Sci. Technol.* **2020**, *50*, 549–611. [CrossRef]
41. Pfaff, K.; Hildebrandt, L.H.; Leach, D.L.; Jacob, D.E.; Markl, G. Formation of the Wiesloch Mississippi Valley-type Zn-Pb-Ag deposit in the extensional setting of the Upper Rhinegraben, SW Germany. Miner. *Deposita* **2010**, *45*, 647–666. [CrossRef]
42. Qasem, N.A.; Mohammed, R.H.; Lawal, D.U. Removal of heavy metal ions from wastewater: A comprehensive and critical review. *Npj Clean Water* **2021**, *4*, 36. [CrossRef]
43. Raaj, S.; Gupta, V.; Singh, V.; Shukla, D.P. A novel framework for peak flow estimation in the Himalayan river basin by integrating SWAT model with machine learning based approach. *Earth Sci. Inform.* **2024**, *17*, 211–226. [CrossRef]
44. Sattari, M.T.; Kamrani, Y.; Javidan, S.; Fathollahzadeh Attar, N.; Apaydın, H. Groundwater quality mapping based on the Wilcox Classification method for agricultural purposes: Qazvin Plain aquifer case. *Turk. J. Eng.* **2025**, *9*, 116–128. [CrossRef]
45. Salamon, J.; Bello, J.P. Deep Convolutional Neural Networks and Data Augmentation for Environmental Sound Classification. *IEEE Signal Process. Lett.* **2017**, *24*, 279–283. [CrossRef]
46. Tchounwou, P.B.; Yedjou, C.G.; Patlolla, A.K.; Sutton, D.J. Heavy metal toxicity and the environment. In *Molecular, Clinical and Environmental Toxicology: Volume 3: Environmental Toxicology*; Springer: Berlin/Heidelberg, Germany, 2012; Volume 3, pp. 133–164. [CrossRef]
47. Verma, S.; Verma, M.K.; Prasad, A.D.; Mehta, D.; Azamathulla, H.M.; Muttil, N.; Rathnayake, U. Simulating the hydrological processes under multiple land use/land cover and climate change scenarios in the mahanadi reservoir complex, Chhattisgarh, India. *Water* **2023**, *15*, 3068. [CrossRef]
48. Verma, S.; Kumar, K.; Verma, M.K.; Prasad, A.D.; Mehta, D.; Rathnayake, U. Comparative analysis of CMIP5 and CMIP6 in conjunction with the hydrological processes of reservoir catchment, Chhattisgarh, India. *J. Hydrol. Reg. Stud.* **2023**, *50*, 101533. [CrossRef]
49. Verma, S.; Verma, M.K.; Prasad, A.D.; Mehta, D.J.; Islam, M.N. Modeling of uncertainty in the estimation of hydrograph components in conjunction with the SUFI-2 optimization algorithm by using multiple objective functions. *Model. Earth Syst. Environ.* **2024**, *10*, 61–79. [CrossRef]
50. Wan Nurazwin Syazwani, R.; Muhammad Asraf, H.; Megat Syahirul Amin, M.A.; Nur Dalila, K.A. Automated image identification, detection and fruit counting of top-view pineapple crown using machine learning. *Alex. Eng. J.* **2022**, *61*, 1265–1276. [CrossRef]
51. Witkowska, D.; Słowik, J.; Chilicka, K. Heavy metals and human health: Possible exposure pathways and the competition for protein binding sites. *Molecules* **2021**, *26*, 6060. [CrossRef] [PubMed]

52. Yalçın, C.; Canlı, H. IP/Resistivity methods for Pb-Zn deposit exploration: A case study in Sudöşeği, (Simav-Kütahya, Türkiye). *Eng. Appl.* **2024**, *3*, 27–35. Available online: https://publish.mersin.edu.tr/index.php/enap/article/view/1503 (accessed on 15 February 2024).
53. Yalçın, C. Geochemical and geological approach to the carbonate-hosted barite deposits in Dadağlı (Kahramanmaraş), Turkey. *Eng. Appl.* **2022**, *1*, 55–62. Available online: https://publish.mersin.edu.tr/index.php/enap/article/view/317 (accessed on 19 June 2022).
54. Yoshioka, K.; Zhdanov, M.S. Three-dimensional nonlinear regularized inversion of the induced polarization data based on the Cole–Cole model. *Phys. Earth Planet. Inter.* **2005**, *150*, 29–43. [CrossRef]
55. Zamora-Ledezma, C.; Negrete-Bolagay, D.; Figueroa, F.; Zamora-Ledezma, E.; Ni, M.; Alexis, F.; Guerrero, V.H. Heavy metal water pollution: A fresh look about hazards, novel and conventional remediation methods. *Environ. Technol. Innov.* **2021**, *22*, 101504. [CrossRef]
56. Zhu, Y.; Fan, W.; Zhou, T.; Li, X. Removal of chelated heavy metals from aqueous solution: A review of current methods and mechanisms. *Sci. Total Environ.* **2019**, *678*, 253–266. [CrossRef]

water

MDPI

Review

An Overview of Evapotranspiration Estimation Models Utilizing Artificial Intelligence

Mercedeh Taheri [1], Mostafa Bigdeli [1,*], Hanifeh Imanian [2] and Abdolmajid Mohammadian [1]

[1] Department of Civil Engineering, University of Ottawa, Ottawa, ON K1N 6N5, Canada; mrc.taheri@gmail.com (M.T.); amohamma@uottawa.ca (A.M.)
[2] Department of Civil and Environmental Engineering, Amirkabir University of Technology, Tehran 15875-4413, Iran; h.imanian@aut.ac.ir
* Correspondence: mbigd068@uottawa.ca

Abstract: Evapotranspiration (ET) has a significant role in various natural and human systems, such as water cycle balance, climate regulation, ecosystem health, agriculture, hydrological cycle, water resource management, and climate studies. Among various approaches that are employed for estimating ET, the Penman–Monteith equation is known as the widely accepted reference approach. However, the extensive data requirement of this method is a crucial challenge that limits its usage, particularly in data-scarce regions. Therefore, as an alternative approach, artificial intelligence (AI) models have gained prominence for estimating evapotranspiration because of their capacity to handle complicated relationships between meteorological variables and water loss processes. These models leverage large datasets and advanced algorithms to provide accurate and timely ET predictions. The current research aims to review previous studies addressing the application of the AI model in ET modeling under four main categories: neuron-based, tree-based, kernel-based, and hybrid models. The results of this study indicated that traditional models like the Penman–Monteith (PM) require extensive input data, while AI-based approaches offer promising alternatives due to their ability to model complex nonlinear relationships. Despite their potential, AI models face challenges such as overfitting, interpretability, inconsistent input variable selection, and lack of integration with physical ET processes, highlighting the need for standardized input configurations, better pre-processing techniques, and incorporation of hydrological and remote sensing data.

Keywords: evapotranspiration; hybrid models; kernel-based models; neuron-based models; tree-based models

1. Introduction

Evapotranspiration (ET), a dual process encompassing evaporation (a physical phase) and transpiration (a biological phase), refers to the phenomenon of water evaporating from the Earth's surface into the atmosphere [1]. It is critically important to estimate evapotranspiration accurately so that water resources can be managed efficiently, irrigation schedules can be devised, and crops can be grown sustainably [2].

A widely used categorization method for determining ET involves two main approaches: direct and indirect. Lysimeter, eddy covariance (EC), sap flow (SF), scintillometer, and Bowen ratio (BR) are techniques that directly measure ET, whereas indirect approaches utilize specific equations to calculate the latent heat flux [3]. Employing direct methods on a large scale for ET mapping faces challenges such as expensive installation and upkeep, disturbances in flow surrounding sensors, and localized measurements [4–6]. In contrast, indirect methods, which utilize theoretical principles to estimate ET, have the capability to reasonably accurately estimate ET over extensive regions. However, some indirect methods have conceptual and mathematical complexities that limit their widespread applications. Among the indirect models, Penman–Monteith (PM), which estimates the reference ET (ET_0) from a reference crop under standard climatic conditions, is considered the most effective approach in both humid and arid climates. This method considers both thermodynamic and aerodynamic factors [1]. By incorporating the crop coefficients suggested by the Food and Agricultural Organization (FAO), the variations between the reference surface and the specific crop's physiology, physics, and morphology are accounted for through the FAO-56 PM model [7–9]. The FAO-56 model has been extensively used to estimate crop ET; however, the blind use of crop coefficients suggested by FAO-56 may cause errors in ET_0 estimates. In addition, applying PM-based models to various locations worldwide, despite their straightforward nature and solid theoretical foundation, poses challenges due to the extensive input data required that are not easily obtained [8,9].

Noteworthy progress in computer technology has resulted in the development of different machine learning (ML) methods, referred to as computational artificial intelligence (AI) models, which employ a dynamic approach to mapping input features to output features. These models consist of mathematical representations derived from training data to predict target variables through the analysis of data characteristics. They possess the ability to approximate intricate nonlinear relationships, including those that are nonlinear, nonmonotonic, and multimodal, without relying heavily on assumptions regarding the distribution of data or predefined conceptual connections between input and output data. ML methods have proven to be effective alternatives for estimating ET_0 using a restricted quantity of meteorological variables, as demonstrated in various studies [10–14]. The early achievement in ET estimation by employing ML algorithms was demonstrated in a previous study [15], in which an artificial neural network (ANN) was employed to estimate ET in California. Their study established the reliability of ML algorithms by comparing their results with lysimetric records and PM-based estimations. Subsequent studies have evaluated the efficiency of ML algorithms in the estimation of ET. During the previous two decades, various ML algorithms, such as extreme learning machine (ELM), support vector machine (SVM), deep learning algorithms, adaptive neuro-fuzzy inference system (ANFIS), and random forest (RF), have been utilized for ET estimation, which has provided the superior capability of AI models in forecasting ET_0, particularly in managing complex relationships and offering flexibility in selecting input data. For example, Ferreira et al. [10] found that both SVM and ANN models were superior compared to the existing equations in the estimation of daily ET_0 across Brazil, even after calibration. Moreover, ELM, as a relatively novel approach, has been confirmed in numerous studies to effectively handle nonlinear problems. For instance, Dou et al. [16] employed the ELM method to simulate ET in various ecosystems and highlighted its fast learning capability, recommending it as an important complement to conventional AI methods because of its flexibility and robustness.

Given that AI methods have demonstrated reliable ET estimations in various areas [17,18], the objective of this study is to review advanced evapotranspiration estimation models derived from AI theories. The assessment was performed in four main categories, namely neuron-based models, kernel-based models, tree-based models, and hybrid models. The studies conducted using these approaches were thoroughly examined. Finally, the limitations and future prospects are deliberated upon.

2. Penman–Monteith (PM) Method

The Penman equation, formulated by Penman [19], is a semi-empirical expression that approximates evaporation occurring from open water bodies. It relies on the principles of mass transfer and energy balance, utilizing monthly meteorological information, such as wind velocity, solar radiation, air temperature, and humidity levels. The fundamental Penman equation [7] is expressed as follows:

$$E = \frac{\Delta \times LE_n + \gamma \times LE_a}{\Delta + \gamma} \quad (1)$$

where E denotes the evaporation rate, Δ is the gradient of the saturation vapor pressure curve at air temperature T, LE_n indicates the evaporation rate due to net radiation, γ shows the psychrometric constant, and LE_a denotes the evaporation rate resulting from mass transfer.

By incorporating physical principles through introducing canopy and aerodynamic resistances, Monteith [20] modified the Penman model, resulting in the development of the well-known PM equation. The PM model is built upon the integration of aerodynamic-based ET, which accounts for mass transfer and surface energy balance (SEB)-based sensible heat flux within a single layer. The PM is represented by the following mathematical formula [7]:

$$ET_0 = \frac{R_n - G}{1 + \frac{\gamma}{\Delta}\left[1 - \frac{(e_s - e_a)}{(e_o - e_a)}\right]\frac{(r_a + r_s)}{r_a}} \quad (2)$$

where ET_0 denotes the reference evapotranspiration, R_n shows the net radiation flux, G is the soil heat flux, r_s represents the surface resistance, r_a is the aerodynamic resistance to heat and/or vapor transport, e_a is the actual vapor pressure of air, e_o is the vapor pressure within the leaf, and e_s is the saturation vapor pressure at the air temperature T.

The model is based on the big canopy assumption, treating it like a single "big leaf" to simplify the parameterization of transpiration via canopy stomatal conductance [19–22]. This assumption, however, becomes problematic when considering changes in canopy structure resulting from differences in the size of the leaf and its density, leaf angle, and the height of the plant. Such changes can lead to unrealistic representations of the canopy configuration [21].

It has been demonstrated that the model adequately simulates ET for densely vegetated canopies [23]. However, the PM model struggles to accurately account for soil evaporation in sparse canopies owing to its big-leaf concept [20]. To address this limitation, Shuttleworth et al. [24] introduced models consisting of two layers. These models incorporate the relationship between vegetation and soil to calculate evaporation from the uncovered soil and transpiration from moist surfaces and plants. While more intricate, these two-layer models provide a better representation of real-world processes. To further enhance accuracy, more complex multi-layer models involving multiple canopy layers have been developed. These models independently compute resistance networks for distinct layers within the canopy, making them highly adaptable and capable of delivering precise ET estimations under varying weather conditions and crop growth stages [22,23]. Despite the detailed structure, the practical use of these models is limited owing to the extensive data requirements. Therefore, FAO further refined and standardized the original PM model for practical applications in agricultural water management. The FAO-56 version provides specific parameter values and recommendations for various situations and crop types by introducing crop coefficients. Although the crop coefficient-driven models decreased the data required for the original PM model, dealing with numerous variables still presents a considerable challenge. Furthermore, blindly adopting coefficients of the crop suggested by FAO-56 can result in errors [25]. The equation below [7] represents the mathematical formula of the PM model:

$$ET_0 = \frac{0.408(R_n - G) + \gamma\left[\frac{900}{T+273}\right]u_2(e_s - e_a)}{\Delta + \gamma[1 + 0.34u_2]} \tag{3}$$

where u_2 is the speed of wind at two meters above the ground.

To overcome the data challenges of PM-based models, researchers have employed alternative approaches such as empirical equations and AI techniques. A growing amount of research has proven the enhanced effectiveness of AI models in predicting ET_0, particularly in handling complex relationships and providing adaptable choices for input selection.

3. AI-Based Models for ET Estimation

3.1. Neuron-Based Models

Neuron-based models have been provided to emulate the structure and function of neural networks of the human brain for various purposes. Popular neuron-based models include ANNs and deep neural networks (DNNs), which are discussed below.

3.1.1. Artificial Neural Networks (ANN)

ANNs are computational models made up of interconnected units called artificial neurons or perceptrons. These networks consist of an input layer, a hidden layer, and an output layer, and the connections between the neurons are associated with weights. By modifying these weights through the training phase, ANNs can learn the patterns among the inputs and outputs to make predictions and decisions [26].

The utilization of ANNs has proven to be effective in simulating the nonlinear patterns of intricate processes (e.g., ET_0), addressing the limitations associated with the use of the PM equation. In this regard, numerous researchers have reported the use of ANNs for ET_0 prediction modeling and have proposed enhancements in the architecture and learning algorithms of ANNs. For instance, Kumar et al. [15] explored the use of a multilayer perceptron (MLP) ANN to estimate ET_0. They compared the performance of the conventional backpropagation algorithms employing learning rates of 0.2 and 0.8, as well as that of backpropagation with momentum, against the traditional PM equation. Daily ET_0 estimation was based on two sets of climatic data for Davis, California, including the maximum and minimum air temperature, maximum and minimum relative humidity, solar radiation, and wind speed. Considering the different architectures of the input, hidden, and output layers, the ANN model consists of six input layers, seven hidden layers, and a single output, exhibiting better efficiency in estimating ET_0 compared to the PM equation against lysimetric measurements.

Kişi [27] evaluated the feasibility of utilizing the generalized regression neural network (GRNN) technique using four different combinations of meteorological records to model ET_0 derived from the PM equation. The findings suggested that by incorporating relative humidity, daily air temperature, solar radiation, and wind speed from the specific station, the GRNN model performed better in modeling ET_0 against the GRNN model trained with data from neighboring stations.

Wang et al. [28] estimated monthly pan evaporation in China using various models, including MLP, GRNN, ANFIS with grid partition (ANFIS-GP), multivariate adaptive regression splines (MARS), fuzzy genetic (FG), and least-squares support vector machine (LSSVM). The effectiveness of these models was evaluated in comparison to two regression techniques: Stephens and Stewart (SS) and multiple linear regression (MLR) models. The meteorological inputs included monthly recorded data of the air temperature, wind speed, sunshine duration, relative humidity, solar radiation, and pan evaporation. The findings indicated that the AI models exhibited higher levels of accuracy than the SS and MLR techniques. According to the statistical criteria, the best models were MLP, GRNN, LSSVM, FG, ANFIS-GP, MARS, and MLR, respectively. It should be noted that the inclusion of more input parameters improved model accuracy.

Kişi [29] evaluated the efficacy of three ANN techniques, i.e., GRNN, MLP, and radial basis function neural network (RBFNN), to simulate PM-based ET_0 by using daily wind speed, air temperature, solar radiation, and relative humidity. They compared the performances of models with empirical models such as the California Irrigation Management Information System (CIMIS), Hargreaves, Penman, and Ritchie. The results indicated that MLP and RBFNN could be effectively utilized in modeling PM ET_0. The Hargreaves model performed the best in estimating ET_0 using data from nearby stations. In a study conducted by [30] in Basque Country (northern Spain), daily estimated ET_0s from empirical and semi-empirical equations, including Turc, Makkink, Priestley–Taylor, Hargreaves and Samani, and ANN, were compared to ET_0 data obtained from the FAO-56 PM equation at four stations. The ANN models, comprising seven different scenarios using various meteorological parameters, including air temperature (minimum, maximum, and mean), wind speed, extra-terrestrial radiation, relative humidity, and solar radiation, outperformed the calibrated equations.

El-Baroudy et al. [31] investigated the performance of data-driven techniques, including ANN, evolutionary polynomial regression (EPR), and genetic programming (GP) models, to model actual ET recorded by the eddy covariance system. They also studied the impact of past states (time lags) of the climatic inputs in two Canadian study areas. The EPR model exhibited performance comparable to that of the GP and ANN models. In addition, this model is simpler. The efficiency of the EPR model was increased by including the previous states, thereby revealing the dynamic nature of the evapotranspiration process.

Antonopoulos et al. [32] compared the performance of ANN with mass transfer models and empirical methods such as Makkink, Priestley–Taylor, and Hargreaves for estimating ET_0 using daily meteorological data, showing the superiority of ANN trained by wind speed and relative humidity over other models.

Kişi [33] made a comparison between the ANN and fuzzy genetic (FG) approaches as well as empirical methods, including the CIMIS, Turc, Ritchie, and Hargreaves methods, in daily ET_0 estimation using daily weather data of relative humidity, solar radiation, wind speed, and air temperature. They reached the conclusion that the FG model surpassed the performance of other models.

Traore et al. [34] investigated the capability of developing an ANN model for estimating ET_0 based on the PM equation to address situations where there might be insufficient climatic data at meteorological stations within the case study. The researchers used the Hargreaves equation, which relies solely on air temperature data, to evaluate the capability of a feedforward backpropagation ANN (FFBPNN) model against standard PM estimations. Results from statistical analysis showed that the FFBPNN model outperformed the Hargreaves equation based on temperature data alone. The introduction of the wind speed parameter significantly increased the R^2 value, whereas the inclusion of relative humidity and sunshine had a lesser impact on the performance of the model. The investigations indicated that wind speed was the most influential parameter in ET_0 simulation.

Nema et al. [35] evaluated ANN capabilities to enhance the accuracy of estimations for monthly evapotranspiration in a sub-humid climatic area located in Dehradun. Different ANN methods, including various combinations of training functions and numbers of neurons, have been developed to model ET_0. During the modeling process, data on monthly relative humidity, temperature, wind speed, rainfall, and sunshine hours were employed. The ANN model that utilized the Levenberg–Marquardt training algorithm along with a single hidden layer containing nine neurons demonstrated the best predictive capabilities for the study station.

Sudheer et al. [36] focused on modeling daily ET_0 specifically for rice cultivation using RBFNNs, experimenting with different sets of climatic inputs including humidity, sunshine duration, wind speed, air temperature, and actual ET measurements. The findings unequivocally indicated the effectiveness of the ANN technique in estimating ET compared with lysimetric ET measurements.

Bruton et al. [37] applied ANN, MLR, and Priestley–Taylor models to model daily pan evaporation by employing daily observations of rainfall, relative humidity, temperature, wind speed, and solar radiation. They reported that the ANN model proved to be the most precise approach, yielding an R^2 value of 0.71 and an RMSE of 1.1 mm.

Sudheer et al. [38] used MLP with a back-propagation algorithm to model daily pan evaporation by considering sunshine hours, temperature, wind speed, and relative humidity. The ANN model exhibited superior performance compared to the SS model, a widely recognized climatic method used for estimating evaporation. Nonetheless, an examination of the discrepancies observed in the residual data of the developed ANN model indicated substantial inaccuracies in the predictions during the validation phase. This suggests that the ability of ANN models to generalize effectively could be compromised unless the training process is meticulously executed. In addition, the research highlighted that the ANN method has the potential to reasonably predict evaporation values based solely on temperature data.

Keskin et al. [39] compared the performance of ANN and Penman approaches in daily pan evaporation retrieval using daily observations of relative humidity, air and water temperature, wind speed, air pressure, solar radiation, and sunshine hours. They indicated that it achieved significantly higher accuracy compared to the Penman equation. An explicit neural network (ENN) was introduced by Aytek et al. [40] to estimate daily ET_0 in California, USA, using relative humidity, air temperature, wind speed, and solar radiation. The findings suggested that while ENN provided an alternative approach for estimating ET_0, the improvement in accuracy was only marginal compared to simpler linear techniques.

Rahimikhoob [41] explored the feasibility of utilizing ANNs to estimate ET_0 in a humid subtropical region along the southern coast of the Caspian Sea in Iran. They compared the findings with the estimations obtained through the Hargreaves equation and the FAO-56 model. The ANNs were trained using input variables including wind speed, daylight hours, minimum and maximum air temperature, and relative humidity. The FAO-56 PM model was employed to evaluate the performance of both approaches. Their findings demonstrated that ANNs successfully estimated ET_0 variability utilizing air temperature data without compromising accuracy.

Parasuraman et al. [42] compared the performance of the GP model with the PM and ANN models for ET_0 prediction as a function of the surface temperature, wind speed, air temperature, relative humidity, and net radiation. The results demonstrated that GP successfully simulated the dynamics of the ET_0 process compared with the other models.

Two types of ANN, MLP and RBFNN, were compared by Tezel and Buyukyildiz [43], along with support vector regression (SVR), in order to estimate monthly pan evaporation using inputs including relative humidity, temperature, wind speed, and total rainfall. The performance of the AI models was further compared to that of two empirical models, the Romanenko and Meyer methods. The findings revealed that ANN outperformed alternative empirical methods and other AI models.

Aghelpour et al. [44] estimated rice evapotranspiration in the southern areas of the Caspian Sea using AIs, such as the group method of data handling neural network (GMDH-

NN), GRNN, MLR, and RBFNN, driven by weather data of relative humidity, air temperature, precipitation, sunshine duration, pan evaporation, and wind speed. They found that these models provided highly accurate estimations of the evapotranspiration rates of rice on a daily basis. The comparison of the models exhibited better performance of GMDH-NN with both NSE and R^2 greater than 0.98, as well as RMSE ranging between 0.214 and 0.234 mm/day.

Kişi [45] examined MLP and RBFNN in modeling monthly pan evaporation using monthly climatic data in California. These models were compared with the MLR and SS methods. It was found that both AI methods exhibited superior performance compared to regression and SS methods.

Three different ANN algorithms, including generalized feedforward (GFF), MLP, and probabilistic neural networks (PNN), as well as linear regression (LR), were compared by Traore et al. [46] to predict short-term ET_0 based on different data combinations, including temperature, extraterrestrial radiation, and net solar radiation in Dallas. The MLP network provided the best results with the scenario of full weather data as inputs. The maximum temperature was considered an important ET_0 predictor, while the estimation accuracy of the net solar radiation determined the rate of model improvement.

Trajkovic [47] highlighted the limitations of the standard PM equation due to the extensive range of weather data required and examined the possibility of achieving reliable outcomes by relying solely on maximum and minimum air temperatures. Four temperature-based methods, including the RBFNN, Hargreaves, Thornthwaite, and reduced-set PM equations, were evaluated and compared to the standard PM equation. The RBFNN performance closely resembled the PM ET_0 at most locations, suggesting that RBFNN can be effective for predicting ET_0 using only air temperature records in locations that suffer from a lack of other climatic inputs, including solar radiation, relative humidity, and wind speed.

Similarly, Tabari and Hosseinzadeh Talaee [48] pointed out that the primary challenge in applying the PM equation lies in the extensive set of meteorological data that are needed as inputs for the ET_0 calculations. Additionally, the nonlinearity of the ET phenomenon poses a challenge in precisely forecasting the impacts of climate change on ET. They employed various MLP networks based on six learning algorithms such as Delta-Bar-Delta, Levenberg–Marquardt, Step, Conjugate Gradient, Momentum, and Quickprop to model daily ET_0 in Hamedan, Iran, as a semiarid region. They explored scenarios involving complete sets of climate variables as well as scenarios where some of these parameters were missing. Different types of learning algorithms have been utilized to optimize the functions of MLP networks. Among the MLP models, the model trained using the conjugate gradient algorithm and full input variables demonstrated the best performance.

Zhu et al. [49] aimed to develop two AI models, the extreme learning machine (ELM) and GRNN, with the objective of ET_0 estimation on a daily basis only based on temperature inputs. ELM is a newly proposed ML technology that has been confirmed to be a promising ET_0 estimation tool because of its higher effectiveness and significantly faster computational speed [49]. The models were evaluated at six meteorological stations in the Sichuan Basin, Southwest China. The historical daily meteorological variables considered in this study were air temperature at a height of 2 m, relative humidity, wind speed at a height of 10 m, and sunshine duration. The effectiveness of the proposed ELM and GRNN models was evaluated against that of FAO-56 PM ET_0 as a benchmark, along with the temperature-based Hargreaves model and calibrated Hargreaves model. Notably, the Hargreaves model was refined using Bayesian theory to account for local conditions at 19 meteorological stations, exhibiting superior performance than the original Hargreaves model [50]. Two scenarios for data management were taken into account for estimating the ET_0. In the first scenario, all models were trained and tested using local data from individual stations. In the second scenario, each model was trained by employing aggregated data from all stations and was subsequently tested independently at all stations. In the initial scenario, the ELM model surpassed the GRNN, Hargreaves model, and its calibrated version, exhibiting an average mean absolute error (MAE) of 0.267 mm/d, relative root mean square error (RRMSE) of 0.198, and Nash–Sutcliffe coefficient (NSE) of 0.891. The GRNN model yielded the most precise results among all considered models, with an average MAE of 0.263 mm/d, RRMSE of 0.194 mm/d, and NSE of 0.895 in the second scenario. Overall, the performance of both the ELM and GRNN models surpassed that of the Hargreaves and calibrated Hargreaves models in both scenarios. Therefore, these models are recommended as suitable alternatives for estimating ET_0 accurately in the Sichuan Basin of southwest China.

In a separate investigation conducted by [51], the authors examined the application of the ELM in predicting weekly evapotranspiration and compared it with ANN, least square support vector machine (LSSVM), and Hargreaves models in Thar Desert, India. They observed that the ELM model exhibited more accurate performance than the ANN and Hargreaves models and similar performance to the LSSVM model.

Heddam et al. [52] suggested two types of ELM, namely optimally pruned ELM (OPELM) and online sequential ELM (OSELM), for predicting the ET_0 on a daily basis in the Mediterranean region of Algeria. Large datasets covering various regions in Algeria from east to west were used to develop the models, employing commonly used climatic variables such as daily relative humidity, minimum and maximum air temperatures, and wind speed as input features. According to the results, the OPELM models exhibited superior overall performance compared with the OSELM models.

Malik et al. [53] predicted daily pan evaporation in India by employing various models, including RBFNN, MLR, Stephens–Stewart, Christiansen, Griffiths, Priestley–Taylor, SOMNN, Jensen–Burman–Allen, and Penman. The climatic inputs employed to train the models were the relative humidity, maximum and minimum air temperatures, sunshine hours, wind speed, and daily pan evaporation. According to their results, the RBFNN model achieved the highest accuracy, with an RMSE of 1.02 mm/day, CE of 0.87, and R of 0.93.

Malik et al. [54] employed several model strategies, including MARS, multi-gene genetic programming (MGGP), support vector machine (SVM), and M5 model tree (M5Tree), to estimate monthly pan evaporation by employing monthly meteorological inputs of sunshine hours, relative humidity, minimum and maximum temperature, pan evaporation, and wind speed. Additionally, this study used a hybrid model known as the multiple model ANN (MM-ANN) to model the evaporation process. It incorporates two tiers of learning, Level 1 and Level 2, within the modeling process. The first level controls the primary learning procedure and involves training potential AI models using input variables (meteorological data) and corresponding output (pan evaporation data) within a single context. Conversely, Level 2, which is based on the results obtained from Level 1, creates a form of binary learning procedure within the ML modeling strategy. The outcomes of the Level 1 learning phase are used as input features, whereas the original output (pan evaporation) serves as the output for the Level 2 stage (Figure 1).

The findings demonstrated that the MGGP and MM-ANN models, with RMSE, NSE, MAPE, WI, and LM equal to 0.536, 0.954, 9.988%, 0.988, and 0.801 for the Pantnagar station and equal to 0.911, 0.975, 0.724, 0.364, and 12.297% for the Ranichauri station, were more successful compared to the other models in simulating pan evaporation in India on a monthly basis.

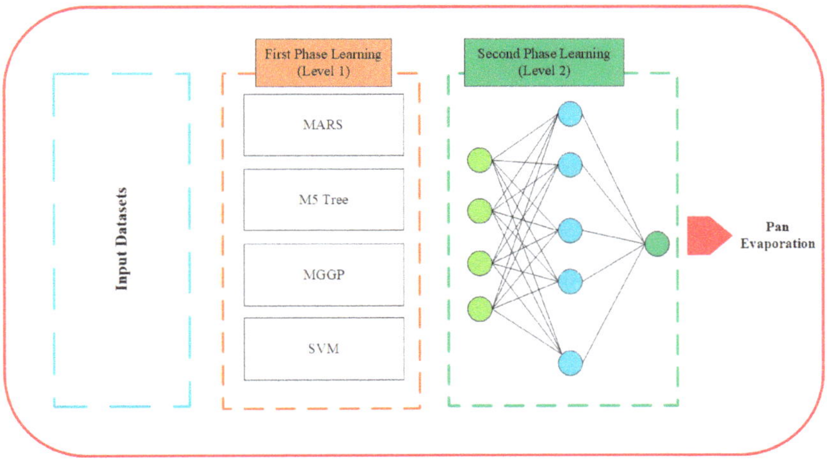

Figure 1. The configuration of the MM-ANN model.

Makwana, Tiwari [55] assessed various AI-based models, including ANN, ELM, M5Tree, and MLR, for the estimation of daily reference evapotranspiration (ET_0) using different combinations of meteorological variables. According to the findings, the best parameters resulting in precise estimation of ET_0 included maximum temperature and wind speed as two significant predictor variables. Additionally, ANN was identified as the best model with R^2, NSE, RMSE, Pdv, and MAE values of 0.98, 98.11%, 0.27 mm day^{-1}, 7.75, and 0.19 mm day^{-1} to calculate ET_0 using limited meteorological data compared to the FAO-56 Penman–Monteith equation used as a benchmark.

In order to estimate daily ET_0, Güzel, Üneş [56] explored the application of AI-based models, including ANN, ANFIS, simple membership functions, the fuzzy rules generation technique (fuzzy-SMRGT), support vector regression (SMOReg), and multivariate regression, including multi-linear regression (MLR), interaction (I-MR), pure quadratic (P-MR), and quadratic (Q-MR), highlighting the novelty of using the fuzzy-SMRGT model in this context. Various soft computing techniques were compared against classical statistical methods using input data of air temperature, wind speed, solar radiation, and relative humidity. In comparison with Penman-based ET_0, ANN achieved the best performance based on R^2, average percentage error (APE), and RMSE. ANFIS ranked second after ANN in terms of these evaluation metrics. Based on R^2, the ranking from best to worst was Q-MR, I-MR, P-MR, fuzzy-SMRGT, MLR, and SMOReg. However, in terms of RMSE, the order changed to I-MR, Q-MR, P-MR, fuzzy-SMRGT, MLR, and SMOReg. When assessed based on APE, the best-performing methods were I-MR, Q-MR, MLR, P-MR, SMOReg, and fuzzy-SMRGT, respectively.

Another study [57] focused on developing a reliable model to estimate ET_0 using climatic predictors, comparing two techniques: stepwise regression and ANNs. The results showed that the ANN model outperformed the stepwise regression model, with an R^2 of 0.99 and a MAPE of 2.7%, compared to an R^2 of 0.95 and a MAPE of 8.06% for the stepwise regression model. Significant climatic factors affecting ET_0 included maximum temperature (100%), wind speed (93%), humidity (83%), and solar radiation (80%). The study also faced limitations, including data availability, model generalization to other regions, and the computational complexity of artificial neural networks, which could hinder practical application. Future research should focus on ensemble modeling, the potential of remotely sensed data as ancillary inputs, and exploring the impact of climate change on ET_0 estimations in arid regions.

In [58], pan evaporation changes using AI techniques driven by daily climate data were modeled. They used all pan evaporation data across the Slovak Republic from 2010 to 2023 to offer insights into (i) which river basin in the Slovak Republic is the most and least impacted by changes in pan evaporation during the given period; (ii) which AI techniques, including ML and deep learning (DL) models, are best suited for modeling pan evaporation in Slovak conditions; and (iii) the key variables that most affect the estimation accuracy. This study utilized linear regression (LR) models, tree-based methods, SVMs, ensemble methods (EM), NNs, and kernels. According to the results, LR models (LR and Interaction LR) provided the best fit, achieving RMSE values between 0.805 and 0.821 and R^2 values above 0.60. The NN models (narrow, medium, wide, and bilayered) yielded similar RMSE and MAE values, not surpassing the traditional regression methods. SVM models exhibited consistent performance, with RMSE values ranging from 0.882 to 0.900 and R^2 values between 0.538 and 0.556. Tree-based models (fine, medium, and coarse) underperformed, with the fine tree model showing the highest RMSE of 1.035. Gaussian process models provided solid RMSE and MAE outcomes, with the best RMSE around 0.877. The results also revealed that evaporation is rising in rivers such as Bodrog, Hornád, and Váh, while rivers like Dunaj and Hron showed a decreasing trend, due to the various ecological factors. Among all models, stepwise linear regression emerged as the most effective, with relative humidity and minimum temperature identified as crucial factors that enhanced the accuracy of evaporation predictions. Regarding the most effective climatic data as inputs, relative humidity followed by minimum temperature showed the highest impact on the target variable. Following them in rank were wind speed, maximum temperature, and average temperature, respectively.

The study in [59] aimed to predict seasonal crop ET (ET_c) and crop water use efficiency (CWUE) by using ANN and MLR under two different scenarios: (i) considering only climatic factors and (ii) incorporating both crop parameters and climatic data. Both models, by using only climatic variables, showed acceptable performance; however, the accuracy of the MLR model significantly improved when physiological parameters were incorporated. It is worth noting that ANN maintained strong predictive capability with or without crop parameters.

The current study [60] sought to assess the effectiveness of ANN model in forecasting ET_0 and ET_c and evaluating the surplus or deficit in cassava's water requirements. ANN predictions were generated using inputs from Blaney–Morin–Nigeria (BMN) and Hargreaves–Samani (HAG) models, referred to as BMN-ANN and HAG-ANN. The ANN model results were then compared to those obtained through the Penman–Monteith method. Based on results, ANN models demonstrated superior predictive accuracy compared to the empirical models. The BMN-ANN model achieved the best performance, with an R^2 of 0.9890, an RMSE of 0.000056 mm/day, and a Willmott's index of agreement (d) of 0.9960. Table 1 summarizes the aforementioned studies on ET estimation using ANN models.

ANNs, particularly models such as the MLP and GRNN, have demonstrated superior performance over conventional empirical methods, including Hargreaves, Penman, and Turc, for estimating ET [28,36,55,56,58]. This superiority is attributed to several key advantages inherent in ANN-based modeling. Notably, ANNs exhibit higher predictive accuracy, enhanced capability in handling multiple input variables, and greater adaptability under data-scarce conditions. Their ability to model nonlinear relationships and their flexible architecture further reinforce their effectiveness across diverse hydrological and climatic settings.

Table 1. A list of studies carried out to estimate ET by ANNs.

Reference	Models	Input	Output	Performance Criteria	Best Model(s)
Kumar et al. [15]	MLP, PM model	Maximum and minimum air temperature, maximum and minimum relative humidity, wind speed, and solar radiation	Daily ET_0	WSEE = 0.3–0.6 mm/day	MLP
Kişi [27]	GRNN, PM model	Air temperature, wind speed, relative humidity, and solar radiation	Daily ET_0	MSE = 0.058 and 0.032 mm^2 day^{-2}, MAE = 0.184 and 0.127 mm day^{-1}, R^2 = 0.985 and 0.986	GRNN
Wang et al. [28]	MLP, GRNN, ANFIS-GP, MARS, FG, LSSVM, MLR, SS	Air temperature, sunshine durations, solar radiation, relative humidity, wind speed, and pan evaporation	Monthly pan evaporation	MAE = 0.2585, RMSE = 0.4668, R^2 = 0.9914, MSE = 0.2214	MLP
Kişi [29]	GRNN, MLP, RBfNN, CIMIS, Hargreaves, Penman, Ritchie	Air temperature, relative humidity, wind speed, and solar radiation	Daily ET_0	MSE = 0.664 and 0.712 mm^2 day^{-2}, MAE = 0.619 and 0.663, R^2 = 0.870 and 0.855	MLP, RBFNN, Hargreaves model
Landeras et al. [30]	ANN, Turc, Makkink, Priestley–Taylor, Hargreaves and Samani, PM model	Air temperatures (minimum, maximum, and mean), relative humidity, wind speed, extraterrestrial radiation, and solar radiation	Daily ET_0	MBE = 0.063–0.048, MAE = 0.174–0.442, RMSE = 0.238–0.646	ANN
El-Baroudy et al. [31]	ANN, EPR, GP	Air temperature, ground temperature, net radiation, relative humidity, wind speed	Actual ET	RMSE = 38.1, MARE = 0.33, R = 0.86	EPR
Antonopoulos et al. [32]	ANN, Makkink, Priestley–Taylor, Hargreaves, mass transfer models	Maximum, minimum, average, and standard deviation values of temperature, relative humidity, wind speed, solar radiation, and ET_0	Daily ET_0	RMSE = 0.574–1.33 mmd^{-1}, R = 0.955–0.986	ANN
Kişi [33]	ANN, FG, CIMIS, Turc, Hargreaves, Ritchie	Air temperature, solar radiation, relative humidity, and wind speed	Daily ET_0	RMSE = 0.138–0.167, MAE = 0.098–0.115, R = 0.998–0.999	FG
Traore et al. [34]	FFBPNN, Hargreaves, PM model	Relative humidity, maximum and minimum air temperature, precipitation, wind velocity, sunshine duration	Daily ET_0	RMSE = 0.048–0.714, MAE = 0.033–0.581, R^2 = 0.693–0.998	FFBPNN
Nema et al. [35]	Different ANNs	Minimum, average, and maximum temperature; relative humidity (minimum and maximum); wind speed; sunshine hours; and rainfall	Monthly ET_0	R = 0.969–0.989, SSE = 1.102–3.047, RMSE = 0.101–0.168, NSE = 0.938–0.978, MAE = 0.843–0.885	ANN with one single hidden layer, nine neurons, and Levenberg–Marquardt training algorithm
Sudheer et al. [36]	Different RBFNNs	Relative humidity, air temperature, relative humidity, sunshine duration, wind speed, and actual ET measurements	Daily ET_0	SEE = 0.030–1.071, RSEE = 0.030–0.945, Efficiency (%) = 98.20–98.40	RBFNN trained with only temperature data
Bruton et al. [37]	ANN, MLR, Priestley–Taylor	Temperature, relative humidity, solar radiation, rainfall, and wind speed	Daily pan evaporation	RMSE = 1.11 mm, R^2 = 0.717	ANN
Sudheer et al. [38]	MLP, SS	Minimum and maximum temperature and relative humidity, wind speed, and sunshine hours	Daily pan evaporation	RMSE = 1.07–2.31, CE = 9.63–70.71, PE = −14.74–11.49, SD = 0.28–0.50, R = 0.54–0.86	ANN
Keskin and Terzi [39]	ANN and Penman models	Air and water temperature, solar radiation, air pressure, sunshine hours, wind speed, and relative humidity	Daily pan evaporation	MSE = 0.007–0.01, R^2 = 0.629–0.778	ANN
Aytek et al. [40]	ENN, MLR	Wind speed, solar radiation, relative humidity, and air temperature	Daily ET_0	MSE = 0.084–0.123, R^2 = 0.983–0.989	ENN

Table 1. *Cont.*

Reference	Models	Input	Output	Performance Criteria	Best Model(s)
Rahimikhoob [41]	ANN, PM, Hargreaves	Wind speed, maximum and minimum air temperature, relative humidity, and daylight hours	Daily ET_0	R^2 = 0.95, R = 1, RMSE = 0.41	ANNs utilizing air temperature data
Parasuraman et al. [42]	GP, ANN,	Surface temperature, air temperature, net radiation, wind speed, and relative humidity	Daily ET_0	RMSE = 38.8–69.8, MARE = 0.34–1.02, R = 0.71–0.85	GP
Tezel et al. [43]	ANN, MLP, RBF, Romanenko, Meyer	Temperature, relative humidity, wind speed, and total rainfall	Monthly pan evaporation	MAE = 0.516–0.671 mm/month, RMSE = 2.419–3.147 mm/month, R^2 = 0.893–0.914	ANN
Aghelpour et al. [44]	GMDH-NN, GRNN, MLR, RBFNN	Minimum, average, and maximum air temperature and relative humidity, sunshine duration, precipitation, wind speed, and pan evaporation	Daily ET_0	RMSE = 0.220–0.234 mm/day, NSE = 0.986–0.990, R^2 = 98.76–99.04, NRMSE = 0.017–0.030, MAE = 0.173–0.613 mm/day	GMDH-NN
Kişi [45]	MLP, RBFNN, SS	Air temperature, solar radiation, wind speed, pressure, and humidity	Monthly evaporation	MSE = 0.009–2.398 mm^2, MARE = 1.778–52.552, R^2 = 0.768–0.999	MLP and RBNN
Traore [46]	GFF, MLP, PNN, LR	Minimum and maximum temperature, net solar radiation, and extraterrestrial radiation	Short-term ET_0	MSE = 1.408–3.197 mm/day, NMSE = 0.262–0.595 mm/day, MAE = 0.944–1.382 mm/day, MSESS = 0.405–0.738, CC = 0.703–0.860	MLP
Trajkovic [47]	RBFNN, Thornthwaite, Hargreaves, PM models	Temperature, wind speed, relative humidity, actual vapor pressure, and sunshine hours	Daily ET_0	MXE = 0.482–0.850, MAE = 0.130–0.193, RMSE = 0.161–0.266	RBFNN
Tabari and Hosseinzadeh Talaee [48]	Different MLPs	Minimum, average, and maximum air temperature, dew point temperature, water vapor pressure, relative humidity, wind speed, precipitation, atmospheric pressure, solar radiation, and sunshine hours	Daily ET_0	RMSE = 0.139–0.698, MAE = 0.117–0.597, R = 0.891–1.065, Ratio = 0.973–0.998	MLP model trained with the Levenberg–Marquardt algorithm
Zhu et al. [49]	ELM, GRNN, PM, Hargreaves, calibrated Hargreaves model	Maximum air temperature, minimum air temperature, mean air temperature at a height of 2 m, mean relative humidity, wind speed at a height of 10 m, and sunshine duration	Daily ET_0	RRMSE = 17.9–21.7%, MAE = 0.445–0.496 mm/d, NSE = 0.907–0.929	ELM trained by local data and GRNN trained using pooled data
Patil et al. [51]	ELM, ANN, LSSVM, Hargreaves	Minimum, average, and maximum air temperature; maximum and minimum relative humidity; solar radiation; wind speed; and ET_0	Weekly ET_0	RMSE = 0.33–0.76 mm/day, NSE = 0.85–0.98, TS = 20.4–94.2	ANN, ELM and LSSVM
Heddam et al. [52]	OPELM, OSELM	Wind speed, maximum and minimum air temperatures, and relative humidity	Daily ET_0	RMSE = 1.267–0.240, MAE = 1.053–0.184, R = 0.668–0.990	OPELM
Malik et al. [53]	RBFNN, MLR, Griffiths, SS, Priestley–Taylor, Christiansen, Penman, SOMNN, and Jensen–Burman–Allen	Relative humidity, minimum and maximum air temperatures, wind speed, sunshine hours, and daily pan evaporation	Daily pan evaporation	RMSE = 1.024 mm/day, CE = 0.874, R = 0.934	RBNN
Malik et al. [54]	MM-ANN, MARS, SVM, MGGP, M5Tree	Sunshine hours, relative humidity, wind speed, maximum and minimum temperature, and pan evaporation	Monthly pan evaporation	MAPE = 9.988–12.297%, WI = 0.975–0.988, RMSE = 0.364–0.536 mm/month, NSE = 0.911–0.954, LM = 0.724–0.801	MM-ANN and MGGP

Table 1. *Cont.*

Reference	Models	Input	Output	Performance Criteria	Best Model(s)
Makwana et al. [55]	ANN, ELM, M5Tree, and MLR	Maximum temperature, minimum temperature, relative humidity, wind speed, and BSS.	Daily ET_0	R^2 = 0.30–0.98, NSE = 28.22–98.11, RMSE = 0.27–1.88, Pdv = 4.70–50.76, MAE = 0.19–1.52	ANN
Güzel et al. [56]	ANN, ANFIS, fuzzy-SMRGT, SMOReg, and multivariate regression models	Air temperature, wind speed, solar radiation, relative humidity, and evapotranspiration	Daily ET_0	R^2 = 0.998, RMSE = 0.075, APE = 3.361%	ANN
Abdel-Fattah et al. [57]	ANN, stepwise regression	Minimum and maximum temperature, humidity, wind speed, sunshine, radiation, ET_0, rain	Monthly ET_0	R^2 = 0.99, MSE = 0.24, RMSE = 0.49, MAPE = 2.7%	ANN
Novotná et al. [58]	LR models, tree-based methods, SVMs, EMs, NNs, Kernels	Pan evaporation, minimum temperature, maximum temperature, relative humidity, average wind speed	Daily pan evaporation	MSE = 0.647–0.674, RMSE = 0.805–0.821, MAE = 0.611–0.625, R^2 = 0.598–0.614	LR and Interaction LR
Faloye et al. [59]	ANN, MLR	Maximum and minimum air temperature, precipitation, wind speed, relative humidity, ET_0, LAI, and plant height	Seasonal ET_c	RMSE = 2.297–15.333, MAE = 0.517–3.049, NRMSE = 0.653–4.367, R^2 = 0.875–0.998	ANN
Eludire et al. [60]	ANN	Minimum air temperature, maximum air temperature, mean relative humidity, solar radiation, wind speed, and rainfall	Daily ET_0 and ET_c	WI = 0.996, R^2 = 0.989, RMSE = 0.000056 mm/day	BMN-ANN

Notes: Abbreviations: weighted standard error of estimate (WSEE), mean absolute errors (MAE), root mean square errors (RMSE), determination coefficient (R^2), Nash–Sutcliffe (NSE), mean square errors (MSE), mean bias error (MBE), mean bias error (MBE), mean absolute relative error (MARE), correlation coefficient (R), sum of squared errors (SSE), standard error of estimates (SEE), raw standard error of estimates (RSEE), percentage error (PE), standard deviation (SD), normalized root mean square error (NRMSE), normalized mean square error (NMSE), mean square error skill score (MSESS), coefficient of efficiency (CE), threshold statistics (TS), relative root mean square error (RRMSE), ratio between average estimated ET_0 values and observed values (Ratio), mean absolute percentage error (MAPE), Willmott's index of agreement (WI), Legate and McCabe's index (LM), maximum absolute error (MXE), bright sunshine hours (BSS), peak percentage deviation (Pdv), average percentage error (APE), leaf area index (LAI).

A critical advantage of ANNs lies in their capacity to integrate a comprehensive range of meteorological variables. including maximum and minimum temperature, solar radiation, relative humidity, and wind speed, significantly enhances model performance compared to traditional approaches that typically rely on a limited set of inputs [26,61]. Furthermore, although many studies do not explicitly address data incompleteness, ANNs inherently manage missing data through implicit interpolation during training. This capacity is especially valuable when using data from multiple stations, where spatial correlations can aid in estimating ET at locations with partial records [61].

Despite these strengths, certain limitations warrant consideration. ANN models are most effective when trained on large, high-quality datasets. Their performance can degrade when confronted with incomplete or poorly representative input data [26,62]. Moreover, successful application requires meticulous optimization of network architecture and hyperparameters to ensure optimal performance [63,64]. Although ANN models outperform conventional methods, they are not exempt from challenges related to model complexity and parameter tuning [61,65].

Another critical concern involves the extrapolation capability of ANN models. Their performance may decline when predicting ET under extreme or previously unseen climatic events, such as severe droughts, heatwaves, or anomalous precipitation patterns, particularly if such events are underrepresented in the training dataset. As climate regimes evolve, models trained solely on historical data risk becoming outdated, necessitating periodic retraining to preserve predictive reliability in dynamically changing environmental conditions [66,67].

3.1.2. DNNs

While regular regression models struggle to handle non-linear patterns in time-series data, ANNs are generally more adept in this regard. However, most ANNs fail to account for seasonality or time dependence. For instance, simple ANN architectures, such as multilayer perceptrons, lack memory blocks to retain previous information, which is crucial for accurate predictions. To address this limitation, DNNs with multiple hidden layers were developed to automatically learn complex features from input data. Convolutional neural networks (CNNs) and recurrent neural networks (RNNs) are popular examples of deep learning architectures.

Convolutional layers are used by CNNs to automatically extract the input data's spatial feature hierarchies. Because of this, CNNs are excellent at tasks like object detection, image creation, and image classification. In contrast, RNNs designed for sequential data incorporate cycles within a network of nodes to capture the dynamics of sequences [68]. They have loops that allow the transfer of information from one step to the subsequent one, enabling RNNs to understand time-based relationships in the data. A significant type of RNN architecture is long short-term memory (LSTM) network, which addresses the issue of vanishing gradients in RNNs and possesses the ability to store important data across extended sequences [69]. LSTMs use memory cells to preserve data across extended sequences, making them more capable of learning long-range dependencies in data.

Although many researchers have addressed the potential assessment of commonly used AI techniques, such as ANN, the most recent ML approaches, for example, RNN and CNN, are less employed for ET_0 forecasting. For example, Afzaal et al. [70] examined the efficiency of conventional LSTM and bidirectional LSTM models in predicting ET_0 obtained from the FAO-56 method on Prince Edward Island, Canada. Nine variables were chosen, including heat degree days, wind speed, air temperature, relative humidity, atmospheric pressure, and dew point temperature as inputs. A novel dataset was generated by computing the average values of the climatic inputs to account for variability across various regions of the study area. Subset regression analysis revealed that maximum air temperature and relative humidity were the most important variables among the selected factors for estimating ET_0. According to the findings, all RNNs utilized in this study, which were trained using the Adam optimizer without applying dropout in the LSTM layers, were able to converge successfully without experiencing overfitting or underfitting, and they achieved comparable training and testing accuracies.

Saggi and Jain [71] conducted a comparative study in India to investigate the effectiveness of generalized linear model (GLM), deep learning-multilayer perceptrons (DLMP), gradient-boosting machine (GBM), and random forest (RF) in daily ET_0 modeling using six attributes: maximum and minimum air temperature, wind speed, relative humidity, sunshine hours, and solar radiation. The analysis of these models was conducted within the H_2O framework, which provided a novel approach for training, validating, testing, and enhancing the effectiveness of classification utilizing ML algorithms. The findings of the analysis demonstrated that all the models exhibited high efficiency when predicting evapotranspiration on a daily basis. However, they concluded that the DLMP demonstrated better performance than the other models.

Ferreira et al. [10] presented a novel evaluation of the application of restricted hourly meteorological data to predict ET_0 in Brazil on a daily basis. Data on the highest and lowest recorded temperatures, highest and lowest relative humidity levels, wind speed, and solar radiation were gathered for analysis. The evaluation was carried out at both regional and local scales using ML models, such as RF, XGBoost, ANN, and CNN. The results revealed that the CNN model achieved higher accuracy compared to the other methods.

Daily ET_0 estimation for a seven-day period was conducted utilizing three contemporary deep learning models: LSTM, ConvLSTM, and ID-CNN trained using daily climatic inputs such as air temperature (maximum, minimum, mean), solar radiation, relative humidity, and wind speed [72]. Although, in both the calibration and validation procedures, the ConvLSTM model showed superior performance compared to both LSTM and CNN by exhibiting the lowest RMSE in the weekly ET_0 estimations, there was not a significant distinction in the daily ET_0 estimations.

The study in [73] investigated the effectiveness of AI-based models to estimate PM-calculated ET_0 by using limited meteorological data. They used an ANN model to estimate long-term ET_0, while LSTM and CNN-LSTM were applied for short-term ET_0 predictions. Compared to traditional models, findings showed the superiority of AI models in ET_0 estimation. ANN, which used temperature, solar radiation, and relative humidity as inputs, yielded the best results with R^2 between 0.992 and 0.998, MAE from 0.07 to 0.16 mm/day, and RMSE between 0.12 and 0.25 mm/day. Conversely, for short-term ET_0 forecasting, the LSTM model using the same inputs was the most effective for predicting ET_0 four days ahead, achieving R^2 ranging from 0.608 to 0.756, MAE from 1.05 to 1.28 mm/day, and RMSE between 1.35 and 1.62 mm/day. This study demonstrated that AI algorithms can effectively estimate and predict ET_0 using limited meteorological data, though their forecasting accuracy could be enhanced by testing over longer periods. While regional AI models, trained with data from multiple meteorological stations, improve generalizability, they tend to be less accurate than location-specific models due to variations in training data. To balance generalizability and performance, grouping stations based on meteorological characteristics rather than geography may lead to more reliable regional models [73].

The study in [74] utilized four deep learning models of CNN, DNN, bidirectional long short-term memory (BiLSTM), and gated recurrent units (GRU) to predict daily ET_0 in Australia. The results demonstrated that deep learning models effectively estimated ET, with GRU and DNN outperforming the other models. During the testing period, the GRU model achieved the highest accuracy, with R^2 = 0.989, RMSE = 0.1794, MSE = 0.0322, and MAE = 0.1417, followed by DNN (R^2 = 0.980, RMSE = 0.185, MSE = 0.0345, MAE = 0.1507). The CNN model had the lowest precision (R^2 = 0.958, RMSE = 0.364, MSE = 0.1330), while BiLSTM performed better than CNN but remained behind GRU and DNN, with R^2 = 0.969 and MSE = 0.0988. According to the sensitivity analysis, radiation with 42% and maximum temperature with 32% had the greatest impact on the evapotranspiration process, while minimum relative humidity contributed the least. Table 2 presents a compilation of research papers that focused on estimating ET using DNN models.

Recent advancements in DL have led to the development of sophisticated models for forecasting ET_0 using a variety of meteorological inputs [10,70,71]. In particular, emerging research has focused on the application of DL techniques for both short-term and long-term ET_0 prediction [72–74]. DL models have demonstrated notable advantages over traditional empirical methods, attributed to their (1) enhanced predictive accuracy, (2) robustness under varying climatic and environmental conditions, (3) ability to capture complex non-linear interactions, (4) improved computational efficiency, and (5) adaptability to newly available data.

Compared to empirical approaches, DL models exhibit reduced sensitivity to missing or incomplete data [75,76]. Furthermore, DL models have demonstrated consistent performance across a wide range of climatic zones, underscoring their applicability in diverse geographic extents [77]. Unlike traditional empirical models that depend on fixed mathematical formulations, DL architectures inherently learn relationships from data without requiring predefined equations or assumptions about the underlying processes [75,78]. Additionally, DL models can be trained on localized datasets and subsequently applied to broader regions using global-scale inputs. This transferability facilitates the accurate upscaling of ET_0 estimates from site-specific to regional and global levels [79]. Once trained, DL models offer computationally efficient ET estimation, in contrast to empirical approaches that often necessitate extensive data inputs [77]. Moreover, DL frameworks support continuous model refinement through retraining or fine-tuning as new data becomes available, thereby maintaining long-term model relevance and accuracy [75,78].

Table 2. Studies related to estimating ET by using DNN models.

Reference	Models	Input	Output	Performance Criteria	Best Model(s)
Afzaal et al. [70]	Different configuration of LSTM, PM	Hourly minimum, mean, and maximum air temperature, heat degree days, relative humidity, wind speed, dew point temperature, and atmospheric pressure	Daily ET_0	MAE = 0.0375–0.0555, RMSE = 0.38–0.58, R^2 = 0.86–0.92	LSTM trained by the Adam optimizer
Saggi and Jain [71]	DLMP, GLM, RF, GBM	Maximum air temperature, minimum air temperature, relative humidity, wind speed, solar radiation, and sunshine hours	Daily ET_0	MSE = 0.0369–0.1215, RMSE = 0.1921–0.2691, NRMSE = 13.90–18.70%, RMSLE = 0.0693–0.1023, R = 0.96–0.98, R^2 = 0.95–0.99, LL, NSE = 0.95–0.98, ACC = 85–95, MCE = 0.042–0.085	DLMP
Ferreira et al. [10]	RF, XGBoost, ANN, and CNN	Minimum and maximum air temperature, minimum and maximum relative humidity, wind speed, and solar radiation	Daily ET_0	RMSE = 0.2–1.2 mm/day, MBE = −1.00–1.00	CNN
Farooque et al. [72]	LSTM, ID-CNN, and ConvLSTM	Minimum, mean, and maximum air temperature, solar radiation, relative humidity, and wind speed	Daily ET_0	RMSE = 0.62–0.95, R^2 = 0.61–0.74, NSE = 0.67–0.75	ConvLSTM
Fang et al. [73]	ANN, LSTM, and CNN-LSTM	Maximum air temperature, minimum air temperature, Relative humidity, solar radiation, wind speed, and mean atmospheric pressure	Daily ET_0	R = 0.992–0.998, MAE = 0.07–0.16 mm/day, RMSE = 0.12–0.25 mm/day	ANN for long-term predictions and LSTM for short-term predictions
Baishnab et al. [74]	CNN, DNN, BiLSTM, GRU	Rainfall, maximum temperature, minimum temperature, solar radiation, maximum relative humidity, minimum relative humidity, and short crop evapotranspiration	Daily ET_0	R^2 = 0.989, RMSE = 0.1794, MAE = 0.1417, MSE = 0.0322	GRU

Notes: Abbreviations: root mean square logarithmic error (RMSLE), log loss (LL), accuracy (ACC), mean per-class error (MCE).

Several critical components, namely data selection, preprocessing, feature selection, and the treatment of missing data, play a pivotal role in the effective application of DL models, as evidenced by the existing literature. Commonly, DL studies utilize key meteorological variables such as air temperature, relative humidity, solar radiation, and wind speed as model inputs. Moreover, hourly data are frequently preferred over daily records due to their higher temporal resolution, which have led to model accuracy [75,79]. To improve compatibility with DL algorithms and to optimize performance, climatic datasets are typically normalized [80]. Feature selection techniques are employed to eliminate redundant or irrelevant variables, thereby reducing computational demands and improving model accuracy [80]. Addressing missing data is also essential for maintaining dataset integrity; this is often achieved through statistical imputation or machine learning-based methods.

The implementation of effective feature selection has been shown to significantly enhance the predictive performance of various models by emphasizing the most informative input variables [80]. In summary, DL-based approaches to ET estimation benefit substantially from the integration of advanced data preprocessing strategies, feature optimization, and robust methods for handling incomplete data, thereby demonstrating strong potential for improved predictive performance.

Previous studies highlight both the strengths and limitations of DL-based ET estimation. DL models are particularly effective in capturing complex and nonlinear interactions among climatic variables; however, their performance is highly dependent on appropriate hyperparameter tuning and rigorous model validation [75]. Despite their predictive capabilities, the interpretability of DL models remains a significant challenge. Nevertheless, the integration of hybrid approaches such as combining DL with gradient boosting techniques like XGBoost offers enhanced scalability and performance [81]. A further concern is the potential for overfitting, particularly in regions with limited training data or low variability in input features [79]. These considerations underscore the need for careful model design and data management when applying DL techniques to ET estimation.

One of the significant DNNs' advantages for ET_0 prediction is in spatiotemporal modeling. Architectures like hybrid CNN-LSTM models excel at capturing temporal dependencies and spatial variability in dynamic weather conditions [82]. However, DNNs face challenges related to high computational cost during training, requiring substantial resources and time, which can limit their use in low-resource settings [83]. Additionally, while transfer learning improves adaptability, DNNs often struggle with generalizability across extreme or sparse climates, such as arid or polar regions. Without targeted data augmentation or domain-specific fine-tuning, DNNs may not perform well in these environments, highlighting the need for careful model adaptation in diverse climatic contexts.

3.2. Tree-Based Models

Tree-based models, which are employed for both regression and classification purposes, construct a hierarchical arrangement of decisions that resembles a tree. These models are specifically useful for capturing complex relationships in data and are widely used in various fields. Typical tree-based models encompass decision trees, random forests, gradient boosting machines, adaptive boosting, gradient boosted regression trees, and extremely randomized trees, among which random forests are more popular for modeling ET_0.

A random forest (RF) model is a collective learning approach that combines multiple decision trees to improve predictive accuracy and reduce overfitting. Every tree was trained on a randomly selected subset of data to make estimations. The final prediction is obtained by averaging (in regression) or using majority voting (for classification) on the predictions from individual trees. Because of its effectiveness across diverse datasets, accurate estimation, a limited set of pre-defined parameters, the capability to predict the relative value of the parameters, and its capacity to prevent overfitting, the RF method has gained significant popularity in recent years [84–87]. For example, Wang et al. [86] assessed the generalization capabilities of the RF algorithm in modeling ET_0 using various input combinations, particularly in scenarios involving missing data. The study also compared the performance of RF with the gene expression programming (GEP) approach by data collected from 24 stations situated in a karst region in China. The findings indicate that the RF-based models successfully modeled ET_0 using both entire and partial meteorological variables. The performance measures, including R^2, NSE, RMSE, and PBIAS, ranged from 0.64 to 0.99, 0.63 to 0.99, 0.11 to 0.56, and −2.92% to 1.57%, respectively. Similarly, the GEP-based models showed promising results, with performance measures of R^2, NSE, RMSE, and PBIAS ranging from 0.64 to 0.94, 0.64 to 0.94, 0.22 to 0.56, and −1.98% to 0.25%, respectively. While models based on RFs demonstrated a slight superiority compared to models based on GEPs, the GEP approach offered the advantage of providing explicit relationships among dependent and independent parameters, enhancing accessibility for irrigators with limited computer proficiency. Therefore, it was recommended to utilize the RF-based models for water balance purposes and the GEP-based models for agricultural irrigation practices. Furthermore, the study observed a decline in model performance over time because of the impact of climate change on ET_0. Finally, both the RF and GEP methods demonstrated the ranking of meteorological variables based on their significance for ET_0, such as air temperature, sunshine duration, wind speed, and relative humidity.

Shi et al. [88] estimated the ET_0 on a daily basis in southeastern Australia using RF-based models and evaluated their efficiency in comparison to empirical methods including Makkink, Abtew, Jensen–Haise, and Hargreaves. Data on daily historical climate conditions were gathered throughout the study period, which included parameters such as relative humidity, temperature, solar radiation, and rainfall. The RF-based models outperformed in estimating ET for all stations with R^2, RMSE, and rMBE values ranging from 0.68 to 0.92, 0.58 mm/day to 1.46 mm/day, and −16.10% to 9.73%, respectively. Subsequently, empirical methods and RF-based models were employed to forecast future ET across eight stations using climatic data from 34 general circulation models (GCMs), which were statistically downscaled on a daily basis. Two distinct representative concentration pathways (RCP4.5 and RCP8.5) were used during the modeling. All models showed an increase in ET under RCP4.5 and RCP8.5 from 33 mm/year to 129 mm/year and from 43 mm/year to 248 mm/year, respectively. Additionally, uncertainties, which were quantified in ET projections, revealed that RCP-related uncertainty was the predominant contributor to the uncertainty in projected ET (approximately 40% for most stations), while uncertainties related to GCMs and ET models contributed approximately equal amounts to the projected ET uncertainty (10–30%).

In the study conducted by Vulova et al. [89], an approach that integrates flux footprint modeling, ML techniques, and a Geographic Information System (GIS) was implemented along by employing data acquired through remote sensing to estimate ET at a half-hourly temporal resolution in a diverse urban area. The hypothesis consisted of incorporating GIS data and remote sensing via footprint modeling, instead of relying merely on meteorological data. Two algorithms, 1D CNN and RF, were compared in the analysis. While RF demonstrated a slightly higher level of accuracy, CNNs also demonstrated promising results.

Feng et al. [90] employed RF and GRNN to predict the daily ET_0 at the Chengdu and Nanchong stations in southwest China. Two different input scenarios were used: a complete dataset including the maximum and minimum air temperature, solar radiation, wind speed, and relative humidity. It was observed that both RF and GRNN exhibited strong performance; however, the RF model exhibited higher accuracy based on statistical performance criteria.

Al-Mukhtar et al. [91] assessed various AI methods, including quick RF (QRF), RF, SVM, and feed-forward ANN (FFANN), as well as multiple regression analysis, to predict pan evaporation rates using monthly data of air temperatures, wind speed, and relative humidity recorded by three meteorological stations in Iraq. Both the QRF and RF methods showed superior performance compared to the other models during both calibration and validation phases in terms of modeling pan evaporation on a monthly basis at all meteorological stations. Overall, the ranking of the investigated models, from best to worst performance, in relation to the observed data, was as follows: QRF > RF > ANN > SVM > MLR.

In the study carried out by [92], aimed to incorporate a robust modeling approach, utilizing diverse sites along Turkey's southern coastline, to develop a general model for predicting monthly ET_0 comprehensively. To this end, a comprehensive investigation was carried out by three distinct approaches: MLR, RF, and ELM, based on 20 different combinations of meteorological and geographical factors, resulting in a total of 60 predictive models. The input data included historical monthly meteorological records, such as humidity, temperature, reference evapotranspiration, and wind speed. The ELM model provided more accurate predictions compared to the MLR and RF models.

Furthermore, other tree-based models, including M5Tree and gradient boosting machines, have been employed by researchers to estimate ET_0 based on different data scenarios. For example, Fan et al. [93] proposed a novel tree-based soft computing model called the light gradient boosting machine (LGBM) to model the daily ET_0 for 49 meteorological stations throughout China.

The LGBM constitutes a gradient-based learning approach built upon decision trees and the boosting concept. The main advantage of the LGBM model over XGBoost lies in its use of histogram-based algorithms to enhance the speed of the training process, curtail memory usage, and implement a leaf-wise growth method with depth constraints. The standard decision tree growth method, known as "level-wise", exhibited weakness because it treats all leaves at the same level, incurring unnecessary memory overhead. Therefore, a more efficient alternative, termed "leaf-wise", was developed to choose leaves that exhibit the greatest branching advantage from the entire leaf pool, leading to superior accuracy with the same number of segmentation attempts. However, it is noteworthy that leaf-wise orientation can lead to deeper decision trees and the possibility of overfitting. To strike a balance between efficiency and overfitting, the LGBM introduces a limit for the maximum depth at the highest leaf level. A visual depiction of level-wise and leaf-wise tree growth methods is presented in Figure 2.

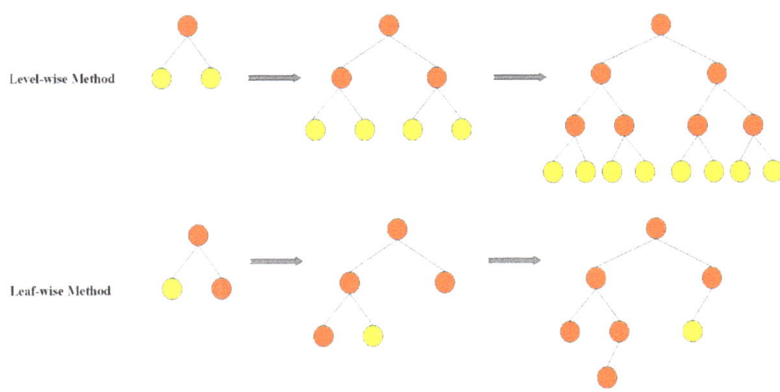

Figure 2. Schematic structure of the growth strategies of leaf-wise and level-wise trees.

The performance of LGBM used in this study was compared with M5Tree, RF, and four empirical models (Tabari, Hargreaves–Samani, Trabert, and Makkink). The LGBM model was found to be the most effective in this comparison. Different combinations of daily meteorological inputs, including minimum and maximum temperature, wind speed at a height of 2 m, relative humidity, global solar, and extraterrestrial solar radiation, were used for developing the models. The results demonstrated that LGBM outperformed RF and M5Tree in both local and external applications across all combinations, with an average RMSE ranging from 0.08 to 0.58 mm/day, 0.11 to 0.62 mm/day, and 0.13 to 0.60 mm/day, respectively. In addition, all three models provided significantly improved estimates of daily ET_0 compared to the corresponding empirical models using the same inputs. Among the meteorological variables, solar radiation was identified as the primary factor affecting the prediction of daily ET_0 estimation in the study area, followed by wind speed, relative humidity, and minimum and maximum temperature.

Huang et al. [11] examined the viability of a novel ML algorithm called CatBoost, which utilizes gradient boosting on decision trees to accurately predict daily ET_0 in humid areas of China using limited meteorological data. Two commonly used ML algorithms, RF and SVM, were also evaluated for comparison. This research utilized different combinations of daily meteorological input data such as solar radiation, minimum and maximum temperatures, wind speed, and relative humidity. According to the results, the SVM model performed the best with limited access to climatic data, whereas the CatBoost model performed the best with full access to climatic data. CatBoost, as a tree-based algorithm, significantly improved stability, accuracy, and computational efficiency compared to RF. It was concluded that CatBoost has great capability for modeling ET_0 in other areas worldwide with similar climates to humid regions of China.

Rahimikhoob [41] investigated the efficiency of ANN and M5Tree models to predict ET_0 using humidity, maximum and minimum air temperature, and extraterrestrial radiation at four meteorological stations in a region with an arid climate. The results demonstrated that the ANN-based ET_0 estimates with an R^2 of 0.98 and an RMSE of 5.6% outperformed the M5Tree model; however, the results obtained from both models were close to those of the PM method.

Ref. [94] introduced an extensive investigation to accurately predict ET_0 across 14 stations situated in China using different types of techniques, namely GRNN, MLP, ANFIS, SVM, MARS, extreme gradient boosting (XGBoost), M5Tree, and a kernel-based nonlinear extension of the Arps decline model (KNEA). The input data included the daily relative humidity, minimum and maximum temperatures, wind speed, solar radiation, and precipitation. The study results demonstrated that ML techniques could offer reliable predictions of ET_0 using merely temperature variables. Additionally, adding precipitation data improved prediction accuracy by 7.6% across China, particularly in humid regions. MARS, KNEA, and SVM generally performed better than the other models.

Wu and Fan [95] utilized the M5Tree, ANFIS, and conjugate gradient (CG) models for pan evaporation simulations using measured daily meteorological input data of sunshine hours, air temperature, relative humidity, wind speed, and pan evaporation. Their results revealed that the CG model outperformed the M5Tree and ANFIS methods.

Three distinct ET models were proposed to predict actual evapotranspiration, with variations in the input variables used [4]. The inputs included net solar radiation, soil moisture content, mean temperature and relative humidity, wind speed, and sensible-heat flux. Each model consisted of four different variants, employing different ML algorithms, namely, the SVR, bagging, RF, and M5P regression tree. According to the results, models using a comprehensive range of climatic input variables offer greater accuracy but require sufficient measuring stations, limiting their applicability. Furthermore, the M5P models exhibited a noticeable rise in result variance when dealing with ET values greater than 3 mm/day. Conversely, the SVR models consistently overestimated ET when it exceeded 3 mm/d. M5P and SVR models showed the largest residuals in the 0–1 mm and 2.5–5 mm ranges, similar to RF models. This suggests that excluding moisture content and sensible-heat flux from the inputs adversely affects the effectiveness of the predictive model. The highest residual values observed across the entire range of experimental measurements were comparable in the RF models.

Wang et al. [96] predicted the pan evaporation on a daily basis in China using the M5Tree, ANFIS-GP, and FG models trained by air pressure and temperature, solar radiation, wind speed, and pan evaporation. Their results demonstrated the considerable capability of the FG model for estimating pan evaporation.

The research in [97] aimed to forecast crop water stress index (CWSI) and crop evapotranspiration (ETc) to design irrigation decision support systems (IDSS) by using different AI algorithms in comparison with traditional methods such as FAO56 ETc and Jackson's theoretical CWSI. The data included different combinations of weather data, soil water content, and ancillary variables. The findings indicated that CatBoost, with RMSE ranging from 0.06 to 0.09, and stacked regression, with RMSE ranging from 0.27 to 0.72 mm d^{-1}, were the best models in prediction of CWSI and estimation of ET_c, respectively.

The study performed in [98] evaluated the effectiveness of machine learning models, including the gradient boosting decision tree (GBDT), partial least square regression (PLSR), random forest regression (RFR), k-nearest neighbors (KNN), backpropagation neural network (BPNN), and SVR, to estimate the spatiotemporal distribution of ET with limited input variables in China. In tandem with machine learning-based models, a feature space method called the long-term sequence feature space method (LTSFSM) was used to improve the estimation accuracy of ET. Among the ML models, PFR demonstrated the highest accuracy with the R, RMSE, MAE, and MBE equal to 0.79, 0.61 mm/day, 0.42 mm/day, and −0.02 mm, respectively. Compared to machine learning models, which generally need a large dataset for training, the feature space method can estimate ET directly without prior data and follows a simpler computational process. As a result, the feature space method offers greater advantages for widespread applications. It is worth mentioning that ET estimates from both types of models showed compatible spatial distribution patterns with MOD16 ET data, confirming the reliability of both methods.

The study performed in [99] endeavored to estimate actual evapotranspiration under semi-arid Mediterranean conditions by combining agro-meteorological data with high-resolution satellite imagery. For this purpose, they used three ML algorithms, including elastic net, random forest, and support vector machine. According to the results, random forest demonstrated the highest predictive performance, achieving an average R^2 of 0.74, an RMSE of 0.577 mm, and an MBE of 0.03 mm. A list of ET estimation studies using tree-based algorithms is presented in Table 3.

Table 3. Studies conducted to estimate ET using tree-based algorithms.

Reference	Models	Input	Output	Performance Criteria	Best Model(s)
Wang et al. [86]	RF, GEP	Minimum and maximum air temperature, duration of sunshine, relative humidity, and wind speed at a height of 2 m	Daily ET_0	R^2 = 0.637–0.987, NSCE = 0.626–0.986, RMSE = 0.107–0.563 mm/day, PBIAS = −2.916–1.571%	RF
Shi et al. [88]	RF, Jensen–Haise, Makkink, Abtew, and Hargreaves	Minimum and maximum air temperature, solar radiation, rainfall, and minimum and maximum relative humidity	Daily ET_0	R^2 = 0.68–0.92, RMSE = 0.58–1.46 mm/day, rMBE = −16.10–9.73%	RF
Vulova et al. [89]	RF, CNN	Air pressure, air temperature, diffuse solar radiation, dry bulb temperature, longwave downward radiation, ET_0, relative humidity, saturated vapor pressure, shortwave downward radiation, soil temperature, solar zenith angle, vapor pressure deficit, wind speed, impervious surface fraction, building height, NDVI, vegetation height, vegetation fraction, water fraction	Half-hourly urban ET	R^2 = 0.840, RMSE = 0.0239 mm/h, MAE = 0.0154, NRMSE = 8.30, PBIAS = 1.80	RF

Table 3. *Cont.*

Reference	Models	Input	Output	Performance Criteria	Best Model(s)
Feng et al. [90]	RF, GRNN	Minimum and maximum air temperature, solar radiation, wind speed, and relative humidity	Daily ET_0	RRMSE = 0.067–0.258, MAE = 0.1–0.4, NSE = 0.834–0.987	RF
Al-Mukhtar et al. [91]	quick RF (QRF), RF, SVM, FFANN	Maximum and minimum air temperatures, relative humidity, and wind speed	Monthly pan evaporation	RMSE = 14.44–23.36 mm, R^2 = 0.98–0.99, NSE = 0.98–0.99	QRF
Hameed et al. [92]	MLR, RF, ELM	Temperature, humidity, wind speed, and ET_0	Monthly ET_0	MAE = 0.946 mm/month, RMSE = 1.155 mm/month, MARE = 0.146, RMSRE = 0.18, RRMSE = 16.544%, MBE = 5.958, erMAX = 1.634	ELM
Fan et al. [93]	LGBM, M5Tree, RF, Hargreaves–Samani, Tabari, Makkink, and Trabert).	Minimum and maximum air temperature, wind speed at 2 m height, relative humidity, and global and extraterrestrial solar radiation	Daily ET_0	RMSE = 0.08–0.58, R^2 = 0.85–1, NRMSE = 0.03–0.24	LGBM
Huang et al. [11]	CatBoost, RF, SVM	Minimum and maximum air temperature, solar radiation, relative humidity, and wind speed	Daily ET_0	RMSE = 13–288, R^2 = 15–287, MBE = 49–221 MAPE = 12–283	SVM with limited access to climatic data, CatBoost with full access to climatic data
Rahimikhoob [41]	ANN, M5Tree	Maximum and minimum air temperature, extraterrestrial radiation, and humidity	ET_0	RMSE = 0.41 mm/day, R^2 = 0.95 R = 1	ANN
Wu et al. [94]	GRNN, MLP, ANFIS, M5Tree, XGBoost, SVM, KNEA, MARS	Minimum and maximum air temperature, wind speed, relative humidity, precipitation, and solar radiation	Daily ET_0	RMSE = 0.718 mm/day, R^2 = 0.829, MAE = 0.508 mm/day NRMSE = 0.250	SVM
Keshtegar et al. [95]	M5Tree, CG, ANFIS	Air temperature, wind speed, sunshine hours, relative humidity, and pan evaporation	Daily pan evaporation	RMSE = 1.944–4.947, EF = 0.840–0.872, d = 0.928–0.964	CG
Granata [4]	M5P Regression Tree, Bagging, RF, SVR	Soil moisture content, net solar radiation, sensible-heat flux, mean relative humidity and temperature, and wind speed	Daily actual evapotranspiration	NSE 0.932–0.987, MAE = 0.14–0.322 mm/day, RMSE = 0.179–0.400 mm/day, RAE = 15.4–35.4%	M5P regression tree trained with net solar radiation, sensible-heat flux, soil moisture, wind speed, relative humidity, and temperature
Wang et al. [96]	M5Tree, FG, ANFIS-GP	Air temperature and pressure, wind speed, solar radiation, and pan evaporation	Daily pan evaporation	RMSE = 0.592 mm/day, MAE = 0.459 mm/day, R^2 = 0.932	FG
Katimbo et al. [97]	CatBoost and Stacked Regression	Minimum, maximum, and mean air temperatures; relative humidity, wind speed, and shortwave incoming solar radiation; soil volumetric water content; ET_0, NDVI; and CGDDs	Daily ET_0	R^2 = 0.40–0.99, RMSE = 0.15–0.25, MAE = 0.10–0.19, MAPE = 7.9 to 15.1%	CatBoost
Sun et al. [98]	GBDT, PLSR, RFR, KNN, BPNN, SVR	NDVI, LST, NSATave, $NSAT_{max}$, $NSAT_{min}$, and ET	Daily ET	R = 0.79, RMSE = 0.61 mm/day, MAE = 0.42 mm/day, MBE = −0.02 mm	PFR
Garofalo et al. [99]	Elastic Net, RF, SVM	Multispectral reflectance data, local meteorological observations	Daily actual ET	R^2 = 0.74, RMSE = 0.577 mm, MBE = 0.03 mm	RF

Notes: Abbreviations: relative mean bias error (Rmbe), root mean square relative error (RMSRE), maximum absolute relative error (erMAX), coefficient of efficiency (CE), agreement index (D), percent bias (PBIAS), relative absolute error (RAE), cumulative growing degree days (CGDDs), normalized difference vegetation index (NDVI), land surface temperature (LST), average near-surface air temperature (NSATave), maximum near-surface air temperature ($NSAT_{max}$), minimum near-surface air temperature ($NSAT_{min}$).

Tree-based machine learning models, particularly RF, have gained substantial recognition for their effectiveness in estimating ET_0 across diverse geographic regions and environmental contexts [86,88]. Numerous studies have demonstrated the successful integration of RF into various methodological frameworks, enhancing its predictive performance not only for ET_0 but also for actual ET estimation under different conditions and data availability scenarios [11,89–92]. In complex and data-scarce environments, such as arid regions, RF's ability to capture nonlinear interactions among climatic variables offers a significant advantage over traditional empirical models [78,100].

The strengths of RF include its superior accuracy and reliability, flexibility in handling diverse meteorological inputs (e.g., temperature, humidity, wind speed, sunshine duration), and robustness against missing data [100,101]. Unlike empirical models that require complete datasets and rely on static equations, RF can maintain predictive performance even in the presence of incomplete data [87]. This resilience is partly due to its ensemble architecture, which also supports the integration of uncertainty analysis and model interpretability through feature importance evaluation [102].

In addition to RF, other tree-based methods, such as regression trees and boosting algorithms (e.g., GBoost and XGBoost), have also been employed for ET prediction. A key advantage of these models lies in their inherent interpretability, enabling identification of the most influential meteorological variables in ET estimation. This feature is critical for both advancing scientific understanding and informing practical water resource management decisions. Furthermore, boosting algorithms offer the unique ability to adapt to new climatic data via incremental learning, thus making them well-suited for modeling under evolving climate regimes.

Despite these advantages, several limitations must be considered. RF and boosting models can be computationally intensive, particularly during feature selection and training phases, and when conducting interpretability or sensitivity analyses [102]. Moreover, like other machine learning techniques, these models are susceptible to overfitting, especially boosting methods, when applied to small or noisy ET datasets unless appropriate regularization strategies are implemented. Additionally, their extrapolation capacity to extreme or previously unobserved weather conditions remains limited unless the ensemble design explicitly promotes model diversity and robustness [103]. RF's integration with remote sensing technologies further enhances its utility in large-scale agricultural water management, offering spatial scalability alongside high predictive accuracy [99,101].

3.3. Kernel-Based Models

Kernel function-embedded methods (known as kernel-based models) are a class of ML algorithms that use kernel functions, including linear and radial basis function (RBF) kernels, sigmoid, and polynomial, to implicitly map data points into an expanded dimensional space, where they might be more easily separable or transformable. The main advantage of kernel-based models is the capability to handle complex relationships among features and capture non-linear patterns. However, they can also be computationally expensive, particularly when dealing with large datasets (i.e., high-dimensional spaces). Additionally, the selection of the kernel function and its associated parameters can significantly affect the performance of the model.

Kernel functions are commonly associated with SVM and SVR for classification and regression purposes, respectively. The SVM aims to discover the most optimal hyperplane for distinguishing between data points of distinct classes, with a focus on maximizing the margin between them. SVR, on the other hand, is used to identify a hyperplane that best fits the data points, aiming to minimize discrepancies between actual and predicted values. Moreover, the relevance vector machine (RVM) is an ML algorithm that falls within the category of kernel methods and is often employed for regression and classification tasks. RVM was considered as a sparser alternative to SVM, meaning that it tends to use fewer support vectors for the final model, which can result in improved interpretability and efficiency. These models have been successfully applied in modeling non-linear systems, including evaporation, evapotranspiration, air temperature, and streamflow. For example, Shrestha and Shukla [104] developed an SVM model to estimate crop coefficient and crop evapotranspiration using a large dataset spanning 10 seasons obtained from lysimeters, specifically focusing on multiple crop-season combinations in the presence of plastic mulch. The meteorological variables included relative humidity, air temperature, wind speed, solar radiation, and rainfall. SVM-based models successfully predicted crop coefficients for bell pepper (R^2 = 0.71) and watermelon (R^2 = 0.82) crops by incorporating factors such as time, rainfall events, and water table depth. It should be noted that the time since transplant was used as an indicator of plant growth and transpiration, whereas depth of water table and rainfall events captured the influence of surface soil moisture on evaporation. Furthermore, the SVM-predicted crop coefficients were statistically similar to the actual crop coefficient values for both the crops. On the other hand, the FAO-56 estimates significantly underestimated the coefficients for bell peppers (p = 0.016) and watermelons (p = 0.025). This model can facilitate the development of region-specific crop coefficients to enhance ET estimates.

Kişi [105] assessed the effectiveness of three heuristic regression approaches, namely the MARS, M5Tree models, and LSSVM, for monthly ET_0 estimation in Turkey by using input data of solar radiation, air temperature, wind speed, and relative humidity. The results showed that the MARS and M5Tree techniques exhibited superior performance under local and external conditions, respectively.

Goyal et al. [106] employed least-squares support vector regression (LSSVR), FL, ANN, and ANFIS techniques to model pan evaporation in India on a daily basis. The measured daily meteorological variables were the minimum and maximum humidity, minimum and maximum air temperatures, sunshine hours, and rainfall. The accuracy of the AI techniques was compared against the SS and Hargreaves–Samani equations. This study demonstrated that ANFIS and LSSVR could be efficiently employed to model daily pan evaporation. Additionally, the AI models outperformed the traditional Hargreaves–Samani and SS empirical methods.

Given the explicit and direct input–output model structure, accounting for the uncertainty in the modeling process becomes crucial. It can become challenging to rely on models in dealing with real-world issues and making informed decisions [107–109]. Furthermore, that research primarily focused on the performance and practicality of AI models in estimating ET_0 by including user-friendly models and optimized combinations of inputs [10,110,111], while the analysis of uncertainty was largely overlooked. Therefore, the objective of the study [112] was to evaluate the uncertainty associated with the modeling process and assess the validity of all potential input combinations of weather variables such as air temperature, solar radiation, relative humidity, wind speed, sunshine duration, and atmospheric pressure for daily ET_0 estimations. By employing ANN, SVM, and ELM models, this study investigated the most effective pattern of input combinations for AI models using the Monte Carlo (MC) approach. The results revealed that wind speed and solar radiation had the greatest effect on variation in ET_0. The maximum temperature exhibited a more significant influence than the minimum temperature, and the relative humidity appeared to have the least influence. Furthermore, ET_0 showed higher sensitivity to higher values of wind speed and solar radiation, whereas it demonstrated sensitivity within certain ranges of minimum and maximum temperature and relative humidity. The outcome of the uncertainty analysis indicated that all three models could be trusted for predicting ET_0 on a daily basis. However, both ELM and ANN models exhibited sensitivity to high ET_0 estimation values, whereas SVM was deemed the most robust model.

Tabari et al. [113] investigated the prediction of monthly ET_0 in Nozheh, western Iran, using ANFIS, SVM, MLR, and multiple non-linear regression (MNLR), as well as some empirical equations. Six different scenarios were considered, incorporating the air temperatures, solar radiation, relative humidity, and wind speed. The findings indicated that the ANFIS and SVM models outperformed the empirical and regression methods for estimating monthly ET_0. Furthermore, these models demonstrated superior accuracy when utilizing mean air temperature, wind speed, relative humidity, and solar radiation as inputs.

Wen et al. [114] utilized ANN, SVM, and three empirical equations of Priestley–Taylor, Hargreaves, and Ritchie for estimating the daily ET_0 in the Ejina Basin in northwestern China. Four scenarios were developed with limited meteorological data, incorporating parameters including the minimum and maximum air temperatures, solar radiation, and wind speed. The SVM model demonstrated superiority over the empirical equations and ANN for modeling daily ET_0.

Sobh et al. [115] employed SVM, GMDH-NN, MARS, DENFIS, and RF to estimate high-resolution daily ET_0 over Egypt. The meteorological data contained daily records of the minimum, mean, and maximum temperature, wind speed, and dew point temperature. The results revealed that wind speed led to an increase in the estimation accuracy of the model more than relative humidity, along with the maximum and minimum temperatures. Among the AI models, the RF models exhibited the highest accuracy, followed closely by SVM, as validated by Mehdizadeh et al. [116], who highlighted the robust performance of SVM-polynomial in predicting monthly ET_0 in Iran. In addition, GMDH-NN exhibited satisfactory results, as reported by [117] in Turkey. In addition, this study identified the DINFIS models as the least effective between the employed algorithms, aligning with the findings of [118] in their modeling of monthly ET_0 in China.

Kişi [119] used the LSSVM, M5Tree, and MARS models to estimate monthly pan evaporation through air temperature, solar radiation, wind speed, humidity, and pan evaporation in Turkey. Their study demonstrated that the MARS model performed better compared to other models, showing significant superiority.

Eslamian et al. [120] employed SVM and MLP models with various input combinations, including humidity, air temperature, solar radiation, wind speed, and precipitation, to improve the accuracy of ET_0 prediction. Although both models performed well in estimating ET, the SVM showed superior performance to the ANN.

Nourani et al. [121] utilized various AI models, including SVR, ANFIS, and FFANN, as well as empirical models such as Hargreaves–Samani, modified Hargreaves–Samani, Makkink, Ritchie, and MLR, to predict daily ET_0 across Iran, Turkey, Iraq, Cyprus, and Libya. They observed that the most accurate predictions were provided by the AI models. The meteorological parameters considered were the surface pressure, relative humidity, precipitation, minimum, mean, and maximum air temperature, minimum, mean, and maximum wind speed, solar radiation, and pan evaporation. The results indicated that the performance of the empirical models varied depending on the stations and the climate of the corresponding regions. However, the empirical models generally yielded acceptable results for ET_0 estimations. Although the MLR model struggled with data nonlinearity, selecting the optimal input variables for the MLR models in the phase of input selection helped to ensure reliable results in both the training and validation stages for ET_0 estimations. Among the AI models, ANFIS demonstrated superior performance across the majority of stations that can be attributed to the combination of the neural network learning capabilities and fuzzy concepts within a unified framework. In addition, the findings highlighted that combining the outputs of multiple models, as ensemble models, achieved greater accuracy in simulating ET_0 than individual models.

Wang et al. [122] assessed the efficiency of various models such as LSSVR, MARS, MLR, FG, and M5Tree in estimating daily pan evaporation in China. The findings revealed that the FG and LSSVR models exhibited the highest accuracies among the employed models. The climatic data in this study included surface and air temperature, wind speed, relative humidity, and sunshine hours. Surface and air temperature and sunshine hours were known as major influencing factors through the nonlinear method of the Gamma Test. Additionally, by considering the wind speed or relative humidity as inputs, the model performance was significantly improved.

Current AI-based models, designed to predict evaporation rates, have not adequately considered the effect of the time increment rate on the accuracy of predictions. Therefore, a previous study evaluated the potential of RBFNN and SVR as AI models in the prediction of evaporation rates [123]. The suggested models were proposed based on the historical time-series data related to evaporation rates to forecast forthcoming evaporation rates. An alternative scenario was examined, which included the average temperature and historical evaporation rate as inputs for the model. Each scenario involved daily, weekly, and monthly time-series intervals. The findings revealed that incorporating temperature data as inputs in the modeling process significantly improved the accuracy of the predictions. The choice of timescale also had a notable impact on model performance. Among the three timescales considered, the utilization of daily evaporation data resulted in a more reliable predictive model for evaporation. The study concluded that the RBFNN model surpassed SVR in predicting evaporation data.

Deo et al. [124] developed the ELM, RVM, and MARS to model pan evaporation in Australia using minimum and maximum temperatures, precipitation, atmospheric vapor pressure, and solar radiation as inputs. The findings revealed that the RVM model demonstrated superior predictive capability when compared to other techniques.

Torres et al. [125] employed RVM and MLP to forecast ET_0 in Central Utah. The general ET_0 model utilized was the 1985 Hargreaves Equation that relies merely on the minimum and maximum daily air temperatures. This makes it suitable for regions with limited climate data. The results demonstrated that the proposed methods in this study could effectively forecast daily ET_0 up to four days ahead with a satisfactory level of accuracy, indicated by a goodness of fit parameter (η) greater than or equal to 0.8. Additionally, applying specified learning machines provided an estimation of the expected variability for each forecast day, resulting in a highly accurate estimation of the forecasted ET_0. Table 4 presents a collection of research endeavors focused on estimating ET using kernel-based models.

By utilizing key meteorological variables such as temperature, wind speed, and solar radiation, SVMs have consistently outperformed empirical equations and other AI techniques like ANNs in daily ET_0 prediction across diverse climatic settings [104–106,113–115]. SVMs have proven effective in capturing complex, nonlinear relationships among climatic variables, and their flexibility and generalizability make them a promising tool for ET estimation in varied environmental conditions. These models also show robustness in data-scarce regions, delivering accurate predictions even when limited climatic data are available, unlike data-intensive empirical models [78,126]. Moreover, SVMs' adaptability to different climates, evidenced by consistently high R^2 and efficiency indices, contrasts with the fixed relationships used by empirical models, which require recalibration for different regions or climates [17].

Table 4. Studies implemented to model ET using kernel-embedded models.

Reference	Models	Input	Output	Performance Criteria	Best Model(s)
Shresthaand Shukla [104]	SVM, PM	Air temperature, wind speed, relative humidity, solar radiation, and rainfall	Actual ET	R^2 = 0.71–0.82, MSE = 0.034–0.116, MAE = 0.026–0.213	SVM
Kişi [105]	LSSVM, MARS, M5Tree	Air temperature, solar radiation, relative humidity, and wind speed	Monthly ET_0	SI = 0.097–0.212, MAE = 0.339–0.783, R^2 = 0.843–0.970	MARS in local conditions and M5Tree in external conditions
Goyal et al. [106]	LSSVR, FL, ANN, ANFIS, SS, Hargreaves–Samani	Minimum and maximum humidity, minimum and maximum air temperatures, sunshine hours, and rainfall	Daily pan evaporation	RMSE = 2.15–2.94, CORR = 0.56–0.73	LSSVR, ANFIS
Yu et al. [112]	ANN, SVM, and ELM	Air temperature, wind speed, solar radiation, relative humidity, sunshine duration, and atmospheric pressure	Daily ET_0	R, RMSE, MAE, NSE	SVR
Tabari et al. [113]	SVM, ANFIS, MLR, MNLR, Blaney–Criddle, Hargreaves, Makkink, Turc, Jensen–Haise, McGuinness–Bordne, Priestley–Taylor, Ritchie, Abtew	Minimum, maximum, and mean air temperatures, solar radiation, relative humidity, and wind speed	Monthly ET_0	R = 0.626–1, RMSE = 0.032–1.511 mm/day, MAE = 0.017–1.130	SVM and ANFIS
Wen et al. [114]	ANN, SVM, Priestley–Taylor, Hargreaves, Ritchie	Minimum and maximum air temperatures, solar radiation, and wind speed	Daily ET_0	R = 0.772–0.950, RMSE = 0.262–0.539 mm/day, MAE = 0.207–0.446 mm/day	SVM
Sobh et al. [115]	SVM, GMDH-NN, MARS, DENFIS, RF	Minimum, mean, maximum, and dewpoint temperatures and wind speed	Daily ET_0	KGE = 0.52–0.75	RF
Kişi [119]	LSSVM, M5Tree, MARS	Air temperature, wind speed, solar radiation, humidity, and pan evaporation	Monthly pan evaporation	RMSE =0.632–1.359, MAE = 0.499–1.370	MARS
Eslamian et al. [120]	SVM, MLP	Humidity, solar radiation, air temperature, precipitation, and wind speed	ET_0	MAE = 0.55, MAXE = 1.95, EF = 0.91, WI = 0.97, R^2 = 0.96	SVM
Nourani et al. [121]	SVR, ANFIS, FFANN, Hargreaves–Samani, modified Hargreaves–Samani, Makkink, Ritchie, MLR	Relative humidity, surface pressure, precipitation, maximum air temperature, minimum air temperature, mean air temperature, minimum wind speed, maximum wind speed, mean wind speed, solar radiation, and pan evaporation	Daily ET_0	R^2 = 0.517–0.918, RMSE = 0.073–0.168	ANFIS
Wang et al. [122]	LSSVR, MARS, MLR, FG, and M5Tree,	Air temperature, surface temperature, wind speed, relative humidity, and sunshine hours	Daily pan evaporation	MAE = 0.54–1.55, RMSE = 0.72–2.03, MBE = −0.18–0.49, R^2 = 0.593–0.928	LSSVR, FG
Allawi et al. [123]	RBFNN, SVR	Pan evaporation and mean air temperature	Daily pan evaporation	MBE = 0.399–0.557, RMSE = 5.549 –11.409 mcm/month, MAE = 3.522–6.598 mcm/month, NSE = 0.447–0.898, SI = 0.746–1.112, WI = 0.804–0.858, Cl = 0.359–0.770, BIAS = 0.849–1.813 mcm/month	RBFNN
Deo et al. [124]	ELM, MARS, RVM	Maximum and minimum temperatures, atmospheric vapor pressure, precipitation, and solar radiation	Daily pan evaporation	R = 0.979, RMSE = 9.306, MAE = 0.034, q = 0.034	RVM
Torres et al. [125]	RVM, MLP	Minimum and maximum air temperatures, crop coefficients (Kc), ET_0, and information about crop distributions and effective agricultural area	Daily ET_0	RMSE = 0.65–0.89 mm/day, NSE = 0.77–0.88%	RVM

Notes: Abbreviations: SI, sensitivity index (SI), relative error (RE), Kling–Gupta efficiency (KGE), relative error (RE), performance index (q), mean bias error (MBE), Willmott index of agreement (WI), confidence index (CI).

Key advantages of SVMs also lie in their ability to integrate optimization algorithms, which fine-tune model parameters and enhance predictive accuracy. This approach addresses the limitations of empirical models, which often rely on iterative calibration [78,126]. In addition, the use of time-series data structures, input variable combinations, and temporal resolution further improves the accuracy of AI-based ET_0 and evaporation rate estimates [123,125]. Furthermore, SVMs can accommodate missing data either through imputation strategies or by training on incomplete datasets, which enhances their applicability in various data availability scenarios [102]. Feature selection techniques can also improve predictive accuracy by reducing the number of input variables, highlighting the importance of data preprocessing in optimizing model performance [80].

However, despite their strengths, SVMs face several challenges. One significant limitation is their sensitivity to kernel function selection and parameter tuning, which are essential for achieving optimal performance. The time-consuming nature of this process, especially when large-scale applications are involved, can limit its practical use [127]. Furthermore, SVMs generally perform less effectively than more advanced models, such as LSTM networks, when addressing highly nonlinear relationships or working with larger datasets [102,128]. Therefore, while SVMs provide promising results in many ET prediction scenarios, their performance depends on careful model tuning and expert knowledge, and in certain conditions, they may be outperformed by alternative AI techniques.

3.4. *Hybrid Models*

Classical neural networks (NNs) are incapable of enhancing the perception of physics related to ET_0 because of their black box nature without explicit expressions. Additionally, it is widely recognized that classical NNs become stuck easily in a local minimum. In this regard, hybrid methods that integrate various modeling approaches have been developed to promote the effectiveness of independent models in estimating ET_0. The subsequent sections explore the various categories of these hybrid models.

3.4.1. Combination of AI Models and Optimization Algorithms

A well-known category of hybrid models is based on combining AI models utilizing optimization algorithms inspired by biological processes, called bio-inspired (Figure 3). AI performance has been notably enhanced in diverse scenarios through the effective application of bio-inspired optimization algorithms [5,129–134]. Mohammadi and Mehdizadeh [135] demonstrated that using the whale optimization algorithm, as a bio-inspired algorithm, improved the AI's accuracy in modeling daily ET_0 values. Other widely recognized bio-inspired algorithms, including genetic and firefly algorithms, have also shown significant improvements in the accuracy of AIs for modeling the evapotranspiration [136–141]. For instance, Maroufpoor et al. [142] investigated the capacity of the ANN-GWO, as a hybrid model, which is a combination of an ANN and grey wolf optimization algorithm, to approximate ET_0 in Iran under different climates. The ANN-GWO model's efficiency was evaluated in comparison to the LSSVR and standalone ANN models. The results demonstrated a reduction in estimation errors by 42% and 30% when compared to the LS-SVR and ANN models, respectively. Among the different scenarios based on different input data combinations, the scenario that utilized only three inputs (i.e., wind speed and minimum and maximum temperature) demonstrated superiority over the other scenarios.

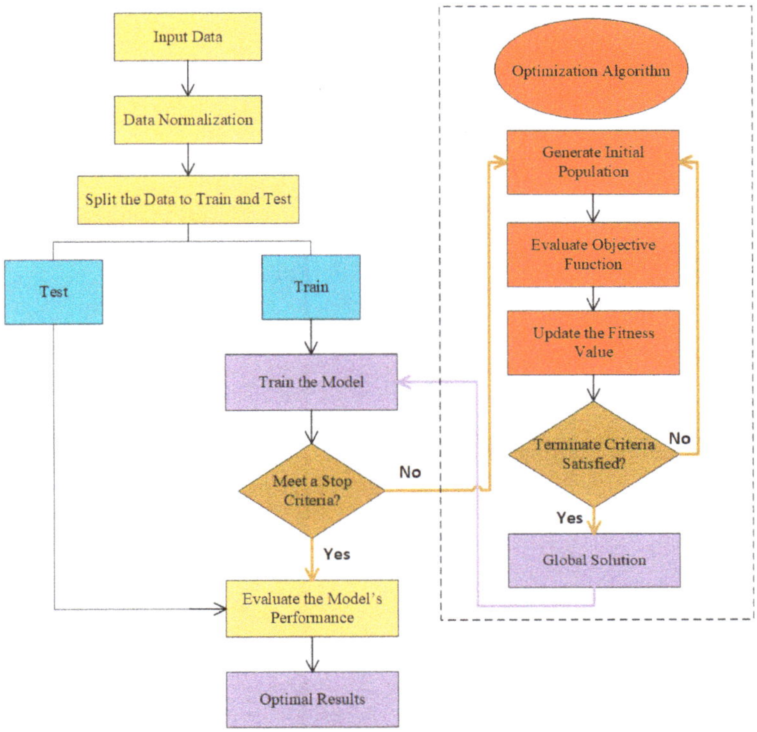

Figure 3. A flowchart of the hybrid AI model boosted with an optimization algorithm.

Wu et al. [143] assessed the performance of combining the ELM model with the flower pollination algorithm (FPA) and whale optimization algorithm (WOA) for monthly pan evaporation prediction. These models were further compared with ANN, improved M5Tree, and ELM optimized by differential evolution (DE), which is known as a floating-point evolutionary optimization algorithm. The climatic factors considered were wind speed, sunshine hours, relative humidity, and minimum and maximum temperatures. They noted that the hybrid ELM model demonstrated superior prediction accuracy against the hybrid M5Tree and ANN models. Additionally, the results explicitly indicate that the performance of the models is noticeably impacted by whether certain inputs are present or absent in the input sets. Among the five predictor variables examined, the maximum temperature and wind speed emerged as the two most effective predictors for accurately estimating ET_0. In the optimal ANN model, all five input variables were deemed significant in contrast to the other models. Additionally, across various performance indices, the ANN model outperformed the ELM and M5Tree models, despite the fact that their performances were generally comparable and closely aligned in most cases. The study carried out by Hadadi et al. [144] aimed to investigate the functionality of the ANFIS model along with its combinations with two optimization algorithms—specifically, the shuffled frog-leaping algorithm (SFLA) and GWO—in predicting monthly actual ET across the Neishaboor watershed located in Iran. The models' input variables were categorized into three groups: (1) meteorological parameters, namely, air temperature (average, minimum, maximum), relative humidity, wind speed, sunshine hours, and dew point; (2) remotely sensed parameters, namely, net radiation, land surface temperature (LST), normalized difference vegetation index (NDVI), soil wetness deficit index (SWDI), and soil adjusted vegetation index (SAVI); and (3) hybrid predictors. The results indicated that the SWDI predictor had an indispensable effect on the estimation of ET in all three models. The estimation error was reduced by 12.5%, 17%, and 26.5% in the testing set and by 18.4%, 25.4%, and 24.5% in the training set in ANFIS, ANFIS-SFLA, and ANFIS-GWO, respectively. The findings also demonstrated the effectiveness of employing either optimization algorithm to enhance ET estimation. The most significant reduction in error rates in the testing set was observed for ANFIS-SFLA compared to ANFIS (12.5%), and ANFIS-GWO exhibited a decrease of 13.7% relative to ANFIS. Tang et al. [6] employed SVM and optimized ANN by a genetic algorithm (GA) to model actual ET in a partial plastic film mulching (MFR) and rainfed maize field under non-mulching (CK) conditions. Meteorological inputs, including minimum, mean, and maximum air temperature; minimum, mean, and maximum relative humidity; wind speed; and solar radiation; as well as crop input data such as plant height and LAI during the seasons in 2011–2012 when maize growing was employed to train the ANN-GA and SVM models. The results revealed that the GANN1 and SVM1 models, incorporating both meteorological and crop data as inputs, successfully predicted ET in maize. This confirmed the effectiveness of the GANN and SVM models in estimating ET in maize. Although the GANN2 and SVM2 models, which just used meteorological inputs, had comparatively lower efficiencies than the GANN1 and SVM1 models, their approximated outcomes remained satisfactory even when just meteorological data were accessible. Notably, the GANN models marginally outperformed the SVM models under both MFR and CK conditions, possibly because of the genetic algorithm optimization capability for ANN, which helps to avoid local minima and overfitting of the neural network.

Tikhamarine et al. [145] employed three different optimization algorithms, GA, GWO, and particle swarm optimizer (PSO), to enhance the precision of SVR in predicting the monthly ET_0 using wind speed, relative humidity, minimum and maximum air temperatures, and solar radiation at three sites in Algeria. The effectiveness of the proposed hybrid AI models was assessed in contrast to traditional ANNs and empirical models, including Ritchie, Turc, three variations of the Valiantzas, and Thornthwaite method. The results demonstrate that the hybrid models incorporating optimization algorithms exhibited superior performance compared to the standalone SVR models. Additionally, the SVR-GWO with five inputs, namely minimum and maximum temperature, wind speed, solar radiation, and relative humidity, exposed a feasible model to estimate ET_0. The GWO algorithm outperformed the GA and PSO algorithms. The conventional empirical methods employed in this study, excluding the Thornthwaite model, could demonstrate reliable estimates for ET_0. Furthermore, it is noteworthy to emphasize that the Valiantzas method performed better than other conventional approaches at the designated study stations.

Eslamian et al. [138] assessed the performance of ANN-GA and ANN models in modeling PM ET_0 using Esfahan, Iran's province, meteorological data. The meteorological inputs included the average, minimum, and maximum air temperature, relative humidity, wind speed, and sunshine during a period from 1951 to 2005. The estimated values of the ANN-GA model outperformed the ANN model predictions when compared with a reference such as PM ET_0.

In a study conducted by Tikhamarine et al. [146], they put forward five hybrid models by combining ANN with five dissimilar optimization algorithms, namely GWO (ANN-GWO), PSO (ANN-PSO), multi-verse optimizer (ANN-MVO), ant lion optimizer (ANN-ALO), and WOA (ANN-WOA), to predict monthly ET_0 at two locations in Algeria and India. The predictions generated by these hybrid machine learning models were contrasted with those derived from the Valiantzas model. The comparative results indicated that the ANN-GWO-1 model was the most effective at both study stations, using five input variables (including maximum and minimum temperature, solar radiation, wind speed, and relative humidity).

The study carried out by Zounemat-Kermani et al. [147] aimed to evaluate the capabilities of three well-known ML approaches, that is, neural network autoregressive with exogenous input (NNARX), ANFIS, and GEP; five metaheuristic algorithms, that is, GA, PSO, firefly algorithm (FA), artificial bee colony (ABC), and continuous ant colony optimization (CACO); and two empirical relationships proposed through algebraic formulas for predicting the pan evaporation on a monthly basis at two meteorological stations. To achieve this goal, solar radiation, measured air temperature, wind speed, and relative humidity data were employed as inputs for the model. A comparison of the total evaporation estimates from the NNARX, GEP, and ANFIS models revealed that the ANFIS and NNARX predictions were closer to the measured one than those of the GEP model. The PSO algorithm performed the best among the five applied metaheuristic algorithms in predicting pan evaporation.

Kişi [148] evaluated the application of ENN in the modeling of pan evaporation on a monthly basis by employing inputs obtained from two stations in the Mediterranean Region of Turkey, including air temperature, pan evaporation, solar radiation, wind speed, and relative humidity. The ENNs used in this study were acquired using the adjustment of the ANN model's weights by employing DE. The performance of ENN models was evaluated against ANFIS, FG, ANN, and SS models, revealing superior results in comparison to the alternatives. Additionally, the results indicated that the proposed models with inputs including full meteorological data outperformed the other models.

Ahmadi et al. [5] employed standalone GEP and SVR models as well as a proposed hybrid model by combining the intelligent water drop (IWD) and SVR algorithms considering different scenarios based on climatic inputs and historical ET_0 data to estimate monthly ET_0. The IWD algorithm acts as a meta-heuristic in order to optimize the weights among the layers of the model and the variables of the membership functions in SVR. The input weather data included the monthly minimum, mean, and maximum air temperature, relative humidity, sunshine duration, and wind speed at 2 m height. The findings indicated that in scenarios based on historical ET_0 data, the GEP models exhibited a slightly better performance against the SVR models. However, the best ET_0 estimates across all study sites were achieved using the hybrid SVR-IWD model.

Recent studies (e.g., Mehdizadeh et al. [135]) have explored the effectiveness of bio-inspired optimization algorithms to improve the accuracy of AI models for estimating ET_0 and pan evaporation. Bio-inspired algorithms like GA and firefly algorithms have also been used to optimize AI models, resulting in better predictions (e.g., Maroufpoor et al. [142]). In addition to these advancements, studies have combined ML models like ELM and SVM with optimization algorithms such as FPA and WOA [143]. Additionally, several studies have shown that incorporating a combination of meteorological and crop-related inputs, along with optimization algorithms, leads to more reliable predictions (e.g., Tang et al. [6] and Tikhamarine et al. [146]). These findings highlight the growing trend of using hybrid AI models and optimization algorithms to enhance the accuracy of ET estimations across diverse regions and climatic conditions.

3.4.2. Gene Expression Programming (GEP)

In accordance with the research conducted, employing function discovery techniques such as symbolic regression (SR) offers several advantages over other data-driven methods. SR, a nonlinear regression technique, aims to uncover suitable mathematical structures and algebraic expressions capable of capturing the nonlinearity problem tailored to the input data. Unlike traditional regression methods that require the user to predefine the model structure, SR automatically generates structures and formulates optimal algebraic expressions [149,150]. Furthermore, SR produces a generalized equation set that enables predictions across a significant spatial range, surpassing the capabilities of black-box data-driven methods [151]. SR primarily involves gene expression programming (GEP), which combines GA and genetic programming (GP). GEP, initially proposed by Ferreira [143], utilizes linear and simple chromosomes with consistent length, resembling the structure of a GA. In addition, it incorporates diverse branch structures along with varying sizes and configurations, comparable to the parsing trees utilized in GP, to construct its structure. Various research endeavors have investigated the predictive precision of GEP for estimating ET_0 [42,152–154]. For instance, Shiri et al. [155] employed the GEP model to predict daily ET_0 in Basque, Spain. They used daily meteorological data gathered from four weather stations, including solar radiation, wind speed, relative humidity, and air temperature, for the period—1999–2003—to estimate the PM ET_0. The efficiency of the GEP model was compared to that of models based on the Priestley–Taylor, ANFIS, and Hargreaves–Samani methods. The results demonstrated that the ANFIS and GEP models performed similarly, but the straightforward mathematical formulation in the GEP model positioned it as the favored option for estimating ET_0 in cases where the required meteorological data for the PM equation were not accessible. [156] conducted a study in Burkina Faso, where they employed GEP for estimating decadal ET_0 at four weather stations. They created a set of 16 distinct GEP models by employing various input parameters, including minimum, mean, and maximum air temperatures, sunshine hours, wind speed, relative humidity, and extra-terrestrial radiation. The findings highlight the reliability of GEP for ET_0 estimation because of its ability to present numerical models.

Wang Sheng et al. [157] employed GEP and ANN models to predict daily ET_0 at four stations in China's southwestern Guangxi Province, known as the karst region. They developed 15 ANN and GEP models utilizing variables such as sunshine duration, maximum and minimum air temperatures, relative humidity, extra-terrestrial radiation, and wind speed. The study indicated that in general, ANN models displayed superior performance compared to their corresponding GEP counterparts. Moreover, the scenarios in which all inputs were considered in both ANN and GEP models were recognized as the most precise in the studied areas. Mehdizadeh et al. [116] carried out research in Iran, estimating the mean ET_0 on a monthly basis at 44 stations using GEP, SVM, MARS, and empirical equations. Sixteen different scenarios were proposed using monthly climate inputs, including the minimum, mean, and maximum temperature, wind speed at 2 m height, relative humidity, solar radiation, vapor pressure deficit, net, and extra-terrestrial radiation. The results showed that the AI methods outperformed the empirical equations, with the MARS model outperforming the GEP and SVM models.

Yassin et al. [158] employed GEP and ANN to determine daily ET_0 in the arid climate of Saudi Arabia using climatic data on minimum, mean, and maximum air temperature and relative humidity, solar radiation, wind speed, and reference crop height. They reported that ANN, with R^2 ranging from 67.6% to 99.8% and RMSE varying from 0.20 to 2.95 mm/day, slightly outperformed the GEP model in ET_0 estimation.

Gavili et al. [159] assessed the effectiveness of three soft computing approaches (GEP, ANFIS, and ANN) to model ET_0 using daily meteorological inputs including minimum and maximum temperature, sunshine hours, wind speed, and relative humidity. In addition to AI models, they used several empirical equations, such as Priestley–Taylor, Hargreaves–Samani, Hargreaves, Makkink, and Makkink–Hansen, and found that the AI models, as the leading approach in which the ANN is on top, outperformed the empirical methods.

The research highlights the advantages of SR over traditional data-driven methods, particularly in its ability to automatically generate optimal mathematical structures for nonlinear problems without requiring predefined models. SR, especially through GEP, combines genetic algorithms and programming to produce predictive models. Studies have demonstrated GEP's effectiveness in estimating ET_0 across various regions and climates, such as Spain, Burkina Faso, China, and Saudi Arabia, comparing it to other models like ANFIS, ANN, and empirical equations. While GEP models often performed comparably or better than other methods, artificial intelligence-based models like ANN were sometimes found to outperform GEP in prediction accuracy, depending on the input variables and scenarios. These findings emphasize the reliability and versatility of GEP for ET_0 estimation, especially when meteorological data are limited.

3.4.3. Adaptive Neuro-Fuzzy Inference Systems (ANFIS)

ANFIS represents another significant subset of hybrid models that blend ANN with FG. These models translate the input parameters into output objectives using a combination of membership functions and conditional rules. Despite the ability of ANFIS to forecast changes in target variables with minimal input parameters and reasonable prediction precision, the enduring adaptability of neural networks is curtailed due to the arbitrary structure of the FG rules. In addition, incorporating further preprocessing entails time-consuming calculations in both frequency domains and time. Nonetheless, ANFIS remains a widely used choice for modeling evapotranspiration. Kişi and Öztürk [160] obtained ET_0 values using the PM equation with four years of climate parameter records on a daily basis and compared the results with empirical methods, ANN, and ANFIS, such as Hargreaves and Ritchie. The comparative analysis demonstrated that ANFIS, with inputs of relative humidity, radiation, temperature, and wind speed, outperformed ANN and empirical methods in modeling daily ET_0.

Dogan [161] investigated the performance of the ANFIS model in daily estimation of ET_0. Four ANFIS models were proposed by employing solar radiation, air temperature, wind speed, and relative humidity as the inputs collected at the Morgan Hill weather station in San Francisco, USA. The results demonstrated the successful implementation of the ANFIS model for estimating daily ET_0. In addition, the scenario incorporating all the input parameters exhibited the best performance.

Dou and Yang [16] utilized four different ML methods, namely ANFIS, ELM, ANN, and SVM, for estimating ET_0 in four different regions. The input data included air temperature, soil temperature, relative humidity, and net radiation. The results demonstrated that all models performed well in modeling daily ET, with high values for R^2 (ranged from 0.9398 to 0.9593) and NSE (ranged from 0.8877 to 0.9147) in the forest ecosystem. In the case of ANFIS, fuzzy c-means clustering and the subtractive clustering algorithms generally outperformed the grid-partitioning algorithm. The hybrid ELM methods, namely, complex ELM, online sequential ELM, and self-adaptive evolutionary ELM, among the ELM models, consistently performed better than the original ELM method across the studied sites, and the computational time required for training these models was significantly decreased. It shows that the advanced ANFIS and ELM models are valuable complements to the SVM and ANN methods because of their robustness and flexibility. Additionally, notable differences in modeling performance were observed between the four major ecosystem types. In forested areas, the models consistently demonstrated superior performance, whereas the lowest performance was observed in cropland ecosystems.

Kişi [162] examined the capabilities of ANN, ANFIS, and the SS models in predicting daily evaporation using different climatic inputs, including daily air temperature, pressure, wind speed, humidity, and solar radiation. Both approaches outperformed the conventional SS method with higher accuracy.

Moghaddamnia et al. [163] employed ANFIS and ANN models to simulate pan evaporation in different areas of Iran. The recorded meteorological data included daily wind speed, air temperature, saturation vapor, pan evaporation, relative humidity, and pressure deficit. Both ANN and ANFIS models demonstrated high capability in estimating pan evaporation values compared to empirical formulas, including Hefner, Lincare, and Marciano. However, ANN slightly outperformed the ANFIS.

del Cerro et al. [164] made a comparison between ANFIS and empirical equations, which are based on radiation and temperature, to model the daily ET_0 in southern India. The results indicated that the ANFIS model trained using meteorological data of temperature, solar radiation, relative humidity, and wind speed provided the most accurate estimates, with a root mean square difference (RMSD) of 0.002 and a correlation coefficient (R) of 0.999. Among the empirical models, the Ritchie method was identified as the most suitable for the region, yielding an RMSD of 0.507 and an R-value of 0.851. Furthermore, a sensitivity analysis was implemented to assess the stability of the best models when introducing errors in the inputs, which indicated the superiority of the empirical models over the ANFIS models.

Pour-Ali Baba et al. [165] investigated the performance of the ANFIS and ANN models in estimating daily ET_0 in Gwangju and Haenam, South Korea. Varied scenarios using sunshine hours, air temperature, relative humidity, and wind speed were considered for both ANN and ANFIS models. The results demonstrated that both the ANN and ANFIS models exhibited capability in the estimation of ET_0, surpassing the accuracy of the empirical equations, including the Priestly–Taylor, Hargreaves– Samani, and PM models.

Citakoglu et al. [166] predicted the mean ET_0 on a monthly basis using ANFIS and ANN models in Turkey at 275 stations based on different scenarios utilizing input parameters including relative humidity, air temperature, wind speed, and solar radiation. It was concluded that the scenario with all inputs yielded the highest accuracy. Additionally, the efficiency of the ANN and ANFIS models was investigated against two empirical equations of Hargreaves and Ritchie, with both models demonstrating higher accuracy than the empirical equations. Notably, the ANFIS model demonstrated a marginally superior performance compared to the ANN model.

Petković et al. [167] investigated the performance of the ANFIS model for estimating ET_0 on a monthly basis at twelve stations in Serbia. They defined 35 different patterns using variables such as minimum and maximum air temperatures, relative humidity, actual vapor pressure, sunshine hours, and wind speed. The results demonstrated that it is possible to estimate monthly ET_0 accurately using the ANFIS model by employing meteorological inputs, with sunshine hours being the most influential factor.

Cobaner [168] compared the performance of two types of ANFIS models, that is, ANFIS with subtractive clustering (ANFIS-SC) and grid partitioning (ANFIS-GP), in estimating ET_0 using daily weather data such as air temperature, solar radiation, wind speed, and relative humidity. In addition, these FIS models were investigated against the MLP and empirical models such as the CIMIS Penman, Ritchie, and Hargreaves methods. The results indicated that ANFIS-SC outperformed ANFIS-GP and other models with fewer computations.

Sanikhani et al. [169] assessed the prediction accuracy of the MLP, GRNN, RBFNN, ANFIS-GP, GEP, ANFIS-SC, Hargreaves–Samani, and calibrated Hargreaves–Samani models by employing monthly data from two stations in Turkey (i.e., Antalya and Isparta). The focus of this study was modeling ET_0 based on the temperature. The climate variables used to predict ET included solar radiation, minimum and maximum air temperatures, relative humidity, and wind speed monthly data over a 25-year period. The ranking of the AI models based on their prediction performance, measured by the RMSE and MAE, was as follows: RBFNN, ANFIS-SC, ANFIS-GP, MLP, GEP, GRNN, calibrated Hargreaves–Samani, and Hargreaves–Samani.

Zakhrouf et al. [170] evaluated the capability of ANFIS-SC, MLR, and ANFIS based on the fuzzy C-means clustering (ANFIS-F) in estimating daily ET_0 by climate data inputs of relative humidity, wind speed, insolation duration, and air temperature on a daily scale. The results revealed that the ANFIS-SC model considerably outperformed the MLR and ANFIS-F models.

The dynamic evolving neural fuzzy inference system (DENFIS) is also a hybrid computational model that employs the elements of NN and FG to create a system that can adapt and evolve over time to handle changing data patterns. The DENFIS is designed to provide accurate predictions in dynamic and non-linear environments. Ref. [171] employed two advanced ML algorithms, namely MARS and DENFIS, for estimating ET_0 on a daily scale by employing daily minimum and maximum temperatures. The hyperparameters of the models were optimized by applying two advanced metaheuristic optimization algorithms, bat Algorithm (BA) and WOA, resulting in four models: MARS-BA, MARS-WOA, DENFIS-BA, and DENFIS-WOA. These optimization algorithms have shown high efficiency in fine-tuning ML-based prediction models for problems in hydrological engineering [18,135,172–175]. By comparing the performances of the two AI-based optimization algorithms employed in this study, it was found that the WOA outperformed the BA. Superior performance of the WOA to evade entrapment in local optima may have participated in the enhanced effectiveness of the DENFIS model since its internal parameters were optimized by employing the WOA. In addition, DENFIS has the capacity to enhance its learning by dynamically developing the neuro-fuzzy functions of inputs and outputs and continuously updating the process of learning. This characteristic likely enables DENFIS to more reliably simulate ET in an environment with complex settings.

Malik et al. [176] assessed the suitability of the co-active neuro-fuzzy inference system (CANFIS) for predicting monthly ET_0 at two locations in India using meteorological data on a monthly basis, including wind speed, minimum and maximum air temperatures, solar radiation, and relative humidity. The effectiveness of the CANFIS models was investigated against that of the MLP, MLR, and RBFNN self-organizing map neural network (SOMNN) models. They concluded that CANFIS provided superior results compared with the other AI techniques used in this study.

Malik et al. [177] applied ANN MLR and CANFIS models to predict daily pan evaporation by employing different combinations of meteorological inputs such as relative humidity, minimum and maximum air temperature, sunshine hours, and wind speed in India at different stations. The findings revealed that the ANN models achieved higher success rates than other models.

In another study, Aytek [178] used CANFIS for daily ET_0 prediction using atmospheric parameters, including solar radiation, relative humidity, air temperature, and wind speed in California. The results indicated that the proposed model provided higher accuracy than the conventional Hargreaves and Turc equations.

This study [179] focused on the capability of three AI models, i.e., ANN, ANFIS, and ANN-gray wolf optimization (ANN-GWO), in estimating ET_0 in arid climates. These models were assessed with various input combinations, including minimum and maximum temperatures, ET_0, and wind speed. In data-scarce conditions, AI modes outperformed the Hargreaves equation, with ANFIS representing the most superior performance.

A substantial body of research has demonstrated the effectiveness of ANFIS and other ML models, as discussed above, in estimating ET_0 using meteorological inputs.

Numerous studies have consistently shown that ANFIS models outperform traditional empirical methods, particularly when key climatic variables such as air temperature, wind speed, relative humidity, and solar radiation are incorporated [160,163,167]. In efforts to further enhance prediction accuracy, several studies have integrated ANFIS with clustering techniques and optimization algorithms (e.g., Ye et al. [171]). Advanced configurations, such as ANFIS-SC, have also shown improved performance compared to empirical models like the Hargreaves–Samani method, particularly across varying climatic regions [169,176]. These findings highlight some advantages of ANFIS over traditional empirical approaches in ET estimation.

ANFIS is particularly adept at modeling the intricate and nonlinear interactions among meteorological variables influencing ET, which are often oversimplified in empirical methods. Its data-driven architecture allows it to adaptively learn from patterns in the input data, enabling better handling of uncertainties and environmental variability [180,181]. Moreover, ANFIS supports hybridization with various metaheuristic optimization algorithms (e.g., Snake Optimizer), which further enhances its predictive capacity. This level of adaptability is lacking in empirical models, which are not designed for dynamic optimization based on dataset characteristics [180]. The application of ANFIS across a range of climatic conditions and datasets has also demonstrated its reliability and consistency, in contrast to empirical approaches that often require regional recalibration due to limited flexibility in response to changing climatic inputs [170,181].

A review of the literature allows for a comprehensive understanding of key methodological aspects such as data selection, preprocessing, feature selection, and the treatment of missing data in the context of ET_0 estimation using ANFIS models. ANFIS has demonstrated the capacity to accurately estimate ET_0 even with a limited set of climatic inputs, emphasizing its adaptability in utilizing alternative variables when conventional meteorological data are incomplete or unavailable [182]. Although the evaluation of acceptable error thresholds often relies on subjective criteria, hybrid adaptive neuro-fuzzy inference systems integrated with heuristic optimization algorithms have shown promise in addressing missing data challenges [183,184]. These hybrid approaches illustrate the potential of soft computing methods to deliver reliable ET_0 estimates across diverse environmental conditions and data availability constraints.

ANFIS represents a robust and effective approach for estimating ET, particularly in contexts characterized by incomplete meteorological data or the need to utilize non-traditional input variables. Its strength lies in its adaptability and flexibility; however, the model's performance is highly contingent upon the quality, relevance, and adequacy of the input data. Despite its potential, the successful application of ANFIS often necessitates considerable domain expertise and computational resources to ensure optimal model configuration and performance.

3.4.4. Combination of Different Models

Another type of hybridization is based on combining different algorithms with AI models, such as wavelet coupled with ANN [185,186], wavelet enhanced extreme learning machine [187], and wavelet transform regression model [188], to increase the estimation precision of evapotranspiration. Falamarzi et al. [185] employed wavelet-ANN (WNN) and ANN for predicting evapotranspiration on a daily scale from wind speed and temperature data. The results revealed that both models provided precise predictions for evapotranspiration.

Partal et al. [187] explored the feasibility of the wavelet extreme learning machine (WELM) model that combines discrete wavelet transform with ELM methods, in order to estimate ET_0 on a daily basis in Ankara and Kirikkale, Turkey. Different combinations of climatic inputs such as temperature, wind speed, solar radiation, and relative humidity were employed to train the models. The performance of the WELM model was investigated against the WANN, single ANN, ELM, and OS ELM models. The findings indicated that models incorporating all four inputs had higher precision than models with fewer input variables. Solar radiation was identified as the most influential input for the ET_0 estimation. Among the models tested, the WELM model emerged as the most accurate for estimating the ET_0. In addition, wavelet-based ensemble models, such as WELM and WANN, exhibit superior accuracy compared with single models.

Mehdizadeh et al. [189] employed GEP and MARS for estimating ET_0 on a daily basis in Iran. This study focused on six different climate stations in Iran, namely Urmia, Isfahan, Tabriz, Shiraz, Zahedan, and Yazd, representing semi-arid, arid, and hyper-arid climates, respectively. Daily weather data, including solar radiation, mean air temperature, wind speed at a height of 2 m, relative humidity, and ET_0 data, were utilized from 2000 to 2014 for this analysis. Two types of scenarios were considered as input: one dependent on weather data and the other on lagged ET_0 data. The findings of the local analysis revealed that both the GEP and MARS approaches were capable of estimating daily ET_0 using meteorological inputs and lagged ET_0 data. However, MARS exhibited the best performance under weather-data-based scenarios. Moreover, there were no significant differences in the models' accuracy by employing data-based scenarios and lagged ET_0. In this study, novel hybrid models in lagged ET_0 data-based scenarios were proposed by combining GEP and MARS models with an autoregressive conditional heteroscedasticity (ARCH) time series model. The findings indicated that these hybrid models, namely GEP-ARCH and MARS-ARCH, promoted the effectiveness of modeling the ET_0 compared with using single GEP or MARS models at all studied stations. Furthermore, an external evaluation of the models' performance at stations exhibiting comparable climatic conditions indicated that data from nearby stations could be applied in order to predict ET_0 on a daily basis at the target station. This finding suggests the potential applicability of neighboring station data for ET_0 estimation in similar climatic regions. According to the sensitivity analysis results, mean air temperature was found to be the best predictor for the accurate estimation of daily ET_0, while wind speed was identified to be the least influential predictor.

In order to optimize irrigation demands and improve crop resilience against water stress, Rahman, Hasan [190] combined AI models, including FFANN, CNN, GRU, and LSTM, with climatic data to predict ET, potential ET (PET), and CWSI in Bangladesh. To depict a wide range of possible future climate scenarios, the study used high-resolution climate data retrieved from ACCESS-ESM. Their findings indicated that the FFNN model, enhanced with a deep learning algorithm (FFNN-DLA), provided the most accurate predictions for representing spatio-temporal distributions of variables in different seasons.

Various statistical and ML models have been developed for ET_0 prediction; however, no single algorithm has proven universally optimal across all conditions. To address this, the current study [191] introduced a novel hybrid modeling approach, a combination of linear mixed effects (LME) model, nonlinear autoregressive neural network (NANN), and support vector machine (SVM) models to predict ET_0. NANN and SVM models are appropriate for capturing nonlinear residuals, while LME is capable of detecting linear patterns. Although both hybrid approaches performed well, the LME-SVM hybrid model demonstrated a slight advantage over the LME-NANN model. A summary of the studies conducted for estimating ET by employing hybrid models is presented in Table 5.

Table 5. Studies conducted to model ET using hybrid techniques.

Reference	Models	Hybrid Model Type	Input	Output	Performance Criteria	Best Model(s)
Maroufpoor et al. [142]	ANN-GWO, LSSVR, ANN	Combination of AI models with optimization algorithms	Maximum and minimum temperatures, relative humidity, wind speed, sunshine hours, and precipitation	Monthly ET_0	ScI = 0.077–0.187, R^2 = 0.890–0.981, MAE = 0.279–0.717 mm/day	ANN-GWO
Wu et al. [143]	ELM-FPA, ELM-WOA, ANN, M5Tree, ELM-DE	Combination of AI models with optimization algorithms	Sunshine hours, wind speed, relative humidity, and maximum and minimum temperature	Monthly pan evaporation	R^2 = 0.853–0.958, NSE = 0.766–0.956, RMSE = 0.2584–0.5032 mm/day, MAE = 0.2041–0.3726 mm/day, MAPE = 0.0928–0.1605	Hybrid ELM
Hadadi et al. [144]	ANFIS, ANFIS-SFLA, ANFIS-GWO	Combination of AI models with optimization algorithms	Meteorological data, including wind speed, relative humidity, air temperature (average, minimum, maximum), dew point, and sunshine hours, remotely sensed data, namely net radiation, NDVI, LST, SAVI, and SWDI	Actual ET	RMSE = 11.06, NSE = 0.74, RRMSE = 0.37	ANFIS-SFLA
Tang et al. [6]	SVM, ANN-GA	Combination of AI models with optimization algorithms	Meteorological data, including minimum, maximum, and mean air temperature; minimum, maximum, and mean relative humidity; solar radiation; and wind speed, as well as crop data like LAI and plant height	Daily actual ET	RMSE = 0.215–0.536 mm/day, MAE = 0.182–0.435 mm/day, NSE = 0.868–0.979	ANN-GA
Tikhamarine et al. [145]	SVR-GWO, SVR-GA, SVR-PSO, ANN, Turc, Ritchie, Thornthwaite, Valiantzas	Combination of AI models with optimization algorithms	Relative humidity, maximum and minimum air temperatures, solar radiation, and wind speed	Monthly ET_0	RMSE = 0.0374–0.0776 mm, NSE = 0.9953–0.9995, PCC = 0.9978–0.9998, WI = 0.9988–0.9999	SVR-GWO
Eslamian et al. [138]	ANN, ANN-GA	Combination of AI models with optimization algorithms	Maximum, minimum, and average air temperature, relative humidity, wind speed, and sunshine duration	ET_0	R^2 = 0.9685, NMSE = 0.0675, MAE = 0.4751, MSE = 0.3693	ANN-GA
Tikhamarine et al. [146]	ANN-GWO, ANN-MVO, ANN-PSO, ANN-WOA, ANN-ALO, Valiantzas	Combination of AI models with optimization algorithms	Minimum and maximum air temperatures relative humidity, wind speed, solar radiation	Monthly ET_0	RMSE = 0.0592–0.0808, NSE = 0.9956–0.9972, PCC = 0.9978–0.9986, WI = 0.9989–0.9993	ANN-GWO trained with minimum temperature, maximum temperature, relative humidity, wind speed, and solar radiation
Zounemat-Kermani et al. [147]	NNARX, GEP, and ANFIS models optimized by GA, PSO, CACO, ABC, and FA	Combination of AI models with optimization algorithms	Air temperature, solar radiation, relative humidity, and wind speed	Monthly pan evaporation	R^2 = 0.959, RMSE = 0.631 mm, MAE = 0.447 mm, Kruskal–Wallis test = 0.0014	ANFIS-PSO
Kişi [148]	Priestley–Taylor, ANFIS, Hargreaves–Samani	Combination of AI models with optimization algorithms	Air temperature, relative humidity, wind speed, and solar radiation	Daily ET_0	MAE = 0.603–1.345 mm, RMSE = 0.789–1.528 mm, R^2 = 0.852–0.935	GEP
Ahmadi et al. [5]	SVR-IWD, GEP	Combination of AI models with optimization algorithms	Solar radiation, relative humidity, air temperature, wind speed, and pan evaporation	Monthly pan evaporation	RMSE = 0.210–1.064 mm/day, R = 0.838–0.996, MAE = 0.160–0.901 mm/day	SVR-IWD
Shiri et al. [155]	GEP	Combination of AI models with optimization algorithms	Minimum air temperature, maximum air temperature, mean air temperature, wind speed at 2 m height, sunshine duration, and relative humidity	Monthly ET_0	RMSE = 0.316–1.159 mm/day, R^2 = 0.881–0.967	GEP
Traore and Guven [156]	GEP, PM	Combination of GP and GA	Minimum, maximum, and mean air temperatures, relative humidity, extraterrestrial radiation, wind speed, and sunshine duration	Decadal ET_0	RMSE = 0.108, R^2 = 0.979	GEP

Table 5. *Cont.*

Reference	Models	Hybrid Model Type	Input	Output	Performance Criteria	Best Model(s)
Wang Sheng et al. [157]	GEP, ANN	Combination of GP and GA	Minimum and maximum air temperatures, extraterrestrial radiation, relative humidity, wind speed, and sunshine duration	Daily ET_0	R^2 = 0.799–0.977, RMSE = 0.225–0.754 mm/day	GEP
Mehdizadeh et al. [116]	GEP, MARS, SVM, and empirical equations	Combination of GP and GA	Minimum temperature, maximum temperature, mean temperature, wind speed at 2 m height, relative humidity, solar radiation. vapor pressure deficit, extraterrestrial radiation, and net radiation	Monthly ET_0	RMSE 0.07–1.75, MAE = 0.05–1.47, R^2 = 0.331–0.999	MARS
Yassin et al. [158]	GEP and ANN	Combination of GP and GA	Air temperature (maximum, minimum, and mean), relative humidity (maximum, minimum, and mean), solar radiation, wind speed, and reference crop height	Daily ET_0	R^2 = 64.4–95.5%, RMSE = 1.13–3.1 mm/day	ANN
Gavili et al. [159]	ANN, ANFIS, GEP, Priestley–Taylor, Hargreaves–Samani, Hargreaves, Makkink, and Makkink–Hansen	Combination of GP and GA	Maximum temperature, minimum temperature, relative humidity, wind speed, and sunshine hours	Daily ET_0	R^2 = 0.9814–0.9873, RMSE = 0.3077–0.3451 mm, MAE = 0.2258–0.2483, NSE = 0.9812–0.9872	ANN
Kişi and Öztürk [160]	ANFIS, ANN, Hargreaves, Ritchie	Combination of NN and FG	Temperature, wind speed, relative humidity, and radiation	Daily ET_0	R^2 = 0.811–0.871, MSE = 0.615–0.712 mm^2day^2, MAE = 0.590–0.850 mm/day	ANFIS
Dogan [161]	ANFIS, PM	Combination of NN and FG	Air temperature, solar radiation, relative humidity, and wind speed	Daily ET_0	AARE = 6.4%, R^2 = 0.996, MSE = 0.016	ANFIS
Dou and Yang [16]	ANFIS, ELM, ANN, SVM	Combination of NN and FG	Air temperature, soil temperature, net radiation, and relative humidity	Daily ET_0	R^2 = 0.9398–0.9593, NSE = 0.8817–0.9147, RMSE = 0.3138–0.6807, MAE = 0.2217–0.4821	Advanced ANFIS and ELM models
Kişi [162]	ANFIS, ANN, SS	Combination of NN and FG	Air temperature, solar radiation, humidity, pressure, wind speed	Daily pan evaporation	MSE = 0.09–6.23 mm^2/day^2, MARE = 3.40–17.9%, R^2 = 0.860–0.998	ANFIS
Moghaddamnia et al. [163]	ANFIS, ANN	Combination of NN and FG	Air temperature, wind speed, saturation vapor pressure deficit, relative humidity, and pan evaporation	Daily pan evaporation	RMSE = 1.51–2.47%, R^2 = 0.92–0.97	ANN, ANFIS
del Cerro et al. [164]	ANFIS, PM, Ritchie, Irmak, Blaney–Criddle, Priestley–Taylor, Hargreaves, Baier–Robertson, McGuiness–Bordne, Jensen–Haise, Turc, Modified Turc, Makkink	Combination of NN and FG	Solar radiation, temperature, wind speed, and relative humidity	Daily ET_0	d = 0.6415–0.9999, MAE = 0.0008–0.7746, SEE = 0.0016–0.5534, RMSD = 0.0016–1.0197	ANFIS, Ritchie
Pour-Ali Baba et al. [165]	ANFIS, ANN, Hargreaves–Samani, Priestly–Taylor, PM	Combination of NN and FG	Air temperature, sunshine hours, wind speed, and relative humidity	Daily ET_0	RMSE = 0.474–0.965 mm, MAE = 0.307–0.659 mm, NSE = 0.764–0.959, R^2 = 0.765–0.959	ANFIS, ANN
Citakoglu et al. [166]	ANFIS, ANN, Hargreaves, Ritchie	Combination of NN and FG	Air temperature, relative humidity, solar radiation, and wind speed	Monthly ET_0	RMSE = 0.198–0.486, MAE = 0.345–0.538, R^2 = 0.960–0.985	ANFIS
Petković et al. [167]	ANFIS	Combination of NN and FG	Maximum and minimum air temperatures, maximum and minimum relative humidity, wind speed, actual vapor pressure, and sunshine hours	Monthly ET_0	RMSE = 0.3148 mm/day, R = 0.9801, R^2 = 0.9607	ANFIS trained by minimum temperature, maximum humidity, and actual vapor pressure
Cobaner [168]	ANFIS-GP, ANFIS-SC, MLP, CIMIS Penman, Hargreaves, Ritchie	Combination of NN and FG	Solar radiation, air temperature, relative humidity, and wind speed	Daily ET_0	RMSE = 0.11–0.66 mm, MAE = 0.06–1.12 mm, R^2 = 0.507–0.995	ANFIS-SC

Table 5. *Cont.*

Reference	Models	Hybrid Model Type	Input	Output	Performance Criteria	Best Model(s)
Sanikhani et al. [169]	MLP, GRNN, RBFNN, ANFIS-GP, ANFIS-SC, GEP, Hargreaves–Samani, and calibrated Hargreaves–Samani	Combination of NN and FG	Solar radiation, minimum and maximum air temperatures, relative humidity, and wind speed	Daily ET_0	RMSE = 0.524 mm/day, MAE = 0.383, R^2 = 0.905	RBNN
Zakhrouf et al. [170]	ANFIS-SC, ANFIS-F, MLR	Combination of NN and FG	Relative humidity, insolation duration, wind speed, and air temperature	Daily ET_0	MARE = 16.43%, MSE = 0.316 mm^2, ME = 0.30 mm, NSE = 94.01%	ANFIS-SC
Ye et al. [171]	DENFIS-WOA, DENFIS-BA, MARS-WOA, and MARS-BA	Combination of NN and FG	Maximum and minimum temperature, ET_0	Daily ET_0	NRMSE = 0.35–0.54, SD = 1.00–1.01, R^2 = 0.85–0.94, md = 0.73–0.82, NSE = 0.71–0.87, KGE = 0.70–0.86	DENFIS-WOA
Malik et al. [176]	CANFIS, MLP, RBFNN, SOMNN, MLR	Combination of NN and FG	Minimum and maximum air temperatures, wind speed, relative humidity, and solar radiation	Monthly ET_0	RMSE = 0.0978–0.1394, ScI = 0.0261–0.0475, CE = 0.9846–0.9963, PCC = 0.9942–0.9982, WI = 0.9959–0.9991	CANFIS
Malik et al. [177]	CANFIS, MLR, ANN	Combination of NN and FG	Relative humidity, air temperature (maximum and minimum), wind speed, and sunshine hours	Daily pan evaporation	RMSE = 1.233–1.406 mm/day, CE = 0.729–0.792, R = 0.887–0.921	ANN
Aytek [178]	CANFIS, Hargreaves, Turc	Combination of NN and FG	Solar radiation, relative humidity, air temperature, and wind speed	Daily ET_0	RMSE = 0.13 mm/day, CE = 0.97%, R^2 = 0.98	CANFIS
Falamarzi et al. [185]	ANN, WNN	Combination of different models	Temperature and wind speed	Daily ET_0	R = 0.82–0.90, RMSE = 1.10–1.54 mm/day, APE = 23–30%, NSE = 0.53–0.76	ANN, WNN
Kişi [187]	WELM, ANN, ELM, and OS-ELM	Combination of different models	Temperature, solar radiation, relative humidity, and wind speed	Daily ET_0	RMSE 0.3590–1.6381 mm/day, R^2 = 0.0.2617–0.9518, MAE = 0.3848–1.3774 mm/day, NSE = 0.2334–0.9492	WELM
Mehdizadeh [189]	MARS-ARCH and GEP-ARCH	Combination of different models	Air temperature, solar radiation, relative humidity, wind speed, ET_0	Daily ET_0	RMSE = 0.40–1.85 mm/day, R^2 = 0.200–0.962, MAE = 0.29–1.64 mm/day, MAPE = 11.20–105.50%	MARS-ARCH and GEP-ARCH
Rahman et al. [190]	FFNN, CNN, GRU, and LSTM	Combination of different models	Air temperature, humidity, solar radiation, wind speed, maximum and minimum temperature, and precipitation	Daily ET and PET	R^2 = 0.94	FFNN-DLA
Bidabadi et al. [179]	ANN, ANFIS, and ANN-GWO	Combination of NN with FG and optimization algorithms	Minimum and maximum air temperature, wind speed, and ET_0	Monthly ET_0	RMSE = 0.335–0.434, R^2 = 0.947–0.970, MSE = 0.112–0.212	ANFIS
Habeeb et al. [191]	LME-SVM, LME-NANN	Combination of different models	Wind speed, maximum and minimum temperatures, average temperature, humidity, and ET_0	Monthly ET_0	MAE = 5.0580–6.9283, MAPE = 1.8713–2.9077, RMSE = 6.1883–9.6620	LME-SVM

Notes: Abbreviations: peak percentage deviation (Pdv), average absolute relative error (AARE), root mean square difference (RMSD), mean error (ME), average performance error (APE), scatter index (ScI), modified index of agreement (md), performance indicator (GPI), uncertainty with a 95% confidence level (U95), linear mixed effects (LME), nonlinear autoregressive neural network (NANN), Pearson correlation coefficient (PCC), coefficient of efficiency (CE).

Recent investigations (e.g., [189–191]) have explored hybrid modeling approaches for ET_0 estimation, reporting consistent—albeit marginal—improvements in predictive performance over standalone models. These hybrid methods offer clear advantages over traditional empirical techniques, including enhanced accuracy, better adaptability to complex datasets, improved interpretability, robustness in data-scarce contexts, more refined estimation of ET components, and reduced uncertainty. Frameworks such as RF-XGBoost and XGBoost-LightGBM have reliably outperformed empirical models by achieving superior statistical metrics—higher KGE and R^2, and lower RMSE—thereby supporting more dependable ET estimates, especially under variable meteorological conditions [78].

Hybrid models demonstrate the capacity to process a wide range of meteorological variables—temperature, relative humidity, solar radiation, and wind speed—addressing a major limitation of empirical approaches and enabling application across diverse climates [78,192]. Models that integrate additive regression with decision tree-based algorithms or ensemble strategies (e.g., RF-M5Tree) strike a favorable balance between predictive accuracy and interpretability. Moreover, their compatibility with remote sensing data and physically based simulations enhances their utility in regions with limited ground-based observations. These capabilities position hybrid models as a significant advancement in ET estimation methodologies, enabling accurate prediction across heterogeneous climatic settings [78,81,193].

A key strength of hybrid models is their ability to integrate multi-source datasets, which helps mitigate issues stemming from sparse observational networks. This often involves combining conventional meteorological data (e.g., air temperature, humidity, wind speed) with remote sensing inputs such as satellite-derived vegetation indices and outputs from eddy covariance systems or physically based models [81,194]. Ensuring computational efficiency and predictive reliability typically requires steps like data normalization and hyperparameter tuning, particularly in models such as XGBoost [78]. Feature selection also plays a crucial role in enhancing performance; in semi-arid regions, temperature-based inputs tend to offer more consistency and reliability than solar radiation variables [78,195]. XGBoost-based hybrids have shown strong performance using combinations of minimum and maximum temperature, solar radiation, humidity, and wind speed [78], reinforcing their adaptability to variable climate conditions and data availability constraints.

Research has underscored the utility of models like RF-XGBoost, which combine high accuracy with moderate interpretability [78]. Deep learning-based hybrids, such as deep long short-term memory (D-LSTM) models, have demonstrated promising results even in data-scarce environments, although often at the expense of transparency and interpretability [196]. Despite their strengths, hybrid models face notable limitations. Equifinality—where multiple parameter sets yield similar outputs—can introduce uncertainty and diminish reliability [197]. These challenges highlight the need for continued refinement to ensure model robustness, particularly when applied across differing climatic regimes or regions.

Hybrid models often integrate metaheuristic optimizers (PSO, GA, Grey Wolf, etc.) to tune parameters, yielding notably better generalization [198]. Modern hybrids not only improve raw predictive accuracy but can also reveal dominant physical drivers: e.g., GEP models in multi-climate tests identified mean relative humidity as the key determinant of ET_0 variability [199]. By combining algorithms (ANN, fuzzy systems, tree ensembles, etc.) with evolutionary optimization, current hybrids provide expressive, flexible models that learn complex ET_0 relationships beyond what classic empirical formulas or simple ML can capture.

On the other hand, one of the challenges is model complexity and computational cost. Advanced hybrids (ensembles of deep nets or many tree models) can require heavy training resources. In one comparative study, RF ensembles gave the best ET_0 estimates but suffered much greater computation than simpler models [200]. Overfitting is another concern: data-driven hybrids like GEP or ANFIS can fit training data too closely, especially with limited records. Ensuring the transferability of a hybrid model to new regions or climate regimes remains an open issue. Recent reviews recommend attention to domain-adaptation strategies (e.g., transfer learning or multi-site training) to improve robustness, since many ML-based ET_0 models fail when relocated.

4. Comparison Between Time Series and AI Models

Comparing time-series and AI models, some studies have reported the advantage of time-series models compared to AI. For example, Aghelpour et al. [201] utilized a hybrid model called ANFIS-DE, a combination of ANFIS and DE, as well as seasonal autoregressive integrated moving average (SARIMA), a classic stochastic time series model, for monthly ET forecasting using input weather data, including air temperature, sunshine duration, and relative humidity. The findings demonstrated the ability of the DE algorithm to significantly promote the accuracy of the simple ANFIS model. However, SARIMA was considered the most suitable model compared to ANFIS and ANFIS-DE for ET_0 predictions on a monthly basis in all study areas. Ashrafzadeh et al. [202] used SVM, GMDH, and SARIMA models to predict ET_0 in Guilan Province, Iran. It was found that the models could predict ET_0 for the subsequent two years; nevertheless, the SARIMA models generally showed better performance than the SVM and GMDH models. Aghelpour et al. [203] compared several stochastic models, such as autoregressive, moving average, autoregressive moving average, and autoregressive integrated moving average (ARIMA), with three AI models (SVM, GRNN, and ANFIS) for daily ET_0 prediction. Although both model types demonstrated high capability in the prediction of daily ET_0 rates for humid regions, the time series models were preferred over AI models, similar to the studies conducted by Ashrafzadeh et al. [202,204], which suggested that linear autocorrelation is more robust compared to nonlinear autocorrelation in the ET_0 time series, explaining the preference for time series models in the study conducted by Aghelpour et al. [201] because of their higher accuracy and ease of use. It should be noted that the limitations of the SARIMA model, which only considers time lags of the evaporation variable as inputs, led to the application of machine learning models using ET_0 time lags to enable a meaningful comparison. Additionally, in Tamil Nadu, India, ref. [205] compared AI and stochastic methods and concluded that stochastic models were more suitable for ET_0 prediction. Conversely, in a study conducted in Brazil [206], AI models provided relatively more accurate predictions of ET_0 compared to time series models, which contradicts the findings of the aforementioned studies. They compared the CNN model with the SARIMA stochastic model for daily ET prediction using meteorological data. This discrepancy could be due to the differences in climatic conditions between the study areas. Landeras et al. [207] conducted a comparison between stochastic and AI methods, that is, ARIMA and ANN, for predicting weekly ET in Spain using meteorological data, including air temperature, relative humidity, wind speed, and solar radiation. According to the results, both types of models improved the weekly ET_0 predictions compared to the mean-based models; however, the AI model slightly outperformed the stochastic model.

5. Challenges to AI Models

Despite the increasing application of AI techniques in estimating ET_0, the use of black-box models without incorporating the fundamental physical principles underlying the ET process can lead to significant inaccuracies [208]. A critical challenge in deploying AI models for ET_0 prediction lies in the quality and quantity of the training data. Small or noisy datasets may prevent these models from learning the true underlying patterns. Specifically, limited data can result in overfitting, where the model captures noise rather than meaningful trends, while excessively noisy data may cause underfitting, particularly when simple learning architectures are used [208]. To mitigate these issues, data preprocessing techniques such as normalization, bias reduction, and expansion of training datasets are essential. Another key consideration is the selection of relevant input features, as the choice of climatic parameters directly influences model accuracy. However, despite their importance, these preprocessing and feature selection methods have been underutilized in many AI-based ET modeling studies. For instance, Goyal et al. [106] evaluated multiple combinations of input variables, ranging from a full set including rainfall, temperature extremes, relative humidity and sunshine hours to reduced sets with fewer variables—using the Gamma test (GT). Their results indicated that including all six climatic parameters yielded the highest accuracy in estimating pan evaporation. Similarly, other studies have employed GT to construct optimal input combinations by minimizing mean square error (MSE). One study [54] found that a combination of relative humidity, minimum and maximum temperatures, sunshine hours, and wind speed provided optimal performance for monthly pan evaporation prediction. Moghaddamnia et al. [163] used the Gamma test across various regions in Iran to assess the impact of input selection on evaporation simulation. While wind speed emerged as the most influential factor based on GT results, this contradicted intuitive assessments that identified air temperature as the most relevant. This inconsistency highlights the complexity and spatial variability in the influence of climatic parameters on ET processes. In another study, Ahmadi et al. [5] employed τ-Kendall and entropy-based methods to identify effective inputs from a dataset comprising minimum, mean, and maximum air temperature, wind speed at 2 m height, relative humidity, and sunshine duration. Their findings showed that models using climatic data outperformed those based solely on antecedent ET_0 values. Air temperature and solar radiation were consistently significant, while wind speed also played a notable role, especially in the τ-Kendall method.

The contribution of wind speed to ET_0 estimation has been further validated in prior research. Although wind speed alone may result in poor model performance, its inclusion alongside other climatic inputs substantially improves estimation accuracy [116,166]. Nevertheless, a persistent challenge in AI-based ET modeling is the variability in optimal input combinations across studies. For example, Wen et al. [114] found that SVM performed well using only minimum and maximum temperatures and solar radiation, while Antonopoulos et al. [32] identified temperature, relative humidity, and wind speed as the best predictors for ANN. Such inconsistencies restrict the generalizability of AI models and underscore the need for location-specific and model-specific input selection strategies.

In addition to the limitations mentioned above, emerging challenges in AI-based ET_0 estimation relate to model generalizability across diverse climatic and geographic regions, the dynamic and nonlinear behavior of environmental variables, and the inadequate temporal adaptability inherent in many existing predictive frameworks. AI models that are predominantly trained on static historical datasets often exhibit poor performance when exposed to non-stationary climate conditions, which are increasingly prevalent due to the compounded effects of global climate change, anthropogenic land-use transformations, and shifting agricultural practices. These evolving conditions introduce variability that traditional machine learning models are not inherently designed to accommodate. Furthermore, the reliance on region-specific datasets for training restricts the spatial transferability of models, as they may fail to capture the localized hydrometeorological characteristics of distinct agroecological or hydrological regimes. Without rigorous retraining, recalibration, or the use of domain adaptation strategies, the applicability of these models across broader regions remains limited [209–211].

6. Conclusions and Future Studies

ET is the combined process of water vapor transfer to the atmosphere through evaporation from surfaces and transpiration from plants. Accurate ET estimation is vital for agriculture, water resource management, climate studies, and ecological assessments. The choice of estimation methods depends on data availability, computational resources, and required accuracy. ET estimation approaches are broadly classified into direct and indirect methods. Direct methods, though accurate, are often limited by high costs, maintenance demands, and localized measurements. Indirect methods, particularly the Penman–Monteith (PM) model, are widely used but require extensive input data, which may not be available globally. As an alternative, AI techniques offer a promising solution due to their ability to model the nonlinear nature of ET processes. This study explores the use of AI in ET estimation—specifically ET_0, actual ET, and pan evaporation—using four model types: neuron-based, tree-based, kernel-based, and hybrid models.

ANN models that fall into the neuron-based category have gained popularity for estimating ET because of their capability to capture complex nonlinear relationships among input variables and output estimates. Although this type of method is well-suited for large datasets with intricate patterns and various data distributions, it suffers from overfitting under limited data conditions and is difficult to interpret, especially in the case of deep architecture. Similarly, tree-based models, especially deep trees, can overfit; however, ensemble methods, such as RF and GB, can mitigate overfitting and generalize effectively. Furthermore, tree-based models can handle non-linear relationships and missing datasets and are more interpretable than neural network-based models. Unlike the aforementioned methods, kernel-based methods are able to capture fewer non-linear relationships and are well-suited for small- to medium-sized datasets. Additionally, selecting an appropriate kernel function is challenging because it requires domain expertise. According to previous studies that estimated ET using AI models, ANNs have been extensively employed due to their flexibility and generalization ability. Another type of neuron-based model, that is, DNN, has rarely been used by researchers, while the superior performance of RNNs compared to simple ANNs is demonstrated in studies. This can be because of additional capabilities in their architecture, like the additional memory units to maintain historical data records. Likewise, CNN can extract valuable features for modeling tasks. The challenges related to data, resources, complexity, and interpretability likely contributed to their reduced use in ET modeling compared to traditional ANNs. Among tree-based models, RF and other ensemble methods have shown acceptable performance in ET modeling. In addition, many investigations concluded that hybrid models, including optimized AI models, ANFIS, GEP, and other combined models, outperformed traditional models due to integrating the strengths of various models to enhance predictive performance.

To further advance the applicability and robustness of AI-based ET prediction, future research should target several strategic areas. A particularly promising direction involves the application of transfer learning, where models initially trained in data-rich regions are adapted for data-scarce areas through fine-tuning or domain adaptation techniques. This not only reduces the reliance on extensive local datasets but also significantly enhances spatial generalizability, making it highly suitable for large-scale ET_0 mapping across diverse climatic zones. In addition, there is increasing recognition of the importance of integrating socioeconomic variables, such as land use change, irrigation intensity, and anthropogenic landscape transformations, into AI modeling frameworks. These factors play a pivotal role in shaping ET dynamics, particularly in rapidly transforming agricultural and peri-urban regions, and their inclusion can lead to more context-sensitive and policy-relevant predictions. Furthermore, the development of climate-adaptive AI frameworks offers a compelling path forward. By incorporating real-time climatic indices (e.g., ENSO events, drought severity indicators) into the model architecture, such systems can dynamically recalibrate in response to evolving environmental conditions. This adaptability is essential for enhancing model resilience under extreme climate variability and can provide critical decision support for sustainable water resource planning and agricultural management in an era of intensifying climate change.

In this regard, there are some concluding remarks outlining potential directions for future studies:

- The integration of physical ET processes into AI models to reduce inaccuracies from the current black-box approaches.
- The development of standardized input variable combinations for AI-based ET_0 estimation to address inconsistencies in variable selection across similar climate conditions.
- The exploration and implementation of advanced pre-processing techniques to improve input variable selection and overall model accuracy.
- A combination of AI models with hydrological modeling and remote sensing data for more accurate, real-time, or near-real-time ET estimations at various spatiotemporal scales.
- The establishment of benchmarks for consistent evaluation of ET_0 models across different geographic and climatic contexts by integrating domain-specific knowledge and multi-source data.
- The transfer learning enhances model generalizability by adapting knowledge from data-rich regions to data-scarce areas.
- Socioeconomic factors, including land use changes and irrigation practices, are crucial for context-sensitive ET_0 predictions.

Author Contributions: Conceptualization, A.M. and M.T.; methodology, A.M., and M.T.; software, M.T.; validation, M.T., and H.I.; formal analysis, H.I.; investigation, M.T., M.B., and H.I.; resources, A.M., and M.T.; data curation, M.B.; writing—original draft preparation, M.T., and M.B.; writing—review and editing, A.M., and H.I.; visualization, M.T.; supervision, A.M.; project administration, A.M.; funding acquisition, A.M. All authors have read and agreed to the published version of the manuscript.

Funding: The research was supported by the Natural Sciences and Engineering Research Council of Canada (NSERC).

Conflicts of Interest: The authors declare no conflict of interest.

References

1. Taheri, M.; Mohammadian, A.; Ganji, F.; Bigdeli, M.; Nasseri, M. Energy-based approaches in estimating actual evapotranspiration focusing on land surface temperature: A review of methods, concepts, and challenges. *Energies* **2022**, *15*, 1264. [CrossRef]
2. Omar, P.J.; Gupta, P.; Wang, Q. Exploring the rise of AI-based smart water management systems. *AQUA Water Infrastruct. Ecosyst. Soc.* **2023**, *72*, iii–iv. [CrossRef]
3. Abdullah, S.S.; Malek, M.A. Empirical Penman-Monteith equation and artificial intelligence techniques in predicting reference evapotranspiration: A review. *Int. J. Water* **2016**, *10*, 55–66. [CrossRef]
4. Granata, F. Evapotranspiration evaluation models based on machine learning algorithms—A comparative study. *Agric. Water Manag.* **2019**, *217*, 303–315. [CrossRef]
5. Ahmadi, F.; Mehdizadeh, S.; Mohammadi, B.; Pham, Q.B.; Doan, T.N.C.; Vo, N.D. Application of an artificial intelligence technique enhanced with intelligent water drops for monthly reference evapotranspiration estimation. *Agric. Water Manag.* **2021**, *244*, 106622. [CrossRef]
6. Tang, D.; Feng, Y.; Gong, D.; Hao, W.; Cui, N. Evaluation of artificial intelligence models for actual crop evapotranspiration modeling in mulched and non-mulched maize croplands. *Comput. Electron. Agric.* **2018**, *152*, 375–384. [CrossRef]
7. Allen, R.G.; Pereira, L.S.; Raes, D.; Smith, M. *Crop Evapotranspiration-Guidelines for Computing Crop Water Requirements-FAO Irrigation and Drainage Paper 56*; FAO: Rome, Italy, 1998; pp. 147–151.
8. Allen, R.G. Using the FAO-56 dual crop coefficient method over an irrigated region as part of an evapotranspiration intercomparison study. *J. Hydrol.* **2000**, *229*, 27–41. [CrossRef]
9. Ding, R.; Kang, S.; Zhang, Y.; Hao, X.; Tong, L.; Du, T. Partitioning evapotranspiration into soil evaporation and transpiration using a modified dual crop coefficient model in irrigated maize field with ground-mulching. *Agric. Water Manag.* **2013**, *127*, 85–96. [CrossRef]
10. Ferreira, L.B.; da Cunha, F.F.; de Oliveira, R.A.; Fernandes Filho, E.I. Estimation of reference evapotranspiration in Brazil with limited meteorological data using ANN and SVM–A new approach. *J. Hydrol.* **2019**, *572*, 556–570. [CrossRef]

11. Huang, G.; Wu, L.; Ma, X.; Zhang, W.; Fan, J.; Yu, X.; Zeng, W.; Zhou, H. Evaluation of CatBoost method for prediction of reference evapotranspiration in humid regions. *J. Hydrol.* **2019**, *574*, 1029–1041. [CrossRef]
12. Qutbudin, I.; Shiru, M.S.; Sharafati, A.; Ahmed, K.; Al-Ansari, N.; Yaseen, Z.M.; Shahid, S.; Wang, X. Seasonal drought pattern changes due to climate variability: Case study in Afghanistan. *Water* **2019**, *11*, 1096. [CrossRef]
13. dos Santos Farias, D.B.; Althoff, D.; Rodrigues, L.N.; Filgueiras, R. Performance evaluation of numerical and machine learning methods in estimating reference evapotranspiration in a Brazilian agricultural frontier. *Theor. Appl. Climatol.* **2020**, *142*, 1481–1492. [CrossRef]
14. Yang, Y.; Sun, H.; Xue, J.; Liu, Y.; Liu, L.; Yan, D.; Gui, D. Estimating evapotranspiration by coupling Bayesian model averaging methods with machine learning algorithms. *Environ. Monit. Assess.* **2021**, *193*, 156. [CrossRef] [PubMed]
15. Kumar, M.; Raghuwanshi, N.; Singh, R.; Wallender, W.; Pruitt, W. Estimating evapotranspiration using artificial neural network. *J. Irrig. Drain. Eng.* **2002**, *128*, 224–233. [CrossRef]
16. Dou, X.; Yang, Y. Evapotranspiration estimation using four different machine learning approaches in different terrestrial ecosystems. *Comput. Electron. Agric.* **2018**, *148*, 95–106. [CrossRef]
17. Ehteram, M.; Singh, V.P.; Ferdowsi, A.; Mousavi, S.F.; Farzin, S.; Karami, H.; Mohd, N.S.; Afan, H.A.; Lai, S.H.; Kisi, O.; et al. An improved model based on the support vector machine and cuckoo algorithm for simulating reference evapotranspiration. *PLoS ONE* **2019**, *14*, e0217499. [CrossRef]
18. Chia, M.Y.; Huang, Y.F.; Koo, C.H. Support vector machine enhanced empirical reference evapotranspiration estimation with limited meteorological parameters. *Comput. Electron. Agric.* **2020**, *175*, 105577. [CrossRef]
19. Penman, H.L. Natural evaporation from open water, bare soil and grass. *Proc. R. Soc. London. Ser. A Math. Phys. Sci.* **1948**, *193*, 120–145.
20. Monteith, J.L. Evaporation and environment. The stage and movement of water in living organisms. In *19th Symp. Society for Experimental Biology*; Cambridge University Press: Cambridge, UK, 1965.
21. Wróblewski, P.; Drożdż, W.; Lewicki, W.; Miązek, P. Methodology for assessing the impact of aperiodic phenomena on the energy balance of propulsion engines in vehicle electromobility systems for given areas. *Energies* **2021**, *14*, 2314. [CrossRef]
22. Luo, X.; Chen, J.M.; Liu, J.; Black, T.A.; Croft, H.; Staebler, R.; He, L.; Arain, M.A.; Chen, B.; Mo, G.; et al. Comparison of big-leaf, two-big-leaf, and two-leaf upscaling schemes for evapotranspiration estimation using coupled carbon-water modeling. *J. Geophys. Res. Biogeosci.* **2018**, *123*, 207–225. [CrossRef]
23. Alves, I.; Perrier, A.; Pereira, L. Aerodynamic and surface resistances of complete cover crops: How good is the "big leaf"? In *Transactions of the ASAE*; ASABE: St Joseph, MI, USA, 1998; Volume 41, pp. 345–351.
24. Shuttleworth, W.J.; Wallace, J. Evaporation from sparse crops-an energy combination theory. *Q. J. R. Meteorol. Soc.* **1985**, *111*, 839–855. [CrossRef]
25. Zhao, W.; Li, A. A review on land surface processes modelling over complex terrain. *Adv. Meteorol.* **2015**, *2015*, 1–17. [CrossRef]
26. Skhiri, A.; Ferhi, A.; Bousselmi, A.; Khlifi, S.; Mattar, M.A. Artificial neural network for forecasting reference evapotranspiration in semi-arid bioclimatic regions. *Water* **2024**, *16*, 602. [CrossRef]
27. Kişi, Ö. Generalized regression neural networks for evapotranspiration modelling. *Hydrol. Sci. J.* **2006**, *51*, 1092–1105. [CrossRef]
28. Wang, X.; Li, W.; Li, Q. A new embedded estimation model for soil temperature prediction. *Sci. Program.* **2021**, *2021*, 1–16. [CrossRef]
29. Kisi, O. The potential of different ANN techniques in evapotranspiration modelling. *Hydrol. Process.* **2008**, *22*, 2449–2460. [CrossRef]
30. Landeras, G.; Ortiz-Barredo, A.; López, J.J. Comparison of artificial neural network models and empirical and semi-empirical equations for daily reference evapotranspiration estimation in the Basque Country (Northern Spain). *Agric. Water Manag.* **2008**, *95*, 553–565. [CrossRef]
31. El-Baroudy, I.; Elshorbagy, A.; Carey, S.; Giustolisi, O.; Savic, D. Comparison of three data-driven techniques in modelling the evapotranspiration process. *J. Hydroinform.* **2010**, *12*, 365–379. [CrossRef]
32. Antonopoulos, V.Z.; Antonopoulos, A.V. Daily reference evapotranspiration estimates by artificial neural networks technique and empirical equations using limited input climate variables. *Comput. Electron. Agric.* **2017**, *132*, 86–96. [CrossRef]
33. Kisi, O. Fuzzy genetic approach for modeling reference evapotranspiration. *J. Irrig. Drain. Eng.* **2010**, *136*, 175–183. [CrossRef]
34. Traore, S.; Wang, Y.-M.; Kerh, T. Artificial neural network for modeling reference evapotranspiration complex process in Sudano-Sahelian zone. *Agric. Water Manag.* **2010**, *97*, 707–714. [CrossRef]
35. Nema, M.K.; Khare, D.; Chandniha, S.K. Application of artificial intelligence to estimate the reference evapotranspiration in sub-humid Doon valley. *Appl. Water Sci.* **2017**, *7*, 3903–3910. [CrossRef]
36. Sudheer, K.; Gosain, A.; Ramasastri, K. Estimating actual evapotranspiration from limited climatic data using neural computing technique. *J. Irrig. Drain. Eng.* **2003**, *129*, 214–218. [CrossRef]
37. Bruton, J.; McClendon, R.; Hoogenboom, G. Estimating daily pan evaporation with artificial neural networks. In *Transactions of the ASAE*; ASABE: St Joseph, MI, USA, 2000; Volume 43, pp. 491–496.

38. Sudheer, K.; Gosain, A.; Mohana Rangan, D.; Saheb, S. Modelling evaporation using an artificial neural network algorithm. *Hydrol. Process.* **2002**, *16*, 3189–3202. [CrossRef]
39. Keskin, M.E.; Terzi, Ö. Artificial neural network models of daily pan evaporation. *J. Hydrol. Eng.* **2006**, *11*, 65–70. [CrossRef]
40. Aytek, A.; Guven, A.; Yuce, M.I.; Aksoy, H. An explicit neural network formulation for evapotranspiration. *Hydrol. Sci. J.* **2008**, *53*, 893–904. [CrossRef]
41. Rahimikhoob, A. Estimation of evapotranspiration based on only air temperature data using artificial neural networks for a subtropical climate in Iran. *Theor. Appl. Climatol.* **2010**, *101*, 83–91. [CrossRef]
42. Parasuraman, K.; Elshorbagy, A.; Carey, S.K. Modelling the dynamics of the evapotranspiration process using genetic programming. *Hydrol. Sci. J.* **2007**, *52*, 563–578. [CrossRef]
43. Tezel, G.; Buyukyildiz, M. Monthly evaporation forecasting using artificial neural networks and support vector machines. *Theor. Appl. Climatol.* **2016**, *124*, 69–80. [CrossRef]
44. Aghelpour, P.; Bahrami-Pichaghchi, H.; Karimpour, F. Estimating daily rice crop evapotranspiration in limited climatic data and utilizing the soft computing algorithms MLP, RBF, GRNN, and GMDH. *Complexity* **2022**, *2022*, 4534822. [CrossRef]
45. Kişi, Ö. Modeling monthly evaporation using two different neural computing techniques. *Irrig. Sci.* **2009**, *27*, 417–430. [CrossRef]
46. Traore, S.; Luo, Y.; Fipps, G. Deployment of artificial neural network for short-term forecasting of evapotranspiration using public weather forecast restricted messages. *Agric. Water Manag.* **2016**, *163*, 363–379. [CrossRef]
47. Trajkovic, S. Temperature-based approaches for estimating reference evapotranspiration. *J. Irrig. Drain. Eng.* **2005**, *131*, 316–323. [CrossRef]
48. Tabari, H.; Hosseinzadeh Talaee, P. Multilayer perceptron for reference evapotranspiration estimation in a semiarid region. *Neural Comput. Appl.* **2013**, *23*, 341–348. [CrossRef]
49. Zhu, B.; Feng, Y.; Gong, D.; Jiang, S.; Zhao, L.; Cui, N. Hybrid particle swarm optimization with extreme learning machine for daily reference evapotranspiration prediction from limited climatic data. *Comput. Electron. Agric.* **2020**, *173*, 105430. [CrossRef]
50. Feng, Y.; Jia, Y.; Cui, N.; Zhao, L.; Li, C.; Gong, D. Calibration of Hargreaves model for reference evapotranspiration estimation in Sichuan basin of southwest China. *Agric. Water Manag.* **2017**, *181*, 1–9. [CrossRef]
51. Patil, A.P.; Deka, P.C. An extreme learning machine approach for modeling evapotranspiration using extrinsic inputs. *Comput. Electron. Agric.* **2016**, *121*, 385–392. [CrossRef]
52. Heddam, S.; Kisi, O.; Sebbar, A.; Houichi, L.; Djemili, L. New formulation for predicting daily reference evapotranspiration (et 0) in the mediterranean region of Algeria country: Optimally pruned extreme learning machine (opelm) versus online sequential extreme learning machine (oselm). In *Water Resources in Algeria-Part I: Assessment of Surface and Groundwater Resources*; Springer Nature: Berlin/Heidelberg, Germany, 2020; pp. 181–199.
53. Malik, A.; Kumar, A.; Kisi, O. Daily pan evaporation estimation using heuristic methods with gamma test. *J. Irrig. Drain. Eng.* **2018**, *144*, 04018023. [CrossRef]
54. Malik, A.; Kumar, A.; Kim, S.; Kashani, M.H.; Karimi, V.; Sharafati, A.; Ghorbani, M.A.; Al-Ansari, N.; Salih, S.Q.; Yaseen, Z.M.; et al. Modeling monthly pan evaporation process over the Indian central Himalayas: Application of multiple learning artificial intelligence model. *Eng. Appl. Comput. Fluid Mech.* **2020**, *14*, 323–338. [CrossRef]
55. Makwana, J.J.; Tiwari, M.K.; Deora, B.S. Development and comparison of artificial intelligence models for estimating daily reference evapotranspiration from limited input variables. *Smart Agric. Technol.* **2023**, *3*, 100115. [CrossRef]
56. Güzel, H.; Üneş, F.; Erginer, M.; Kaya, Y.Z.; Taşar, B.; Erginer, İ.; Demirci, M. *A Comparative Study on Daily Evapotranspiration Estimation by Using Various Artificial Intelligence Techniques and Traditional Regression Calculations*; AIMS Press: Springfield, MO, USA, 2023.
57. Abdel-Fattah, M.K.; Kotb Abd-Elmabod, S.; Zhang, Z.; Merwad, A.R.M. Exploring the Applicability of Regression models and Artificial neural networks for calculating reference evapotranspiration in arid regions. *Sustainability* **2023**, *15*, 15494. [CrossRef]
58. Novotná, B.; Cviklovič, V.; Chvíla, B.; Minárik, M. Application of Developing Artificial Intelligence (AI) Techniques to Model Pan Evaporation Trends in Slovak River Sub-Basins. *Sustainability* **2025**, *17*, 526. [CrossRef]
59. Faloye, O.T.; Ajayi, A.E.; Babalola, T.; Omotehinse, A.O.; Adeyeri, O.E.; Adabembe, B.A.; Ogunrinde, A.T.; Okunola, A.; Fashina, A. Modelling crop evapotranspiration and water use efficiency of maize using artificial neural network and linear regression models in biochar and inorganic fertilizer-amended soil under varying water applications. *Water* **2023**, *15*, 2294. [CrossRef]
60. Eludire, O.O.; Faloye, O.T.; Alatise, M.; Ajayi, A.E.; Oguntunde, P.; Badmus, T.; Fashina, A.; Adeyeri, O.E.; Olorunfemi, I.E.; Ogunrinde, A.T. Evaluation of Evapotranspiration Prediction for Cassava Crop Using Artificial Neural Network Models and Empirical Models over Cross River Basin in Nigeria. *Water* **2025**, *17*, 87. [CrossRef]
61. Diamantopoulou, M.; Georgiou, P.; Papamichail, D. Performance evaluation of artificial neural networks in estimating reference evapotranspiration with minimal meteorological data. *Glob. NEST J.* **2011**, *13*, 18–27.
62. Genaidy, M. Estimating of evapotranspiration using artificial neural network. *Misr J. Agric. Eng.* **2020**, *37*, 81–94. [CrossRef]
63. Kartal, V. Prediction of monthly evapotranspiration by artificial neural network model development with Levenberg–Marquardt method in Elazig, Turkey. *Environ. Sci. Pollut. Res.* **2024**, *31*, 20953–20969. [CrossRef]

64. Dimitriadou, S.; Nikolakopoulos, K.G. Artificial neural networks for the prediction of the reference evapotranspiration of the Peloponnese Peninsula, Greece. *Water* **2022**, *14*, 2027. [CrossRef]
65. Sattari, M.T.; Apaydin, H.; Band, S.S.; Mosavi, A.; Prasad, R. Comparative analysis of kernel-based versus ANN and deep learning methods in monthly reference evapotranspiration estimation. *Hydrol. Earth Syst. Sci.* **2021**, *25*, 603–618. [CrossRef]
66. Shi, H.; Cai, X. Extrapolability improvement of machine learning-based evapotranspiration models via domain-adversarial neural networks. *Environ. Model. Softw.* **2025**, *187*, 106383. [CrossRef]
67. Hettiarachchi, P.; Hall, M.; Minns, A. The extrapolation of artificial neural networks for the modelling of rainfall—Runoff relationships. *J. Hydroinform.* **2005**, *7*, 291–296. [CrossRef]
68. Lipton, Z.C.; Berkowitz, J.; Elkan, C. A critical review of recurrent neural networks for sequence learning. *arXiv* **2015**, arXiv:1506.00019.
69. Hochreiter, S.; Schmidhuber, J. Long short-term memory. *Neural Comput.* **1997**, *9*, 1735–1780. [CrossRef]
70. Afzaal, H.; Farooque, A.A.; Abbas, F.; Acharya, B.; Esau, T. Computation of evapotranspiration with artificial intelligence for precision water resource management. *Appl. Sci.* **2020**, *10*, 1621. [CrossRef]
71. Saggi, M.K.; Jain, S. Reference evapotranspiration estimation and modeling of the Punjab Northern India using deep learning. *Comput. Electron. Agric.* **2019**, *156*, 387–398. [CrossRef]
72. Farooque, A.A.; Afzaal, H.; Abbas, F.; Bos, M.; Maqsood, J.; Wang, X.; Hussain, N. Forecasting daily evapotranspiration using artificial neural networks for sustainable irrigation scheduling. *Irrig. Sci.* **2022**, *40*, 55–69. [CrossRef]
73. Fang, S.L.; Lin, Y.S.; Chang, S.C.; Chang, Y.L.; Tsai, B.Y.; Kuo, B.J. Using artificial intelligence algorithms to estimate and short-term forecast the daily reference evapotranspiration with limited meteorological variables. *Agriculture* **2024**, *14*, 510. [CrossRef]
74. Baishnab, U.; Hossen Sajib, M.S.; Islam, A.; Akter, S.; Hasan, A.; Roy, T.; Das, P. Deep learning approaches for short-crop reference evapotranspiration estimation: A case study in Southeastern Australia. *Earth Sci. Inform.* **2025**, *18*, 4. [CrossRef]
75. Ba-ichou, A.; Zegoumou, A.; Benhlima, S.; Bekr, M.A. Daily reference evapotranspiration estimation utilizing deep learning models with varied combinations of weather data. In Proceedings of the E3S Web of Conferences, 9–20 May 2023; EDP Sciences: Kenitra, Morocco; Volume 492, p. 01002.
76. Treder, W.; Klamkowski, K.; Wójcik, K.; Tryngiel-Gać, A. Evapotranspiration estimation using machine learning methods. *J. Hortic. Res.* **2023**, *31*, 35–44. [CrossRef]
77. Ravindran, S.M.; Bhaskaran, S.K.M.; Ambat, S.K.N. A deep neural network architecture to model reference evapotranspiration using a single input meteorological parameter. *Environ. Process.* **2021**, *8*, 1567–1599. [CrossRef]
78. Acharki, S.; Raza, A.; Vishwakarma, D.K.; Amharref, M.; Bernoussi, A.S.; Singh, S.K.; Al-Ansari, N.; Dewidar, A.Z.; Al-Othman, A.A.; Mattar, M.A. Comparative assessment of empirical and hybrid machine learning models for estimating daily reference evapotranspiration in sub-humid and semi-arid climates. *Sci. Rep.* **2025**, *15*, 2542. [CrossRef] [PubMed]
79. Kraft, B.; Nelson, J.A.; Walther, S.; Gans, F.; Weber, U.; Duveiller, G.; Reichstein, M.; Zhang, W.; Rußwurm, M.; Tuia, D.; et al. On the added value of sequential deep learning for upscaling evapotranspiration. *EGUsphere* **2024**, *2024*, 1–30.
80. Caminha, H.D.; da Silva, T.L.C.; da Rocha, A.R.; Lima, S.C.R.V. Estimating Reference Evapotranspiration using Data Mining Prediction Models and Feature Selection. In Proceedings of the ICEIS (1), Porto, Portugal, 26–29 April 2017; pp. 272–279.
81. Chen, J.; Dafflon, B.; Tran, A.P.; Falco, N.; Hubbard, S.S. A deep learning hybrid predictive modeling (HPM) approach for estimating evapotranspiration and ecosystem respiration. *Hydrol. Earth Syst. Sci.* **2021**, *25*, 6041–6066. [CrossRef]
82. Farhangmehr, V.; Imanian, H.; Mohammadian, A.; Cobo, J.H.; Shirkhani, H.; Payeur, P. A spatiotemporal CNN-LSTM deep learning model for predicting soil temperature in diverse large-scale regional climates. *Sci. Total Environ.* **2025**, *968*, 178901. [CrossRef]
83. Im, M.S.; Dasari, V.R. Computational complexity reduction of deep neural networks. *arXiv* **2022**, arXiv:2207.14620.
84. Karimi, S.; Shiri, J.; Marti, P. Supplanting missing climatic inputs in classical and random forest models for estimating reference evapotranspiration in humid coastal areas of Iran. *Comput. Electron. Agric.* **2020**, *176*, 105633. [CrossRef]
85. Shiri, J. Improving the performance of the mass transfer-based reference evapotranspiration estimation approaches through a coupled wavelet-random forest methodology. *J. Hydrol.* **2018**, *561*, 737–750. [CrossRef]
86. Wang, S.; Lian, J.; Peng, Y.; Hu, B.; Chen, H. Generalized reference evapotranspiration models with limited climatic data based on random forest and gene expression programming in Guangxi, China. *Agric. Water Manag.* **2019**, *221*, 220–230. [CrossRef]
87. Gonzalo-Martin, C.; Lillo-Saavedra, M.; Garcia-Pedrero, A.; Lagos, O.; Menasalvas, E. Daily evapotranspiration mapping using regression random forest models. *IEEE J. Sel. Top. Appl. Earth Obs. Remote Sens.* **2017**, *10*, 5359–5368. [CrossRef]
88. Shi, L.; Feng, P.; Wang, B.; Li Liu, D.; Cleverly, J.; Fang, Q.; Yu, Q. Projecting potential evapotranspiration change and quantifying its uncertainty under future climate scenarios: A case study in southeastern Australia. *J. Hydrol.* **2020**, *584*, 124756. [CrossRef]
89. Vulova, S.; Meier, F.; Rocha, A.D.; Quanz, J.; Nouri, H.; Kleinschmit, B. Modeling urban evapotranspiration using remote sensing, flux footprints, and artificial intelligence. *Sci. Total Environ.* **2021**, *786*, 147293. [CrossRef] [PubMed]
90. Feng, Y.; Cui, N.; Gong, D.; Zhang, Q.; Zhao, L. Evaluation of random forests and generalized regression neural networks for daily reference evapotranspiration modelling. *Agric. Water Manag.* **2017**, *193*, 163–173. [CrossRef]

91. Al-Mukhtar, M. Modeling the monthly pan evaporation rates using artificial intelligence methods: A case study in Iraq. *Environ. Earth Sci.* **2021**, *80*, 39. [CrossRef]
92. Hameed, M.M.; AlOmar, M.K.; Mohd Razali, S.F.; Kareem Khalaf, M.A.; Baniya, W.J.; Sharafati, A.; AlSaadi, M.A. Application of artificial intelligence models for evapotranspiration prediction along the southern coast of Turkey. *Complexity* **2021**, *2021*, 1–20. [CrossRef]
93. Fan, J.; Ma, X.; Wu, L.; Zhang, F.; Yu, X.; Zeng, W. Light Gradient Boosting Machine: An efficient soft computing model for estimating daily reference evapotranspiration with local and external meteorological data. *Agric. Water Manag.* **2019**, *225*, 105758. [CrossRef]
94. Wu, L.; Fan, J. Comparison of neuron-based, kernel-based, tree-based and curve-based machine learning models for predicting daily reference evapotranspiration. *PLoS ONE* **2019**, *14*, e0217520. [CrossRef]
95. Keshtegar, B.; Piri, J.; Kisi, O. A nonlinear mathematical modeling of daily pan evaporation based on conjugate gradient method. *Comput. Electron. Agric.* **2016**, *127*, 120–130. [CrossRef]
96. Wang, L.; Kisi, O.; Zounemat-Kermani, M.; Li, H. Pan evaporation modeling using six different heuristic computing methods in different climates of China. *J. Hydrol.* **2017**, *544*, 407–427. [CrossRef]
97. Katimbo, A.; Rudnick, D.R.; Zhang, J.; Ge, Y.; DeJonge, K.C.; Franz, T.E.; Shi, Y.; Liang, W.-z.; Qiao, X.; Heeren, D.M.; et al. Evaluation of artificial intelligence algorithms with sensor data assimilation in estimating crop evapotranspiration and crop water stress index for irrigation water management. *Smart Agric. Technol.* **2023**, *4*, 100176. [CrossRef]
98. Sun, D.; Zhang, H.; Qi, Y.; Ren, Y.; Zhang, Z.; Li, X.; Lv, Y.; Cheng, M. A Comparative Analysis of Different Algorithms for Estimating Evapotranspiration with Limited Observation Variables: A Case Study in Beijing, China. *Remote Sens.* **2025**, *17*, 636. [CrossRef]
99. Garofalo, S.P.; Ardito, F.; Sanitate, N.; De Carolis, G.; Ruggieri, S.; Giannico, V.; Rana, G.; Ferrara, R.M. Robustness of Actual Evapotranspiration Predicted by Random Forest Model Integrating Remote Sensing and Meteorological Information: Case of Watermelon (Citrullus lanatus,(*Thunb*) Matsum. & Nakai, 1916). *Water* **2025**, *17*, 323. [CrossRef]
100. Abed, M.; Imteaz, M.A.; Ahmed, A.N.; Huang, Y.F. Modelling monthly pan evaporation utilising Random Forest and deep learning algorithms. *Sci. Rep.* **2022**, *12*, 13132. [CrossRef] [PubMed]
101. Wu, M.; Feng, Q.; Wen, X.; Deo, R.C.; Yin, Z.; Yang, L.; Sheng, D. Random forest predictive model development with uncertainty analysis capability for the estimation of evapotranspiration in an arid oasis region. *Hydrol. Res.* **2020**, *51*, 648–665. [CrossRef]
102. Rai, P.; Kumar, P.; Al-Ansari, N.; Malik, A. Evaluation of machine learning versus empirical models for monthly reference evapotranspiration estimation in Uttar Pradesh and Uttarakhand States, India. *Sustainability* **2022**, *14*, 5771. [CrossRef]
103. Ayaz, A.; Rajesh, M.; Singh, S.K.; Rehana, S. Estimation of reference evapotranspiration using machine learning models with limited data. *AIMS Geosci.* **2021**, *7*, 268–290. [CrossRef]
104. Amani, S.; Shafizadeh-Moghadam, H. A review of machine learning models and influential factors for estimating evapotranspiration using remote sensing and ground-based data. *Agric. Water Manag.* **2023**, *284*, 108324. [CrossRef]
105. Shrestha, N.; Shukla, S. Support vector machine based modeling of evapotranspiration using hydro-climatic variables in a sub-tropical environment. *Agric. For. Meteorol.* **2015**, *200*, 172–184. [CrossRef]
106. Kisi, O. Modeling reference evapotranspiration using three different heuristic regression approaches. *Agric. Water Manag.* **2016**, *169*, 162–172. [CrossRef]
107. Goyal, M.K.; Bharti, B.; Quilty, J.; Adamowski, J.; Pandey, A. Modeling of daily pan evaporation in sub tropical climates using ANN, LS-SVR, Fuzzy Logic, and ANFIS. *Expert Syst. Appl.* **2014**, *41*, 5267–5276. [CrossRef]
108. Cherkassky, V.; Krasnopolsky, V.; Solomatine, D.P.; Valdes, J. Computational intelligence in earth sciences and environmental applications: Issues and challenges. *Neural Netw.* **2006**, *19*, 113–121. [CrossRef]
109. Elshorbagy ACorzo, G.; Srinivasulu, S.; Solomatine, D.P. Experimental Investigation of the Predictive Capabilities of Data Driven Modeling Techniques in Hydrology-Part 1: Concepts and methodology. *Hydrol. Earth Syst. Sci.* **2010**, *14*, 1931–1941. [CrossRef]
110. Jain, S.; Nayak, P.; Sudheer, K. Models for estimating evapotranspiration using artificial neural networks, and their physical interpretation. *Hydrol. Process. Int. J.* **2008**, *22*, 2225–2234. [CrossRef]
111. Abdullah, S.S.; Malek, M.A.; Abdullah, N.S.; Kisi, O.; Yap, K.S. Extreme learning machines: A new approach for prediction of reference evapotranspiration. *J. Hydrol.* **2015**, *527*, 184–195. [CrossRef]
112. Güçlü, Y.S.; Subyani, A.M.; Şen, Z. Regional fuzzy chain model for evapotranspiration estimation. *J. Hydrol.* **2017**, *544*, 233–241. [CrossRef]
113. Yu, H.; Wen, X.; Li, B.; Yang, Z.; Wu, M.; Ma, Y. Uncertainty analysis of artificial intelligence modeling daily reference evapotranspiration in the northwest end of China. *Comput. Electron. Agric.* **2020**, *176*, 105653. [CrossRef]
114. Tabari, H.; Kisi, O.; Ezani, A.; Talaee, P.H. SVM, ANFIS, regression and climate based models for reference evapotranspiration modeling using limited climatic data in a semi-arid highland environment. *J. Hydrol.* **2012**, *444*, 78–89. [CrossRef]
115. Wen, X.; Si, J.; He, Z.; Wu, J.; Shao, H.; Yu, H. Support-vector-machine-based models for modeling daily reference evapotranspiration with limited climatic data in extreme arid regions. *Water Resour. Manag.* **2015**, *29*, 3195–3209. [CrossRef]

116. Sobh, M.T.; Nashwan, M.S.; Amer, N. High-resolution reference evapotranspiration for arid Egypt: Comparative analysis and evaluation of empirical and artificial intelligence models. *Int. J. Climatol.* **2022**, *42*, 10217–10237. [CrossRef]
117. Mehdizadeh, S.; Behmanesh, J.; Khalili, K. Using MARS, SVM, GEP and empirical equations for estimation of monthly mean reference evapotranspiration. *Comput. Electron. Agric.* **2017**, *139*, 103–114. [CrossRef]
118. Adnan, R.M.; Heddam, S.; Yaseen, Z.M.; Shahid, S.; Kisi, O.; Li, B. Prediction of potential evapotranspiration using temperature-based heuristic approaches. *Sustainability* **2020**, *13*, 297. [CrossRef]
119. Muhammad Adnan, R.; Chen, Z.; Yuan, X.; Kisi, O.; El-Shafie, A.; Kuriqi, A.; Ikram, M. Reference evapotranspiration modeling using new heuristic methods. *Entropy* **2020**, *22*, 547. [CrossRef] [PubMed]
120. Kisi, O. Pan evaporation modeling using least square support vector machine, multivariate adaptive regression splines and M5 model tree. *J. Hydrol.* **2015**, *528*, 312–320. [CrossRef]
121. Eslamian, S.; Abedi-Koupai, J.; Amiri, M.; Gohari, S. Estimation of Daily Reference Evapotranspiration Using Support Vector. *Res. J. Environ. Sci.* **2009**, *3*, 439–447.
122. Nourani, V.; Elkiran, G.; Abdullahi, J. Multi-station artificial intelligence based ensemble modeling of reference evapotranspiration using pan evaporation measurements. *J. Hydrol.* **2019**, *577*, 123958. [CrossRef]
123. Wang, L.; Niu, Z.; Kisi, O.; Yu, D. Pan evaporation modeling using four different heuristic approaches. *Comput. Electron. Agric.* **2017**, *140*, 203–213. [CrossRef]
124. Allawi, M.F.; Jaafar, O.; Hamzah, F.M.; El-Shafie, A. Novel reservoir system simulation procedure for gap minimization between water supply and demand. *J. Clean. Prod.* **2019**, *206*, 928–943. [CrossRef]
125. Deo, R.C.; Samui, P.; Kim, D. Estimation of monthly evaporative loss using relevance vector machine, extreme learning machine and multivariate adaptive regression spline models. *Stoch. Environ. Res. Risk Assess.* **2016**, *30*, 1769–1784. [CrossRef]
126. Torres, A.F.; Walker, W.R.; McKee, M. Forecasting daily potential evapotranspiration using machine learning and limited climatic data. *Agric. Water Manag.* **2011**, *98*, 553–562. [CrossRef]
127. Tausif, M.; Iqbal, M.W.; Bashir, R.N.; AlGhofaily, B.; Elyassih, A.; Khan, A.R. Federated learning based reference evapotranspiration estimation for distributed crop fields. *PLoS ONE* **2025**, *20*, e0314921. [CrossRef]
128. Chia, M.Y.; Huang, Y.F.; Koo, C.H.; Fung, K.F. Recent advances in evapotranspiration estimation using artificial intelligence approaches with a focus on hybridization techniques—A review. *Agronomy* **2020**, *10*, 101. [CrossRef]
129. Ahmadianfar, I.; Shirvani-Hosseini, S.; He, J.; Samadi-Koucheksaraee, A.; Yaseen, Z.M. An improved adaptive neuro fuzzy inference system model using conjoined metaheuristic algorithms for electrical conductivity prediction. *Sci. Rep.* **2022**, *12*, 4934. [CrossRef] [PubMed]
130. Mehdizadeh, S.; Mohammadi, B.; Ahmadi, F. Establishing coupled models for estimating daily dew point temperature using nature-inspired optimization algorithms. *Hydrology* **2022**, *9*, 9. [CrossRef]
131. Babanezhad, M.; Behroyan, I.; Nakhjiri, A.T.; Marjani, A.; Rezakazemi, M.; Heydarinasab, A.; Shirazian, S. Investigation on performance of particle swarm optimization (PSO) algorithm based fuzzy inference system (PSOFIS) in a combination of CFD modeling for prediction of fluid flow. *Sci. Rep.* **2021**, *11*, 1505. [CrossRef]
132. Mohammadi, B.; Guan, Y.; Moazenzadeh, R.; Safari, M.J.S. Implementation of hybrid particle swarm optimization-differential evolution algorithms coupled with multi-layer perceptron for suspended sediment load estimation. *Catena* **2021**, *198*, 105024. [CrossRef]
133. Aghelpour, P.; Varshavian, V. Forecasting different types of droughts simultaneously using multivariate standardized precipitation index (MSPI), MLP neural network, and imperialistic competitive algorithm (ICA). *Complexity* **2021**, *2021*, 1–16. [CrossRef]
134. Deo, R.C.; Ghorbani, M.A.; Samadianfard, S.; Maraseni, T.; Bilgili, M.; Biazar, M. Multi-layer perceptron hybrid model integrated with the firefly optimizer algorithm for windspeed prediction of target site using a limited set of neighboring reference station data. *Renew. Energy* **2018**, *116*, 309–323. [CrossRef]
135. Mohammadi, B.; Mehdizadeh, S. Modeling daily reference evapotranspiration via a novel approach based on support vector regression coupled with whale optimization algorithm. *Agric. Water Manag.* **2020**, *237*, 106145. [CrossRef]
136. Roy, D.K.; Lal, A.; Sarker, K.K.; Saha, K.K.; Datta, B. Optimization algorithms as training approaches for prediction of reference evapotranspiration using adaptive neuro fuzzy inference system. *Agric. Water Manag.* **2021**, *255*, 107003. [CrossRef]
137. Tao, H.; Diop, L.; Bodian, A.; Djaman, K.; Ndiaye, P.M.; Yaseen, Z.M. Reference evapotranspiration prediction using hybridized fuzzy model with firefly algorithm: Regional case study in Burkina Faso. *Agric. Water Manag.* **2018**, *208*, 140–151. [CrossRef]
138. Eslamian, S.S.; Gohari, S.A.; Zareian, M.J.; Firoozfar, A. Estimating Penman–Monteith reference evapotranspiration using artificial neural networks and genetic algorithm: A case study. *Arab. J. Sci. Eng.* **2012**, *37*, 935–944. [CrossRef]
139. Aghajanloo, M.-B.; Sabziparvar, A.-A.; Hosseinzadeh Talaee, P. Artificial neural network–genetic algorithm for estimation of crop evapotranspiration in a semi-arid region of Iran. *Neural Comput. Appl.* **2013**, *23*, 1387–1393. [CrossRef]
140. Yin, Z.; Wen, X.; Feng, Q.; He, Z.; Zou, S.; Yang, L. Integrating genetic algorithm and support vector machine for modeling daily reference evapotranspiration in a semi-arid mountain area. *Hydrol. Res.* **2017**, *48*, 1177–1191. [CrossRef]

141. Gocić, M.; Motamedi, S.; Shamshirband, S.; Petković, D.; Ch, S.; Hashim, R.; Arif, M. Soft computing approaches for forecasting reference evapotranspiration. *Comput. Electron. Agric.* **2015**, *113*, 164–173. [CrossRef]
142. Maroufpoor, S.; Bozorg-Haddad, O.; Maroufpoor, E. Reference evapotranspiration estimating based on optimal input combination and hybrid artificial intelligent model: Hybridization of artificial neural network with grey wolf optimizer algorithm. *J. Hydrol.* **2020**, *588*, 125060. [CrossRef]
143. Wu, L.; Huang, G.; Fan, J.; Ma, X.; Zhou, H.; Zeng, W. Hybrid extreme learning machine with meta-heuristic algorithms for monthly pan evaporation prediction. *Comput. Electron. Agric.* **2020**, *168*, 105115. [CrossRef]
144. Hadadi, F.; Moazenzadeh, R.; Mohammadi, B. Estimation of actual evapotranspiration: A novel hybrid method based on remote sensing and artificial intelligence. *J. Hydrol.* **2022**, *609*, 127774. [CrossRef]
145. Tikhamarine, Y.; Malik, A.; Souag-Gamane, D.; Kisi, O. Artificial intelligence models versus empirical equations for modeling monthly reference evapotranspiration. *Environ. Sci. Pollut. Res.* **2020**, *27*, 30001–30019. [CrossRef]
146. Tikhamarine, Y.; Malik, A.; Kumar, A.; Souag-Gamane, D.; Kisi, O. Estimation of monthly reference evapotranspiration using novel hybrid machine learning approaches. *Hydrol. Sci. J.* **2019**, *64*, 1824–1842. [CrossRef]
147. Zounemat-Kermani, M.; Kisi, O.; Piri, J.; Mahdavi-Meymand, A. Assessment of artificial intelligence–based models and metaheuristic algorithms in modeling evaporation. *J. Hydrol. Eng.* **2019**, *24*, 04019033. [CrossRef]
148. Kişi, Ö. Evolutionary neural networks for monthly pan evaporation modeling. *J. Hydrol.* **2013**, *498*, 36–45. [CrossRef]
149. Ferreira, C. Gene expression programming: A new adaptive algorithm for solving problems. *arXiv* **2001**, arXiv:cs/0102027.
150. Schmidt, M.; Lipson, H. Distilling free-form natural laws from experimental data. *Science* **2009**, *324*, 81–85. [CrossRef] [PubMed]
151. Razaq, S.A.; Shahid, S.; Ismail, T.; Chung, E.S.; Mohsenipour, M.; Wang, X.J. Prediction of flow duration curve in ungauged catchments using genetic expression programming. *Procedia Eng.* **2016**, *154*, 1431–1438. [CrossRef]
152. Shiri, J. Evaluation of FAO56-PM, empirical, semi-empirical and gene expression programming approaches for estimating daily reference evapotranspiration in hyper-arid regions of Iran. *Agric. Water Manag.* **2017**, *188*, 101–114. [CrossRef]
153. Shiri, J.; Zounemat-Kermani, M.; Kisi, O.; Mohsenzadeh Karimi, S. Comprehensive assessment of 12 soft computing approaches for modelling reference evapotranspiration in humid locations. *Meteorol. Appl.* **2020**, *27*, e1841. [CrossRef]
154. Jovic, S.; Nedeljkovic, B.; Golubovic, Z.; Kostic, N. Evolutionary algorithm for reference evapotranspiration analysis. *Comput. Electron. Agric.* **2018**, *150*, 1–4. [CrossRef]
155. Shiri, J.; Kişi, Ö.; Landeras, G.; López, J.J.; Nazemi, A.H.; Stuyt, L.C. Daily reference evapotranspiration modeling by using genetic programming approach in the Basque Country (Northern Spain). *J. Hydrol.* **2012**, *414*, 302–316. [CrossRef]
156. Traore, S.; Guven, A. New algebraic formulations of evapotranspiration extracted from gene-expression programming in the tropical seasonally dry regions of West Africa. *Irrig. Sci.* **2013**, *31*, 1–10. [CrossRef]
157. Wang, S.; Fu, Z.-y.; Chen, H.-s.; Nie, Y.-p.; Wang, K.-l. Modeling daily reference ET in the karst area of northwest Guangxi (China) using gene expression programming (GEP) and artificial neural network (ANN). *Theor. Appl. Climatol.* **2016**, *126*, 493–504. [CrossRef]
158. Yassin, M.A.; Alazba, A.; Mattar, M.A. Artificial neural networks versus gene expression programming for estimating reference evapotranspiration in arid climate. *Agric. Water Manag.* **2016**, *163*, 110–124. [CrossRef]
159. Gavili, S.; Sanikhani, H.; Kisi, O.; Mahmoudi, M.H. Evaluation of several soft computing methods in monthly evapotranspiration modelling. *Meteorol. Appl.* **2018**, *25*. 128–138. [CrossRef]
160. Kişi, Ö.; Öztürk, Ö. Adaptive neurofuzzy computing technique for evapotranspiration estimation. *J. Irrig. Drain. Eng.* **2007**, *133*, 368–379. [CrossRef]
161. Dogan, E. Reference evapotranspiration estimation using adaptive neuro-fuzzy inference systems. *Irrig. Drain.* **2009**, *58*, 617–628. [CrossRef]
162. Kişi, Ö. Daily pan evaporation modelling using a neuro-fuzzy computing technique. *J. Hydrol.* **2006**, *329*, 636–646. [CrossRef]
163. Moghaddamnia, A.; Gousheh, M.G.; Piri, J.; Amin, S.; Han, D. Evaporation estimation using artificial neural networks and adaptive neuro-fuzzy inference system techniques. *Adv. Water Resour.* **2009**, *32*, 88–97. [CrossRef]
164. del Cerro, R.T.G.; Subathra, M.; Kumar, N.M.; Verrastro, S.; George, S.T. Modelling the daily reference evapotranspiration in semi-arid region of South India: A case study comparing ANFIS and empirical models. *Inf. Process. Agric.* **2021**, *8*, 173–184.
165. Pour-Ali Baba, A.; Shiri, J.; Kisi, O.; Fard, A.F.; Kim, S.; Amini, R. Estimating daily reference evapotranspiration using available and estimated climatic data by adaptive neuro-fuzzy inference system (ANFIS) and artificial neural network (ANN). *Hydrol. Res.* **2013**, *44*, 131–146. [CrossRef]
166. Citakoglu, H.; Cobaner, M.; Haktanir, T.; Kisi, O. Estimation of monthly mean reference evapotranspiration in Turkey. *Water Resour. Manag.* **2014**, *28*, 99–113. [CrossRef]
167. Petković, D.; Gocic, M.; Trajkovic, S.; Shamshirband, S.; Motamedi, S.; Hashim, R.; Bonakdari, H. Determination of the most influential weather parameters on reference evapotranspiration by adaptive neuro-fuzzy methodology. *Comput. Electron. Agric.* **2015**, *114*, 277–284. [CrossRef]

168. Cobaner, M. Evapotranspiration estimation by two different neuro-fuzzy inference systems. *J. Hydrol.* **2011**, *398*, 292–302. [CrossRef]
169. Sanikhani, H.; Kisi, O.; Maroufpoor, E.; Yaseen, Z.M. Temperature-based modeling of reference evapotranspiration using several artificial intelligence models: Application of different modeling scenarios. *Theor. Appl. Climatol.* **2019**, *135*, 449–462. [CrossRef]
170. Zakhrouf, M.; Bouchelkia, H.; Stamboul, M. Neuro-fuzzy systems to estimate reference evapotranspiration. *Water SA* **2019**, *45*, 232–238. [CrossRef]
171. Ye, L.; Zahra, M.M.A.; Al-Bedyry, N.K.; Yaseen, Z.M. Daily scale evapotranspiration prediction over the coastal region of southwest Bangladesh: New development of artificial intelligence model. *Stoch. Environ. Res. Risk Assess.* **2022**, *36*, 451–471. [CrossRef]
172. Niu, T.; Wang, J.; Zhang, K.; Du, P. Multi-step-ahead wind speed forecasting based on optimal feature selection and a modified bat algorithm with the cognition strategy. *Renew. Energy* **2018**, *118*, 213–229. [CrossRef]
173. Bui, D.T.; Hoang, N.-D.; Nguyen, H.; Tran, X.-L. Spatial prediction of shallow landslide using Bat algorithm optimized machine learning approach: A case study in Lang Son Province, Vietnam. *Adv. Eng. Inform.* **2019**, *42*, 100978.
174. Han, Y.; Wu, J.; Zhai, B.; Pan, Y.; Huang, G.; Wu, L.; Zeng, W. Coupling a bat algorithm with xgboost to estimate reference evapotranspiration in the arid and semiarid regions of china. *Adv. Meteorol.* **2019**, *2019*, 9575782. [CrossRef]
175. Tikhamarine, Y.; Malik, A.; Pandey, K.; Sammen, S.S.; Souag-Gamane, D.; Heddam, S.; Kisi, O. Monthly evapotranspiration estimation using optimal climatic parameters: Efficacy of hybrid support vector regression integrated with whale optimization algorithm. *Environ. Monit. Assess.* **2020**, *192*, 696. [CrossRef]
176. Malik, A.; Kumar, A.; Ghorbani, M.A.; Kashani, M.H.; Kisi, O.; Kim, S. The viability of co-active fuzzy inference system model for monthly reference evapotranspiration estimation: Case study of Uttarakhand State. *Hydrol. Res.* **2019**, *50*, 1623–1644. [CrossRef]
177. Malik, A.; Kumar, A. Pan evaporation simulation based on daily meteorological data using soft computing techniques and multiple linear regression. *Water Resour. Manag.* **2015**, *29*, 1859–1872. [CrossRef]
178. Aytek, A. Co-active neurofuzzy inference system for evapotranspiration modeling. *Soft Comput.* **2009**, *13*, 691–700. [CrossRef]
179. Bidabadi, M.; Babazadeh, H.; Shiri, J.; Saremi, A. Estimation reference crop evapotranspiration (ET0) using artificial intelligence model in an arid climate with external data. *Appl. Water Sci.* **2024**, *14*, 3. [CrossRef]
180. Akiner, M.E.; Ghasri, M. Comparative assessment of deep belief network and hybrid adaptive neuro-fuzzy inference system model based on a meta-heuristic optimization algorithm for precise predictions of the potential evapotranspiration. *Environ. Sci. Pollut. Res.* **2024**, *31*, 42719–42749. [CrossRef] [PubMed]
181. Mohammadrezapour, O.; Piri, J.; Kisi, O. Comparison of SVM, ANFIS and GEP in modeling monthly potential evapotranspiration in an arid region (Case study: Sistan and Baluchestan Province, Iran). *Water Supply* **2019**, *19*, 392–403. [CrossRef]
182. Gökkuş, M.K. Anfis Based Reference Evapotranspiration (ET0) Estimation Using Limited and Different Climate Parameters. *ISPEC J. Agric. Sci.* **2024**, *8*, 1022–1033.
183. Córdova, M.; Carrillo-Rojas, G.; Crespo, P.; Wilcox, B.; Célleri, R. Evaluation of the Penman-Monteith (FAO 56 PM) method for calculating reference evapotranspiration using limited data. *Mt. Res. Dev.* **2015**, *35*, 230–239. [CrossRef]
184. Adnan, R.M.; Mostafa, R.R.; Islam, A.R.M.T.; Kisi, O.; Kuriqi, A.; Heddam, S. Estimating reference evapotranspiration using hybrid adaptive fuzzy inferencing coupled with heuristic algorithms. *Comput. Electron. Agric.* **2021**, *191*, 106541. [CrossRef]
185. Falamarzi, Y.; Palizdan, N.; Huang, Y.F.; Lee, T.S. Estimating evapotranspiration from temperature and wind speed data using artificial and wavelet neural networks (WNNs). *Agric. Water Manag.* **2014**, *140*, 26–36. [CrossRef]
186. Partal, T. Modelling evapotranspiration using discrete wavelet transform and neural networks. *Hydrol. Process. Int. J.* **2009**, *23*, 3545–3555. [CrossRef]
187. Kisi, O.; Alizamir, M. Modelling reference evapotranspiration using a new wavelet conjunction heuristic method: Wavelet extreme learning machine vs wavelet neural networks. *Agric. For. Meteorol.* **2018**, *263*, 41–48. [CrossRef]
188. Kişi, Ö. Evapotranspiration modeling using a wavelet regression model. *Irrig. Sci.* **2011**, *29*, 241–252. [CrossRef]
189. Mehdizadeh, S. Estimation of daily reference evapotranspiration (ETo) using artificial intelligence methods: Offering a new approach for lagged ETo data-based modeling. *J. Hydrol.* **2018**, *559*, 794–812. [CrossRef]
190. Rahman, M.; Hasan, M.M.; Hossain, M.A.; Das, U.K.; Islam, M.M.; Karim, M.R.; Faiz, H.; Hammad, Z.; Sadiq, S.; Alam, M. Integrating deep learning algorithms for forecasting evapotranspiration and assessing crop water stress in agricultural water management. *J. Environ. Manag.* **2025**, *375*, 124363. [CrossRef] [PubMed]
191. Habeeb, R.; Almazah, M.M.; Hussain, I.; Al-Rezami, A.; Raza, A.; Ray, R.L. Improving Reference Evapotranspiration Predictions with Hybrid Modeling Approach. *Earth Syst. Environ.* **2025**, 1–18. [CrossRef]
192. Guo, N.; Chen, H.; Han, Q.; Wang, T. Evaluating data-driven and hybrid modeling of terrestrial actual evapotranspiration based on an automatic machine learning approach. *J. Hydrol.* **2024**, *628*, 130594. [CrossRef]
193. Lu, H.; Liu, T.; Yang, Y.; Yao, D. A hybrid dual-source model of estimating evapotranspiration over different ecosystems and implications for satellite-based approaches. *Remote Sens.* **2014**, *6*, 8359–8386. [CrossRef]

194. Koppa, A.; Rains, D.; Hulsman, P.; Poyatos, R.; Miralles, D.G. A deep learning-based hybrid model of global terrestrial evaporation. *Nat. Commun.* **2022**, *13*, 1912. [CrossRef]
195. Tosan, M.; Gharib, M.R.; Attar, N.F.; Maroosi, A. Enhancing Evapotranspiration Estimation: A Bibliometric and Systematic Review of Hybrid Neural Networks in Water Resource Management. *Comput. Model. Eng. Sci. (CMES)* **2025**, *142*, 1109–1154. [CrossRef]
196. Xing, L.; Cui, N.; Guo, L.; Du, T.; Gong, D.; Zhan, C.; Zhao, L.; Wu, Z. Estimating daily reference evapotranspiration using a novel hybrid deep learning model. *J. Hydrol.* **2022**, *614*, 128567. [CrossRef]
197. ElGhawi, R.; Kraft, B.; Reimers, C.; Reichstein, M.; Körner, M.; Gentine, P.; Winkler, A.J. Hybrid modeling of evapotranspiration: Inferring stomatal and aerodynamic resistances using combined physics-based and machine learning. *Environ. Res. Lett.* **2023**, *18*, 034039. [CrossRef]
198. Heramb, P.; Ramana Rao, K.; Subeesh, A.; Srivastava, A. Predictive modelling of reference evapotranspiration using machine learning models coupled with grey wolf optimizer. *Water* **2023**, *15*, 856. [CrossRef]
199. Raza, A.; Vishwakarma, D.K.; Acharki, S.; Al-Ansari, N.; Alshehri, F.; Elbeltagi, A. Use of gene expression programming to predict reference evapotranspiration in different climatic conditions. *Appl. Water Sci.* **2024**, *14*, 152. [CrossRef]
200. Aly, M.S.; Darwish, S.M.; Aly, A.A. High performance machine learning approach for reference evapotranspiration estimation. *Stoch. Environ. Res. Risk Assess.* **2024**, *38*, 689–713. [CrossRef]
201. Aghelpour, P.; Varshavian, V.; Khodamorad Pour, M.; Hamedi, Z. Comparing three types of data-driven models for monthly evapotranspiration prediction under heterogeneous climatic conditions. *Sci. Rep.* **2022**, *12*, 17363. [CrossRef]
202. Ashrafzadeh, A.; Kişi, O.; Aghelpour, P.; Biazar, S.M.; Masouleh, M.A. Comparative study of time series models, support vector machines, and GMDH in forecasting long-term evapotranspiration rates in northern Iran. *J. Irrig. Drain. Eng.* **2020**, *146*, 04020010. [CrossRef]
203. Aghelpour, P.; Norooz-Valashedi, R. Predicting daily reference evapotranspiration rates in a humid region, comparison of seven various data-based predictor models. *Stoch. Environ. Res. Risk Assess.* **2022**, *36*, 4133–4155. [CrossRef]
204. Ashrafzadeh, A.; Kişi, O.; Aghelpour, P.; Mostafa Biazar, S.; Askarizad Masouleh, M. Closure to "comparative study of time series models, support vector machines, and gmdh in forecasting long-term evapotranspiration rates in northern Iran" by Afshin Ashrafzadeh, Ozgur Kişi, Pouya Aghelpour, Seyed Mostafa Biazar, and Mohammadreza Askarizad Masouleh. *J. Irrig. Drain. Eng.* **2021**, *147*, 07021006.
205. Kishore, V.; Pushpalatha, M. Forecasting evapotranspiration for irrigation scheduling using neural networks and ARIMA. *Int. J. Appl. Eng. Res.* **2017**, *12*, 10841–10847.
206. e Lucas, P.d.O.; Alves, M.A.; e Silva, P.C.d.L.; Guimaraes, F.G. Reference evapotranspiration time series forecasting with ensemble of convolutional neural networks. *Comput. Electron. Agric.* **2020**, *177*, 105700. [CrossRef]
207. Landeras, G.; Ortiz-Barredo, A.; López, J.J. Forecasting weekly evapotranspiration with ARIMA and artificial neural network models. *J. Irrig. Drain. Eng.* **2009**, *135*, 323–334. [CrossRef]
208. Kumar, M.; Raghuwanshi, N.; Singh, R. Artificial neural networks approach in evapotranspiration modeling: A review. *Irrig. Sci.* **2011**, *29*, 11–25. [CrossRef]
209. Materia, S.; García, L.P.; van Straaten, C.; O, S.; Mamalakis, A.; Cavicchia, L.; Coumou, D.; de Luca, P.; Kretschmer, M.; Donat, M. Artificial intelligence for climate prediction of extremes: State of the art, challenges, and future perspectives. *Wiley Interdiscip. Rev. Clim. Change* **2024**, *15*, e914. [CrossRef]
210. Ye, Y.; González-Vidal, A.; Zamora-Izquierdo, M.A.; Skarmeta, A.F. Transfer and deep learning models for daily reference evapotranspiration estimation and forecasting in Spain from local to national scale. *Smart Agric. Technol.* **2025**, *11*, 100886. [CrossRef]
211. Banerjee, D.; Ganguly, S.; Tsai, W.-P. A novel hybrid machine learning framework for spatio-temporal analysis of reference evapotranspiration in India. *J. Hydrol. Reg. Stud.* **2025**, *58*, 102271. [CrossRef]

water

MDPI

Article

Time-Lag Effects of Winter Arctic Sea Ice on Subsequent Spring Precipitation Variability over China and Its Possible Mechanisms

Hao Wang, Wen Wang * and Fuxiong Guo

The National Key Laboratory of Water Disaster Prevention, Hohai University, Nanjing 210098, China; 221301010089@hhu.edu.cn (H.W.); guofuxiong1995@gmail.com (F.G.)

* Correspondence: wangwen@hhu.edu.cn

Abstract: Arctic sea ice variations exhibit relatively strong statistical associations with precipitation variability over northeastern and southern China. Using Arctic Ocean reanalysis data from the EU Copernicus Project, this study examines the time-lagged statistical relationships between winter Arctic sea ice conditions and subsequent spring precipitation variability over China through wavelet analysis and Granger causality tests. Singular value decomposition (SVD) identifies the Barents, Kara, East Siberian, and Chukchi Seas as key regions exhibiting strong associations with spring precipitation anomalies. Increased winter sea ice in the East Siberian and Chukchi Seas generates positive geopotential height anomalies over the Arctic and negative anomalies over Northeast Asia, adjusting upper-level jet streams and influencing precipitation patterns in Northeast China. Conversely, increased sea ice in the Barents–Kara Seas leads to persistent negative geopotential height anomalies simultaneously occurring over both the Arctic and South China regions, enhancing southern jet stream activity and intensifying warm-moist airflow at the 850 hPa level, thus favoring precipitation in southern China. Compared to considering only climate factors such as the Pacific Decadal Oscillation (PDO), El Niño–Southern Oscillation (ENSO), and Arctic Oscillation (AO), the inclusion of Arctic sea ice significantly enhances the influence of multiple climate factors on precipitation variability in China.

Keywords: Arctic sea ice; precipitation; atmospheric circulation; climate change; multi-wavelet coherence (MWC); SVD

1. Introduction

Arctic sea ice is a "regulator" of the Arctic climate and an "indicator" of global change [1]. With the intensification of global warming, the Arctic region has experienced the most significant temperature rise in the world, exceeding twice the global average, which is also known as "Arctic amplification" [2,3]. From the late 20th century to the 21st century, the coverage, thickness, and surface snow depth of Arctic sea ice have decreased, and the speed of sea ice's movement and the sea temperature have increased, with longer melting seasons [4–6]. These trends are expected to continue, and seasonal ice-free conditions in the Arctic Ocean could occur by the middle of this century or even earlier [7].

Variations in Arctic sea ice are linked to changes in China's climate and extreme weather events, influencing temperature [8] and precipitation patterns in various regions. There have been extensive studies on the correlation between Arctic sea ice and precipitation in Northeast China. Guo et al. [9] found that the total Arctic sea ice extent in spring (February–April) is positively correlated with summer precipitation in the Yangtze River Basin but negatively correlated with summer precipitation in Northeast China, the

Indochina Peninsula, and the South China Sea. Zhang et al. [10] identified a significant negative correlation between spring precipitation in Northeast China and the preceding winter's Barents Sea ice concentration. Similarly, Li et al. [11] reported a negative correlation between the intensity of Barents Sea ice and precipitation in Northeast China during May. Furthermore, Han et al. [12], through observational analysis and atmospheric simulation experiments, found that an increase in Barents Sea ice in early spring leads to increased rainfall in northern Northeast China while causing an abnormal decrease in precipitation in the southern part of the region.

Changes in Arctic sea ice also influence precipitation in southern China. Zhao et al. [13] found that a reduction in spring sea ice extent in the Bering Sea and the Sea of Okhotsk leads to an increase in summer monsoonal precipitation in southeastern China. Similarly, Wu et al. [14] demonstrated that a decline in the spring sea ice concentration in the Arctic Ocean and Greenland Sea is associated with a reduction in summer precipitation in southern China. Additionally, Wu et al. [15] discovered that the variability of Arctic sea ice in the Norwegian Sea and the Barents Sea during the preceding winter is closely linked to the interannual variation in East Asian spring precipitation, showing a significant negative correlation.

Arctic sea ice variability is closely linked to atmospheric circulation changes over the Arctic, with the Arctic Oscillation (AO) and North Atlantic Oscillation (NAO) acting as key intermediaries connecting sea ice conditions to the Northern Hemisphere's climate patterns [16]. He et al. [17] found that the decline in Arctic sea ice in June triggers an anomalous Rossby wave train, which propagates downstream and impacts the circulation over East Asia. This leads to the formation of a tripole precipitation pattern in August, characterized by negative precipitation anomalies in northern and southern China, while Japan, Korea, and central China experience positive anomalies. Wu et al. [15] demonstrated that Arctic sea ice anomalies persist from winter into the following spring, continuously generating Rossby wave trains across the Eurasian continent and inducing downstream circulation anomalies. This wave train pattern is often associated with an anomalous low-pressure system over the Mongolian Plateau, which enhances the East Asian subtropical westerly jet stream The strengthening of this jet stream, accompanied by low-level convergence and upper-level divergence, enhances local convection, thereby favoring increased spring precipitation in East Asia. Physical analyses based on observations and numerical simulations [11] indicate that a decrease in the Barents Sea ice concentration may trigger a Rossby wave train that propagates southeastward, crossing the Mongolia–Baikal region and reaching Northeast China. This process establishes a zonal anticyclone–cyclone–anticyclone teleconnection pattern over the mid-to-high latitudes of Eurasia. Under the influence of this pattern, southerly winds prevail over Northeast China, transporting warm and moist air into the region. Additionally, increased local atmospheric instability and upward motion further enhance favorable moisture and dynamic conditions, leading to increased precipitation in Northeast China during May.

Although previous studies have explored the potential link between Arctic sea ice melting and East Asian precipitation, their explanations regarding the key regions and underlying mechanisms remain inconsistent. The objective of this study is to conduct a comprehensive analysis of the spatiotemporal characteristics of Arctic sea ice variations based on the sea ice reanalysis dataset from the European Union's Copernicus Program. By examining different sea ice regions, this study explores the spatial and temporal relationships between Arctic sea ice anomalies and the precipitation variability over northeastern and southern China. Furthermore, from the perspective of atmospheric circulation, we investigate the possible mechanisms through which Arctic sea ice anomalies are statistically linked to spring precipitation patterns in these selected regions. The remainder of this paper is organized as follows: Section 2 introduces the datasets and methods used in this study. Section 3 presents the results of the spatiotemporal characteristics of Arctic sea ice and its statistical associations with precipitation variability. Section 4 discusses the possible physical mechanisms and further analyzes the influences of multiple climate factors, including ENSO, PDO, AO, and Arctic sea ice, on precipitation variability in China. Section 5 concludes the paper with a summary of the key findings and suggestions for future research.

2. Datasets and Methods

2.1. List of Abbreviations

Throughout this paper, a number of abbreviations and acronyms are employed. To enhance clarity and facilitate comprehension, the definitions of key terms and their corresponding abbreviations are presented in Table 1. This table encompasses all relevant abbreviations that are frequently referenced throughout the study.

Table 1. List of abbreviations used in this paper.

Term	Definition
ENSO	El Niño-Southern Oscillation
PDO	Pacific Decadal Oscillation
AO	Arctic Oscillation
SIC	Sea Ice Concentration
SIT	Sea Ice Thickness
EOF	Empirical Orthogonal Function
SVD	Singular Value Decomposition
MWC	Multiple Wavelet Coherence
AWC	Average Wavelet Coherence
PASC	Percentage of Area with Significant Coherence
Z200	200 hPa Geopotential Height
Z500	500 hPa Geopotential Height
SLP	Sea Level Pressure
U200	200 hPa Zonal Wind
UV850	850 hPa Horizontal Wind

2.2. Study Area

The study area is confined to the ocean above the latitude of 66.5° N on the Atlantic side, with Fury and Hecla Strait as the southern boundary of the Canadian Archipelago. We divided the area into the following 5 sub-regions, as shown in Figure 1: Barents–Kara Seas, Arctic Marginal Seas (Beaufort–Chukchi–East Siberian–Laptev Seas), Baffin–Archipelago, Central Arctic, and Greenland–Norwegian Seas.

2.3. Datasets

The Arctic Ocean Physics Reanalysis dataset "ARCTIC_MULTIYEAR_PHY_002_003" is a reanalysis product provided by the Arctic Monitoring and Forecasting Centre (ARC MFC), operated by the Nansen Environmental and Remote Sensing Center (NERSC) in Bergen, Norway, through the TOPAZ4b system, and is available from the Copernicus Marine Service (http://marine.copernicus.eu/) (accessed on 21 March 2024). It employs advanced sequential data assimilation methods and mixed coordinate ocean models, integrating remote sensing data for enhanced accuracy and comprehensiveness. Observations that are assimilated by TOPAZ4 include sea surface temperature from the Operational Sea Surface Temperature and Sea Ice Analysis (OSTIA), sea ice concentration from OSI-SAF, and CS2SMOS sea-ice thickness data from AWI. The ice thickness product is the weekly merged CS2SMOS dataset from AWI [18], combining thin sea ice measurements from the ESA SMOS mission to the thick sea ice retrievals from another ESA mission CryoSAT2.

The spatial resolution of the reanalysis datasets is 12.5 km, and the temporal resolution is monthly.

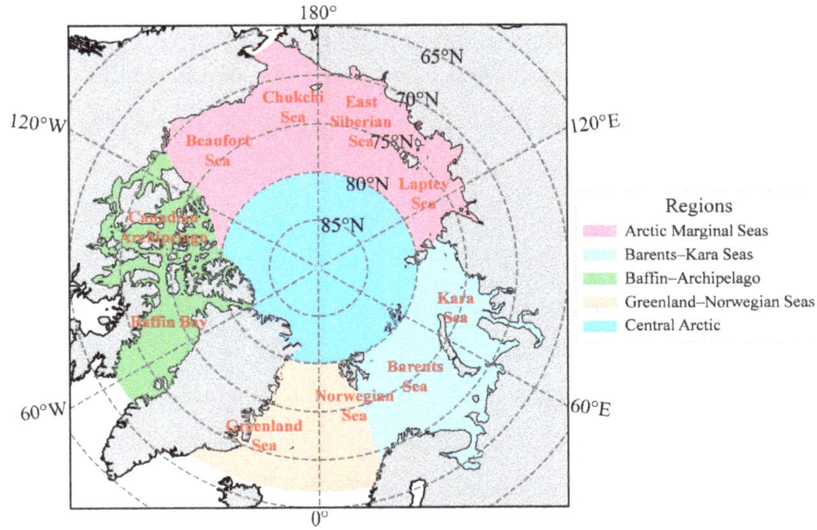

Figure 1. Location map of the Arctic regions.

The precipitation dataset used in this study is the CHM_PRE dataset provided by the National Tibetan Plateau Data Center [19], operated in Beijing, China, and is available at https://data.tpdc.ac.cn/ (accessed on 11 December 2024). This dataset is constructed based on daily observational data collected from 2839 stations, including 2419 stations across mainland China and 420 stations in surrounding regions. The dataset covers the period from 1961 to 2022, with a spatial resolution of 0.1° and a geographical range of 18° N–54° N and 72° E–136° E.

The atmospheric circulation factors used in this study are derived from ERA5 (https://cds.climate.copernicus.eu/), a widely used atmospheric reanalysis dataset produced by the European Centre for Medium-Range Weather Forecasts (ECMWF), located in Reading, United Kingdom. ERA5 is generated using 4D-Var data assimilation and model forecasts from cycle CY41R2 of the ECMWF Integrated Forecast System (IFS) (accessed on 2 December 2024) [20]. The ERA5 dataset contains more than 200 parameters and variables with global coverage, spanning from 1940 to the present date in near real time. ERA5 provides a large number of atmospheric, land, and oceanic climate variables. In this study, we selected the 200 hPa geopotential height (Z200), 500 hPa geopotential height (Z500), sea level pressure (SLP), 200 hPa zonal wind (U200), and 850 hPa horizontal wind (UV850).

The Niño 3.4 index, PDO index, and AO index were obtained from the Earth System Research Laboratory of the National Oceanic and Atmospheric Administration (NOAA), located in Boulder, Colorado, United States, and available at http://www.esrl.noaa.gov/ (accessed on 2 December 2024).

2.4. Methods

This study adopts a multi-method, multi-scale, and integrated analytical framework to reveal the spatiotemporal coupling mechanisms and dominant variability patterns between precipitation in China and Arctic sea ice. First, empirical orthogonal function (EOF) analysis is employed to reduce the dimensionality of high-dimensional spatiotemporal precipitation data, extracting the dominant variability structures and their evolution patterns within the region. Secondly, singular value decomposition (SVD) is used to construct a cross-covariance matrix between the precipitation and Arctic sea ice fields, extracting paired spatial modes with the highest covariance to analyze their teleconnection and co-evolution patterns. For dynamical mechanisms, both regression analysis and composite analysis are applied. Regression analysis quantifies the statistical relationships between precipitation and circulation fields, identifying key large-scale circulation anomalies. Composite analysis further verifies and attributes these atmospheric circulation and moisture transport patterns by comparing climate field characteristics during different sea-ice anomaly years, ensuring the robustness of the identified mechanisms. Additionally, the T–N three-dimensional wave-activity flux theory is introduced to quantitatively diagnose the energy transmission process of quasi-stationary Rossby waves, exploring the physical link between precipitation anomalies and Arctic sea ice variations. Furthermore, bivariate wavelet coherence (WTC) and partial wavelet coherence (PWC) analyses are used to assess the influence of multiple climate factors on precipitation in China across different timescales. Statistical significance is determined using the Student's t-test, with the degrees of freedom set to $n-2$, where n represents the number of rows in the variable dataset. The technical roadmap of this study is shown in Figure 2.

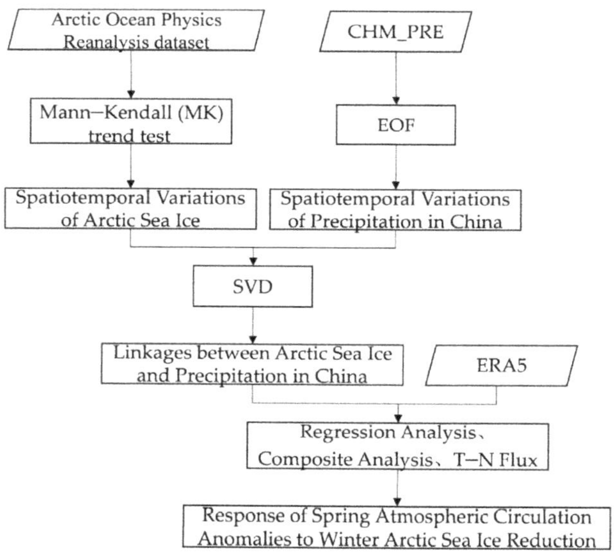

Figure 2. Technical flowchart.

2.4.1. Empirical Orthogonal Function (EOF)

EOF is a dimensionality reduction method for climate fields based on the orthogonal decomposition of the covariance matrix [21]. Its core idea is to decouple high-dimensional spatiotemporal datasets into a linear combination of spatial modes and temporal coefficients, thereby extracting dominant spatiotemporal variability signals. Given the observational data of a climate variable field in the matrix form:

$$X = \begin{bmatrix} x_{11} & \cdots & x_{1n} \\ \vdots & \ddots & \vdots \\ x_{m1} & \cdots & x_{mn} \end{bmatrix} \tag{1}$$

where m represents spatial points, typically observation stations or grid points, and n denotes temporal points, i.e., the number of observations; x_{mn} represents the observation value at the m-th station or grid point during the n-th observation. The EOF decomposition expresses X as:

$$X = VZ\# \tag{2}$$

where

$$V = \begin{bmatrix} v_{11} & \cdots & v_{1m} \\ \vdots & \ddots & \vdots \\ v_{m1} & \cdots & v_{mm} \end{bmatrix}, Z = \begin{bmatrix} z_{11} & \cdots & z_{1n} \\ \vdots & \ddots & \vdots \\ z_{m1} & \cdots & z_{mn} \end{bmatrix} \tag{3}$$

These are referred to as the spatial function matrix and the temporal coefficient matrix, respectively. Given the orthogonality properties of V and Z, the spatial function matrix, V, is obtained from the eigenvectors of XX^T, while the temporal coefficient matrix, Z, is derived as V^TX. The variance contribution of each eigenvector is calculated as follows:

$$R_k = \frac{\lambda_k}{\sum_{i=1}^{m} \lambda_i} (k = 1, 2, \ldots, p(p < m)) \tag{4}$$

where λ_k is the k-th eigenvalue. The cumulative variance contribution of the first p eigenvectors is given by the following:

$$G = \frac{\sum_{i=1}^{p} \lambda_i}{\sum_{i=1}^{m} \lambda_i (p < m)} \tag{5}$$

The leading eigenvectors obtained from EOF decomposition represent the dominant variability structure of the climate variable field across the entire region. Given the advantages of EOF in extracting the primary variability patterns of climate fields, this study applies it to analyze the spatiotemporal evolution of precipitation in China. The first three modes are used to reveal the dominant spatial patterns of precipitation variability and their temporal evolution.

2.4.2. Singular Value Decomposition (SVD)

SVD [22] is an orthogonal decomposition method based on the cross-covariance matrix of two variable fields. By applying SVD, the original high-dimensional spatiotemporal correlation problem can be transformed into a limited number of time-varying coupling modes, effectively extracting the coupled coherent patterns between two meteorological fields. Mathematically, SVD performs spectral decomposition on the cross-covariance matrix to isolate paired spatial modes with the highest covariance, thereby revealing the spatiotemporal teleconnection characteristics between the two fields. In this study, SVD is used to analyze the cross-regional teleconnection mechanism between precipitation in China and Arctic sea ice, focusing on the first two modes with significant covariance contributions to uncover key co-evolution patterns.

2.4.3. T–N Three-Dimensional Wave-Activity Flux

Takaya and Nakamura [23], building upon Plumb's theoretical foundation, developed the formulation for the three-dimensional propagation of quasi-stationary waves, known as the T–N flux (also referred to as wave-activity flux), under conditions where the basic flow is variable. This formulation allows for the diagnosis of Rossby wave perturbation energy propagation and the significance of nonlinear interactions, providing a means to describe the energy propagation characteristics of quasi-stationary Rossby waves. Under the Wentzel–Kramers–Brillouin (WKB) approximation, this flux is independent of the wave phase and aligns with the local group velocity of stationary Rossby wave trains in the horizontal direction. In the logarithmic pressure coordinate system, it can be expressed as follows:

$$W = \frac{\text{pcos}\varphi}{2|U|}\begin{pmatrix} \frac{U}{a^2\cos^2\varphi}\left[\left(\frac{\partial\psi'}{\partial\lambda}\right)^2 - \psi'\frac{\partial^2\psi'}{\partial\lambda^2}\right] + \frac{V}{a^2\cos\varphi}\left[\frac{\partial\psi'}{\partial\lambda}\frac{\partial\psi'}{\partial\varphi} - \psi'\frac{\partial^2\psi'}{\partial\lambda\partial\varphi}\right] \\ \frac{U}{a^2\cos\varphi}\left[\frac{\partial\psi'}{\partial\lambda}\frac{\partial\psi'}{\partial\varphi} - \psi'\frac{\partial^2\psi'}{\partial\lambda\partial\varphi}\right] + \frac{V}{a^2}\left[\left(\frac{\partial\psi'}{\partial\varphi}\right)^2 - \psi'\frac{\partial^2\psi'}{\partial\varphi^2}\right] \\ \frac{f_0^2}{N^2}\left\{\frac{U}{a\cos\varphi}\left[\frac{\partial\psi'}{\partial\lambda}\frac{\partial\psi'}{\partial z} - \psi'\frac{\partial^2\psi'}{\partial\lambda\partial z}\right] + \frac{V}{a}\left[\frac{\partial\psi'}{\partial\varphi}\frac{\partial\psi'}{\partial z} - \psi'\frac{\partial^2\psi'}{\partial\varphi\partial z}\right]\right\} \end{pmatrix} \tag{6}$$

In Equation (6), p denotes the pressure normalized by 1000 hPa; φ and λ represent the latitude and longitude, respectively; f is the Coriolis parameter; a is the Earth's radius; U and V are the zonal and meridional components of the basic-state flow field; ψ' is the quasi-geostrophic perturbation stream function; and N^2 represents the buoyancy frequency (Brunt–Väisälä frequency). This formulation couples dynamical and thermodynamic perturbation terms to describe the three-dimensional propagation of Rossby wave energy, where the horizontal component reflects barotropic energy transmission. In this study, this method is employed to analyze the dynamical linkage between Arctic sea ice anomalies and precipitation variability in China.

2.4.4. Multi-Wavelet Coherence (MWC)

MWC is an effective method for investigating the scale-dependent relationships between two variables [24]. MWC can unravel a series of multivariate relationships, determine the multivariate relationships of variables in space or time scales, and also identify the proportion of variance related to predictive variables and response variables [25]. In wavelet analysis, a 95% confidence level is obtained based on the first-order autocorrelation coefficient through Monte Carlo simulations (1000 repetitions) [26]. The resonance period refers to a strong coupling relationship that occurs on a specific time scale. The quantitative assessment of the predictive variables' explanatory power for the response variable is conducted by calculating the percentage area of significant coherence outside the cone of influence (percent area of significant coherence, PASC), as well as the average wavelet coherence (AWC) [24,27]. The AWC was calculated by averaging the wavelet coherence produced over all scales according to the coherence values produced in the PWC computation [26]. The PASC was obtained by calculating the ratio of the number of significant values of power over the total number of values of the power. The larger the PASC, the higher the AWC, indicating that the specific predictor variable can explain more of the sea ice variation.

2.4.5. Mann–Kendall (MK) Trend Test

Mann–Kendall (MK) trend test, which is a non-parametric test method recommended by the World Meteorological Organization [28], is used to test the temporal changes in sea ice. The specific principles of the MK are as follows:

Assuming that the samples in the sequence $\{x_1, \ldots, x_N\}$ are randomly independent, the statistic S is constructed as follows:

$$S = \sum_{i=1}^{N-1}\sum_{j=i+1}^{N} sgn(x_j - x_i) \tag{7}$$

among

$$sgn(x_j - x_i) = \begin{cases} +1, & x_j - x_i > 0 \\ 0, & x_j - x_i = 0 \\ -1, & x_j - x_i < 0 \end{cases} \tag{8}$$

where $S > 0$ indicates that the sequence has an increasing trend, and $S < 0$ indicates that the sequence has a decreasing trend.

The statistic τ measures the strength of a continuous trend, as follows:

$$\tau = \frac{2S}{N(N-1)} \tag{9}$$

Mann and Kendall proved that when $n \geq 8$, the statistic S roughly follows a normal distribution, and its variance is as follows:

$$\sigma_s^2 = \frac{1}{18}\left[N(N-1)(2N+5) - \sum_{i=1}^{m} t_i(t_i-1)(2t_i+5)\right] \tag{10}$$

where m is the number of groups, $Var(S)$ is equal to x in the dataset, and t_i is the number of data points in group i equal group.

The hypothesis test Z is defined below and approximates the standard normal distribution:

$$z = \begin{cases} (S-1)/\sigma_s & \text{if } S > 0 \\ 0 & \text{if } S = 0 \\ (S+1)/\sigma_s & \text{if } S < 0 \end{cases} \tag{11}$$

When $Z > 0$, there is an upward trend; when $Z < 0$, there is a decreasing trend.

For a given significance level α, if $|Z| > Z_{1-\alpha/2}$, this indicates a significant upward or downward trend in the time series. In this paper, the p-value is used to represent Z. When the p-value is less than a pre-defined significance level (typically 0.05), the data are considered to exhibit a significant trend. In this study, the p-values are categorized into three levels, as follows: a trend with a p-value between 0.05 and 0.1 is classified as slightly significant (indicated by +1 or −1 for positive or negative trends, respectively); a trend with a p-value between 0.01 and 0.05 is considered moderately significant (indicated by +2 or −2); and a trend with a p-value less than 0.01 is regarded as highly significant (indicated by +3 or −3).

Computational resources: All data processing and statistical analyses in this study were conducted using MATLAB R2022a and Python 3.9 on a personal workstation equipped with an AMD Ryzen 7 6800H CPU (8 cores, 3.20 GHz), 16 GB RAM, and a 477 GB SSD. The datasets processed in this study (including precipitation observations, sea ice reanalysis, and atmospheric circulation data) totaled approximately 35 GB. The computational requirements for wavelet analysis, singular value decomposition (SVD), and Granger causality testing were moderate, and all analytical tasks could be completed within several hours. The available computational resources were sufficient to ensure efficient data processing and analysis without compromising the accuracy and reliability of the results.

3. Results

3.1. Spatiotemporal Variations in Arctic Sea Ice

Sea ice concentration (SIC) refers to the percentage of sea ice area in each grid, the sea ice extent (SIE) refers to the total area of all grid cells where the SIC is not less than a certain threshold, usually 15%. In this study, a grid cell is considered ice-free if its SIC is less than 15%, following the threshold commonly used in previous studies by Parkinson and Cavalieri [29,30]. The spatial resolution of the grid is 12.5 × 12.5 km. The average SIC within the study area is calculated using a weighted method. Overall, SIC reached its annual maximum in March and minimum in September. In March, the Central Arctic, Arctic Marginal Seas, and Baffin–Archipelago were mostly covered by sea ice (Figure 3a). In September, the SIC was the highest in the Central Arctic, and decreased toward the marginal areas of the Arctic Ocean, with almost no sea ice coverage in the Barents–Kara Seas and Greenland–Norwegian Seas (Figure 3b).

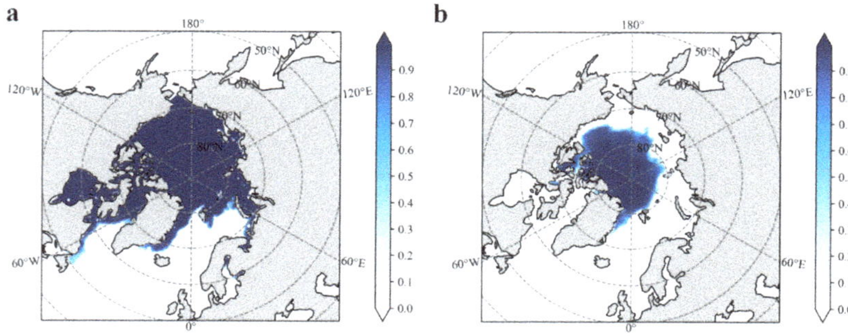

Figure 3. Seasonal distribution of sea ice variables: average SIC in March (**a**) and September (**b**) during 1991–2022.

The analysis of the annual SIC variations in Arctic sub-regions (Figure 4a) reveals a significant Arctic sea ice decline. The most pronounced ice loss occurred in the Barents Sea, Chukchi Sea, Beaufort Sea, Labrador Sea, and Greenland Sea, consistent with findings from previous studies [4,31]. Regional decline rates exhibit spatial heterogeneity, as follows: the Central Arctic (0.0017%/yr), Arctic marginal seas (0.0047%/yr), and Barents–Kara Sea (0.0055%/yr) form a distinct descending gradient in annual ice loss. The grid-scale Mann–Kendall trend tests (Figure 4b) confirm significant seasonal disparities in ice retreat. Summer–autumn (June–November) SIC fluctuations intensified by 42% compared to winter–spring (December–May). The record minimum SIC occurred in September 2012 due to extreme storm activity [32], with secondary lows observed in September 2016 and 2020.

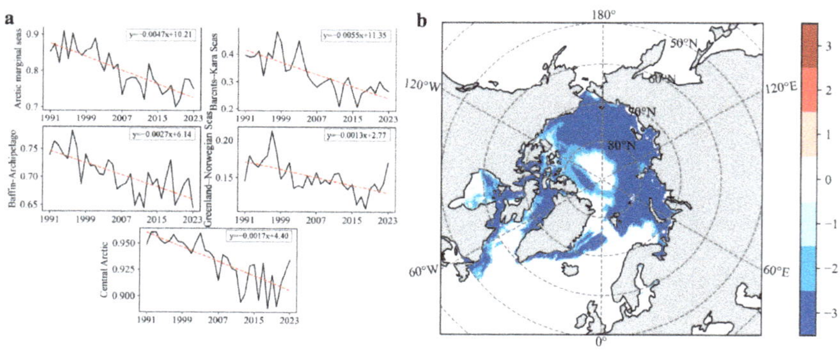

Figure 4. Spatiotemporal variations in SIC during 1991–2022: (**a**) variation in annual average SIC in Arctic sub-regions; (**b**) grid-based MK trend test results for SIC.

Spatiotemporal analysis of sea ice thickness (SIT) (Figure 5) demonstrates that the "perennial multiyear ice zone" along the Canada–Greenland coast maintains maximum thickness, while the Barents–Kara Seas and Greenland–Norwegian Seas, influenced by the warm North Atlantic Current, remain persistently thin. The pan-Arctic annual mean SIT decreased by 0.13 m overall (1991–2023), with the Arctic marginal seas exhibiting the fastest thinning rate (0.0238 m/yr), followed by the Baffin–Archipelago region (0.0096 m/yr). Notably, localized thickening occurred in northern Greenland Sea and the Arctic Basin–Greenland Sea junction (Figure 5b). SIT seasonal cycles exhibited phase delay characteristics relative to SIC, peaking in May and reaching the minima in October. Although thinning trends persist across all seasons, winter–spring periods show relative stability, while summer–autumn (June–November) account for 76% of annual thickness loss.

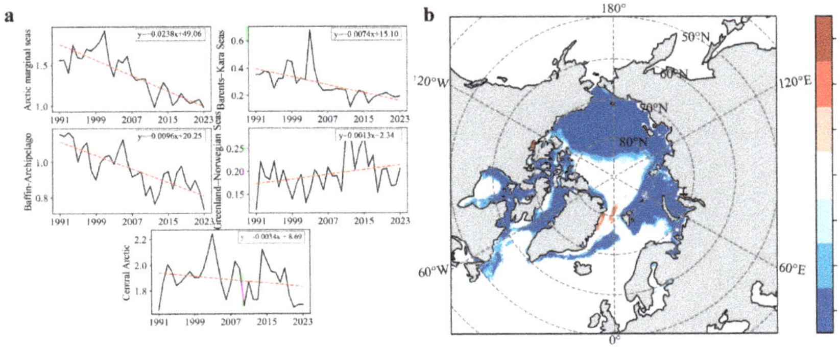

Figure 5. Spatiotemporal variations in SIT during 1991–2022: (**a**) variation in annual average SIT in Arctic sub-regions; (**b**) grid-based MK trend test results for SIT.

The seasonal cycle of SIT lagged behind that of SIC, with a maximum value in May and a minimum value in October. SIT showed a general declining trend across all four seasons. However, it remained relatively stable during winter and spring, while experiencing significant declines in summer and autumn.

3.2. Spatiotemporal Variations in Precipitation in China

Figure 6 presents the spatial distribution of total precipitation and the interannual variation trends in China from 1991 to 2022. The precipitation trend is obtained by performing linear regression between the grid-based precipitation data and time. The total precipitation distribution (Figure 6a) exhibited a typical monsoon climate pattern; over the past 32 years, the Yangtze River Basin and areas to the south have received an average annual precipitation exceeding 1500 mm, with some regions experiencing over 2000 mm due to orographic uplift. In contrast, the North China Plain and the Loess Plateau received less than 400 mm, while the arid inland regions of Northwest China recorded even lower values, below 250 mm. The interannual variation trend (Figure 6b) highlights the significant wetting in the Northeast China Plain, Sichuan Basin, Hainan Island, and areas surrounding the Yangtze River Delta. Conversely, drying trends were observed in the eastern Yunnan–Guizhou Plateau and along the Fujian coast.

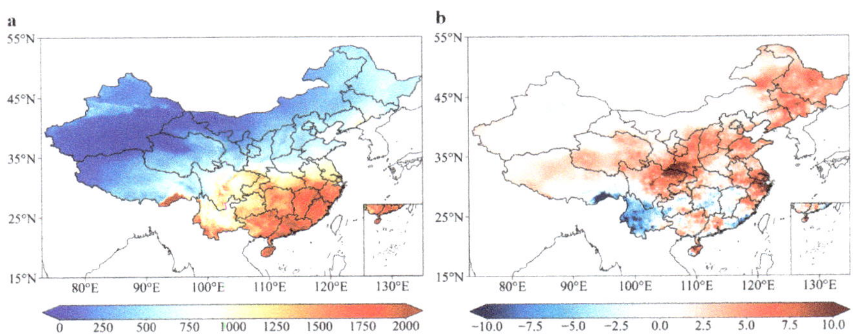

Figure 6. Spatial distribution (**a**) and trends (**b**) in the precipitation in China from 1991 to 2022.

Figure 7 presents the EOF analysis results of the precipitation in China from 1991 to 2022, providing insight into its spatial distribution and temporal variations. The percentage of variance explained by each mode is indicated in the upper right corner of each subplot. EOF1 (Figure 6a) explains 24.2% of the total variance, representing the primary mode of precipitation variability, particularly the north–south contrast. The spatial pattern exhibits a "− + − +" distribution from north to south, with positive anomalies centered in northern Northeast China and the region south of the Yangtze River, while negative anomalies extend across the Xinjiang–Inner Mongolia–Jilin belt and the Tibetan Plateau to the Loess Plateau. This mode reflects the typical out-of-phase relationship between northern and southern China's precipitation. The corresponding PC1 time series (Figure 7b) shows complex fluctuations, with multiple alternating positive and negative values. EOF2 (Figure 7c) explains 17.4% of the variance and exhibits a uniform negative phase across most of China, indicating synchronous precipitation variations (either increasing or decreasing). However, Hainan Island and its surrounding coastal areas display a significant positive anomaly, forming a local inverse-phase relationship. EOF3 (Figure 7e) accounts for 9.7% of the variance, highlighting a positive phase over the Yunnan–Guizhou Plateau and southern Tibet, while Northeast China exhibits a distinct negative phase. The corresponding PC3 time series (Figure 7f) remains mostly in a positive phase from 1991 to 2010 but undergoes a phase shift to negative after 2010.

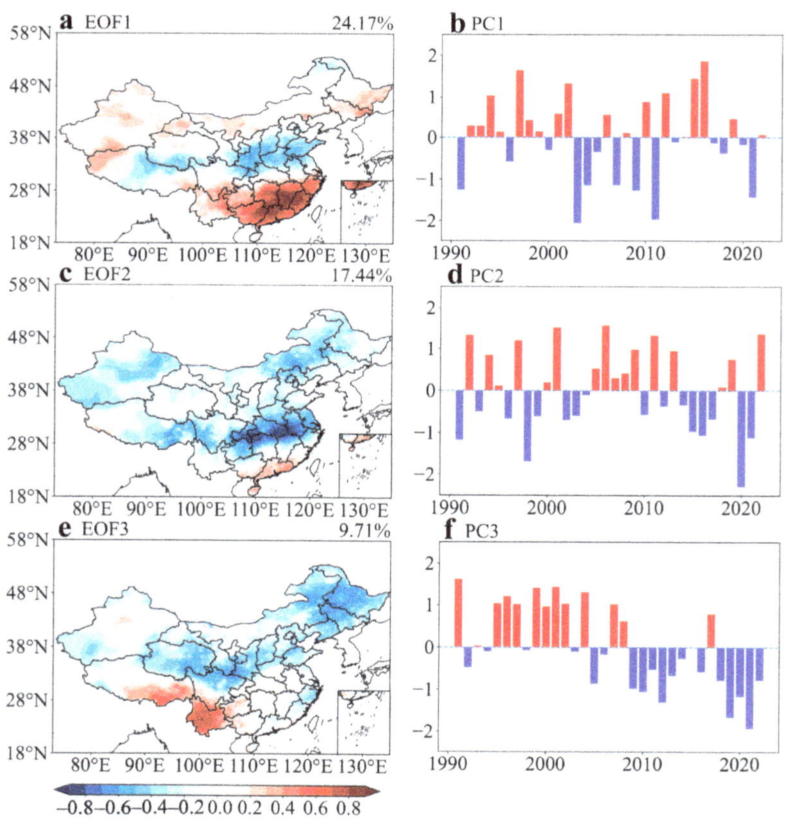

Figure 7. Dominant EOF modes of precipitation in China (**a**,**c**,**e**) and their corresponding standardized time series (**b**,**d**,**f**). Red/blue indicate positive/negative values.

3.3. Time-Lag Effects of Arctic Sea Ice on Precipitation in China and Key Sea Area

To identify the key time periods with strong correlation between Arctic sea ice and precipitation in China, this study employed multiple statistical methods, including wavelet analysis, lagged correlation analysis, and Granger causality testing, for cross-validation. Given the seasonal fluctuations in the monthly data, all variables were deseasonalized before analysis. In Figure 8, cross-wavelet transform analysis is used to examine the phase relationship and significance of the two variables across different timescales. The bold contours indicate the 95% confidence level against yellow noise. The scale is measured in months, and the shaded region represents the cone of influence (COI), where edge effects may distort the results. Rightward arrows indicate a positive phase relationship, while leftward arrows indicate a negative phase relationship. The results (Figure 8) indicate a significant correlation between sea ice and precipitation within the 4–20-month timescale, with an estimated 1–4-month lag of precipitation behind SIC based on the phase relationships. Lagged correlation analysis and Granger causality testing show peak values when precipitation lags behind sea ice by one month, with correlations gradually weakening in the subsequent months. These statistical analyses suggest that the impact of sea ice on precipitation is most pronounced on a seasonal timescale. Therefore, further correlation analysis is conducted between spring sea ice and summer precipitation, summer sea ice and autumn precipitation, autumn sea ice and winter precipitation, and winter sea ice and spring precipitation of the following year. The results indicate that winter sea ice has the strongest correlation with spring precipitation in the following year. Thus, the subsequent analyses primarily focus on these two seasons.

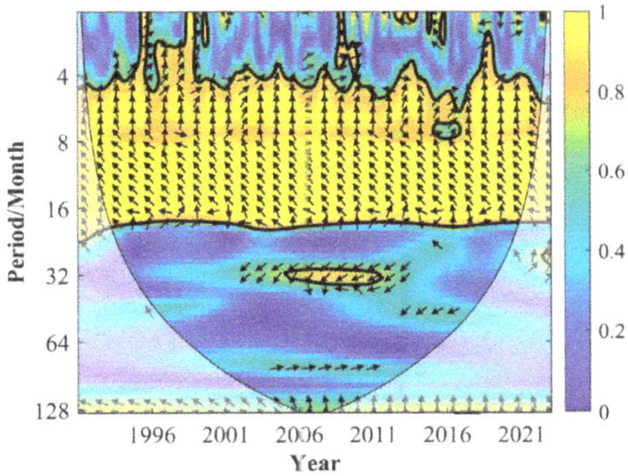

Figure 8. WTC spectrum between area-weighted Arctic sea ice and precipitation in China.

To further identify the key Arctic sea ice regions influencing precipitation distribution in China, this study applies SVD. For clarity, we define September–November (SON), December–February (DJF), March–May (MAM), and June–August (JJA) as autumn, winter, spring, and summer, respectively. The study analyzes the co-variability patterns between winter Arctic SIC anomalies and spring precipitation in China (CHM) from 1991 to 2022. Heterogeneous correlation coefficients are calculated to characterize the spatial relationships between the two fields. The key response regions across fields are identified by computing the correlation between raw data and the opposite-field mode time series, with dotted areas indicating statistically significant regions.

The first SVD mode ($n = 0$) (Figure 9a,b) and the second SVD mode ($n = 1$) (Figure 9c,d) explain 31.9% and 18.8% of the covariance, respectively, collectively accounting for 50.6% of the total variance. This suggests that the primary Arctic sea ice–China precipitation coupling relationships can be well represented by the first two modes, indicating a strong teleconnection signal between Arctic sea ice and East Asian precipitation. In the first mode, the Barents Sea, Kara Sea, and Bering Strait exhibit significant negative correlations in the left field (Figure 9a), corresponding to a spatially coherent negative correlation pattern in southern China in the right field (Figure 9b). The high mode correlation coefficients pass the significance test, indicating that when winter sea ice concentration in the Barents Sea, Kara Sea, and Bering Strait decreases (increases), spring precipitation in the lower-latitude regions of China also decreases (increases). In the second mode, the East Siberian Sea, Chukchi Sea, and parts of the Arctic Basin near the Bering Strait exhibit negative values in the left field, while northern Northeast China, parts of southern Tibet, and the Sichuan Basin show negative precipitation anomalies in the right field. This indicates that when winter sea ice concentration in the East Siberian Sea and Chukchi Sea decreases (increases), precipitation in northern Northeast China, southern Tibet, and the Sichuan Basin also decreases (increases). A linear regression analysis of the SVD mode time coefficients shows a significant downward trend in the first mode (slope: −0.0847/year), suggesting that the impact of Arctic sea ice anomalies on East Asian precipitation is gradually intensifying. The second mode also shows a decreasing trend (slope: −0.0464/year), though with a relatively smaller magnitude of change.

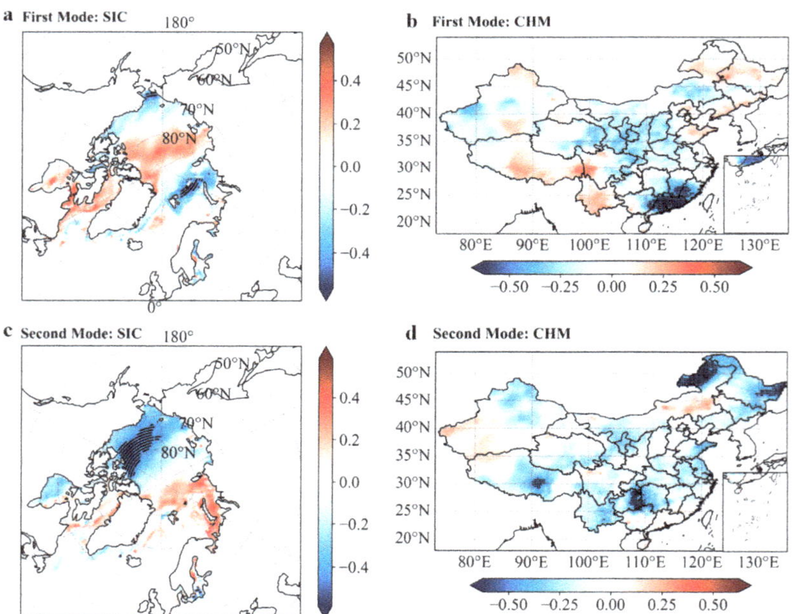

Figure 9. Heterogeneous correlation coefficients of SVD between the winter Arctic sea ice concentration and spring precipitation in China for the following year.

3.4. Response of Spring Atmospheric Circulation Anomalies to Winter Arctic Sea Ice Reduction

Previous studies have shown that sea ice melting significantly reduces surface albedo, leading to an increase in surface heat flux anomalies over the Arctic [33–35]. This diabatic heating effect causes substantial polar warming, weakens the meridional temperature gradient, and subsequently reduces the strength of the westerly jet stream while enhancing meridional atmospheric activity [36]. The thermal anomalies induced by sea ice loss can also trigger Rossby wave source anomalies, which propagate eastward toward East Asia in the form of wave trains, leading to significant adjustments in the mid-to high-latitude circulation system over Eurasia [37,38]. Meanwhile, Arctic warming decreases atmospheric static stability, promoting the development of baroclinic instability, which further amplifies circulation anomalies through enhanced dynamic instability [39].

To further analyze whether the impact of winter Arctic sea ice on spring precipitation in China aligns with the mechanisms mentioned above, key regions with high correlations in Figure 9a,c, namely, the Barents–Kara Seas and Arctic Marginal Seas—are identified as critical areas where sea ice influences precipitation. The area-weighted SIC indices for these key regions are defined as BSIC and ASIC, respectively. When investigating the possible mechanisms of sea ice's influence on precipitation, it is essential to examine multiple atmospheric levels and key circulation factors. In this study, the selected circulation indices include: Z200 (200 hPa geopotential height) and Z500 (500 hPa geopotential height) to represent the upper and mid-tropospheric geopotential height distributions, which help identify wave train structures and ridge-trough locations; SLP (sea level pressure) to reflect the intensity and spatial distribution of cyclonic and anticyclonic activities near the surface; U200 (200 hPa zonal wind) to determine the position and strength of the upper-level westerly jet stream; UV850 (850 hPa horizontal wind) to analyze low-level moisture transport pathways and wind convergence/divergence characteristics.

Regression analysis is performed between ASIC and the selected circulation indices. Additionally, ASIC is ranked, and high-value years and low-value years are identified for composite difference analysis using the same circulation indices. For ASIC, the high-value years are 1995, 1999, 2000, 1991, and 2003, while the low-value years are 2008, 2018, 2020, 2017, and 2007. The results show that the spatial patterns and signs in the key regions (such as the mid-to-high latitudes and East Asia) are consistent between the regression analysis maps and composite analysis maps, indicating that the detected signals are stable and reliable.

The left panel of Figure 10 illustrates that ASIC sea ice anomalies primarily influence precipitation in northern China, particularly in Northeast China, through a "Polar–Siberia–Northeast Asia" wave train propagation pattern. At the 200 hPa geopotential height level (Figure 10a), when winter ASIC is high, a significant positive geopotential height anomaly appears over the Arctic, extending toward the high latitudes of Eurasia, with an opposite-sign anomaly center over Northeast Asia. This pattern resembles a wave train (Rossby wave) propagating from the polar region to mid-latitudes. The position and intensity of the upper-level jet stream over the North Pacific and Northeast Asia also change, leading to north–south jet stream oscillations, which in turn influence frontal activity and interactions between warm and cold air masses over East Asia. The composite analysis (Figure 10b) aligns well with the regression results, showing that in high-ASIC years, positive height anomalies are observed near the Arctic and northern Eurasia, while negative anomalies appear over the North Pacific and parts of East Asia. The opposite pattern occurs in low-ASIC years. At the 500 hPa level (Figure 10c), a series of alternating positive and negative anomaly centers extends from the polar region to East Asia, reflecting mid-tropospheric wave train structures that further regulate convective conditions in Northeast China, influencing regional precipitation processes. Meanwhile, SLP anomaly distributions (Figure 10e) suggest that increased sea ice stabilizes the polar low-pressure system. The regression and composite analysis of U200 (Figure 10g,h) indicate that when winter ASIC is high, the upper-level westerlies in this region are significantly strengthened, potentially enhancing upper-level divergence. This, in turn, promotes lower-level convergence and ascending motion, creating favorable conditions for moisture convergence and precipitation.

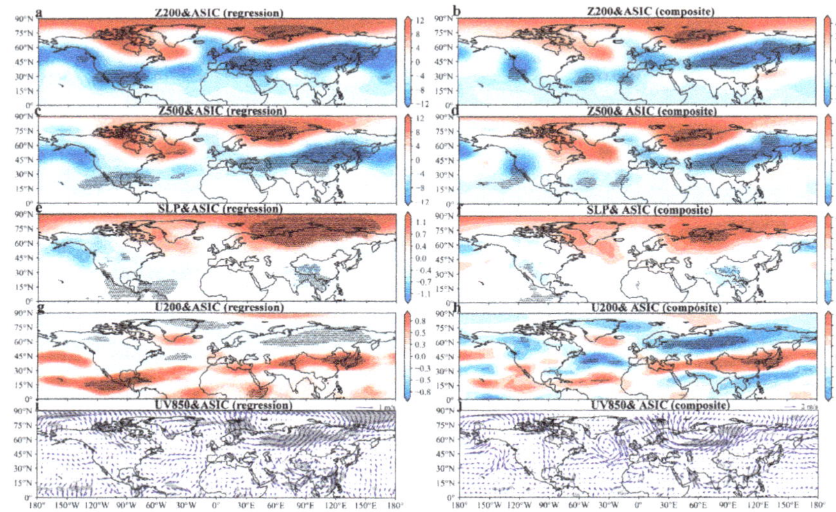

Figure 10. Relationship between winter ASIC anomalies and spring atmospheric circulation: (**a**,**c**,**e**,**g**,**i**) standardized regression responses; (**b**,**d**,**f**,**h**,**j**) composite differences between high- and low-ASIC years. (**a**,**b**) 200 hPa geopotential height (unit: gpm); (**c**,**d**) 500 hPa geopotential height; (**e**,**f**) sea level pressure (unit: hPa); (**g**,**h**) 200 hPa zonal wind (unit: m/s); (**i**,**j**) 850 hPa horizontal wind field. Shaded areas and dotted regions indicate statistical significance at the 95% confidence level, tested using Student's *t*-test.

A similar analysis is conducted for BSIC, including regression analysis with circulation indices and ranking BSIC values to identify high-value and low-value years for composite difference analysis. The high-value years for BSIC are 1994, 1998, 1999, 2003, and 2004, while the low-value years are 2012, 2013, 2016, 2017, and 2018. The 200 hPa geopotential height regression map (Figure 11a) shows that when winter Barents–Kara Seas ice increases (BSIC is high), negative geopotential height anomalies tend to appear over high-latitude regions (Arctic–North Atlantic) and mid- to high-latitude Eurasia, forming a wave train pattern propagating from the polar region to Eurasia. At the same time, the jet stream near South China strengthens, improving upper-level divergence conditions, which favors convection initiation or intensified frontal activity. At the 500 hPa level (Figure 11c,d), a wave train signal can also be observed extending from the Arctic–Eurasia region to South Asia. When the BSIC increases, negative height anomalies appear over the Barents Sea and parts of Siberia, while a negative anomaly center forms near South China. The U200 regression and composite maps (Figure 11g,h) show positive anomalies over South China, indicating that the negative geopotential height anomalies are coupled with the subtropical jet stream, which intensifies over South China, further enhancing upper-level divergence and atmospheric instability. The 850 hPa wind field maps (Figure 11i,j) reveal a clear northward transport of warm and moist air over South China, providing ample moisture supply for local convective activity, thereby facilitating convection development and precipitation formation.

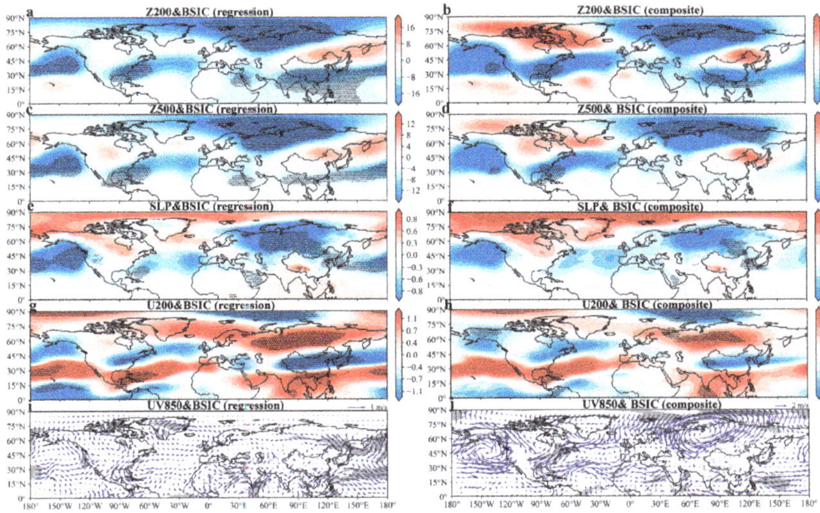

Figure 11. Relationship between winter BSIC anomalies and spring atmospheric circulation: (**a**,**c**,**e**,**g**,**i**) standardized regression responses; (**b**,**d**,**f**,**h**,**j**) composite differences between high- and low-ASIC years. Subplots correspond to those in Figure 10.

Overall, an increase in sea ice induces multi-level circulation adjustments from the upper atmosphere (Z200, Z500, and U200) to the near-surface (SLP, UV850), generating different teleconnection wave trains. The ASIC anomalies primarily led to positive geopotential height anomalies over the Arctic, which propagate toward Northeast Asia, forming negative geopotential height anomalies in this region. This process is accompanied by upper-level jet adjustments, influencing the interaction of cold and warm air masses and frontal activity in Northeast China, thereby regulating precipitation in northern regions. In contrast, the BSIC anomalies exhibit a continuous negative geopotential height anomaly pattern extending from the Barents Sea to South China, which is associated with jet stream intensification over South China and enhanced warm-moist air transport at the 850 hPa level, creating more favorable conditions for precipitation in southern regions. Both the ASIC and BSIC influence precipitation by altering large-scale circulation patterns, but their effects are observed in different regions.

To further verify the reasonableness of the mechanism analysis, this study conducts composite analysis of the 300 hPa stream function and the corresponding wave-activity flux in response to winter sea ice anomalies. This flux intuitively reflects the propagation path and intensity of quasi-stationary Rossby wave energy in the atmosphere. The distribution of the stream function in Figure 11 largely coincides with the anomaly centers in the regression and composite analyses. Both subplots in Figure 12 reveal a clear propagation path, illustrating a strong wave perturbation originating from the Barents–Kara Seas, which propagates southeastward, eventually reaching East Asia, forming a stable "Barents–Kara Seas–East Asia" teleconnection mode.

For ASIC (Figure 12a), winter sea ice anomalies generate significant positive geopotential height anomalies over the Arctic, forming an anticyclonic region. This anomaly triggers planetary wave energy perturbations, which extend from Siberia to Northeast Asia. As energy propagates downstream, the disturbance transitions into negative geopotential height anomalies, forming an upper-level trough, which adjusts the position and strength of the upper-level jet stream, modifies the interaction of cold and warm air masses, and influences frontal activity, ultimately regulating precipitation distribution and intensity in northern

China. For BSIC (Figure 12b), winter sea ice anomalies over the Barents–Kara Seas generate a pronounced cyclonic negative geopotential height anomaly at high latitudes, which serves as a source region for planetary wave energy. The wave-activity flux propagates southeastward from Siberia to East Asia, while an additional propagation pathway extends from the North Pacific, traveling from high-latitude Pacific regions to southern China, thereby modulating precipitation in South China. Overall, the two T–N wave-activity flux diagrams provide a clear visualization of how Arctic sea ice anomalies remotely modulate East Asian atmospheric circulation by exciting and adjusting the planetary wave energy propagation paths. These mechanisms exert distinct influences on precipitation patterns in northern and southern China through different pathways.

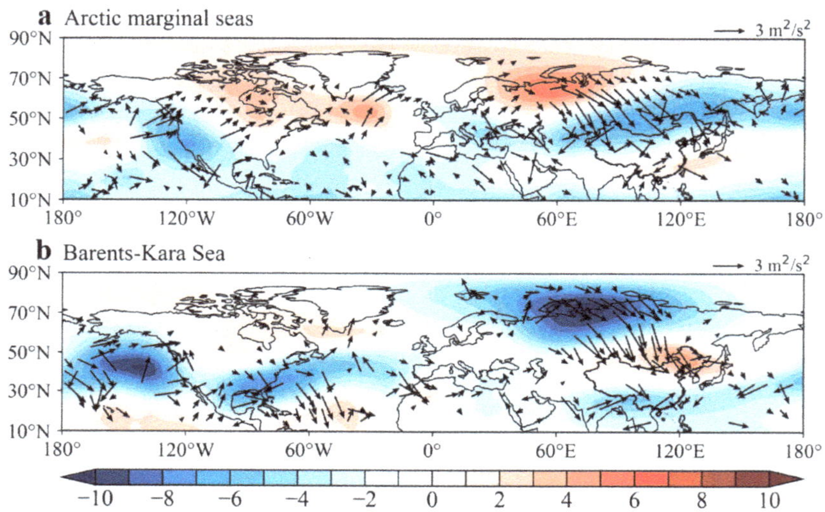

Figure 12. Composite analysis of 300 hPa stream function and wave-activity flux (unit: m^2/s^2) for spring in high and low winter ASIC (**a**) and BSIC (**b**) years.

4. Discussion

The interdecadal variability of precipitation in China is influenced by multiple factors, including El Niño-Southern Oscillation (ENSO), Pacific Decadal Oscillation (PDO), Arctic Oscillation (AO), Antarctic Oscillation (AAO), and Arctic sea ice (SIC) [40–42]. In this section, the Niño 3.4 index is selected to represent the impact of ENSO, combined with the PDO index, AO index, and SIC for analysis. The multi-wavelet coherence (MWC) method is employed to reveal the scale-dependent effects and synergistic relationships of different climate factors on precipitation in China.

Figure 13 presents the results of the MWC analysis between precipitation in China and major climate factors at the monthly timescale. Specifically, Figure 13a illustrates the time–frequency relationships between precipitation and ENSO, PDO, and AO. Significant resonance regions are observed within the 4–16-month band, although they are relatively scattered and do not fully cover all time periods. After 2011, coherence gradually emerges in the 32–64-month mid-range scale, suggesting that the synergistic influence of ENSO, PDO, and AO on China's precipitation may have strengthened in recent years. Overall, the average wavelet coherence between precipitation and these three climate factors is 0.54, with significant coherence regions accounting for 18.61% of the total area.

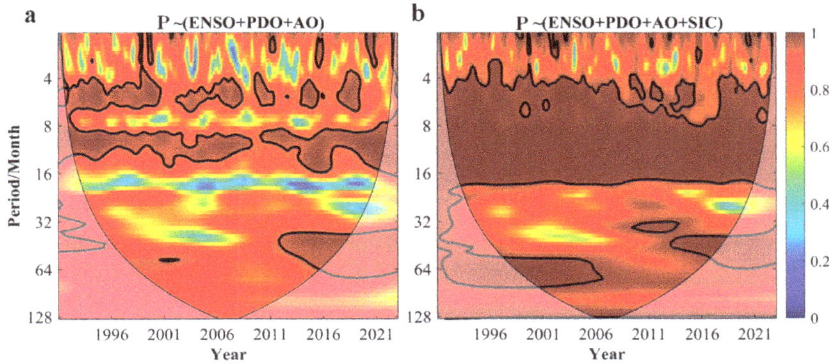

Figure 13. MWC analysis between precipitation in China and multiple climate factors: (**a**) precipitation and ENSO, PDO, and AO; (**b**) precipitation and ENSO, PDO, AO, and SIC.

Based on this, Figure 13b further incorporates the influence of SIC. The results show a notable enhancement of coherence within the 4–20-month low- to mid-frequency ranges, covering a broader time span. The proportion of significant coherence area increases to 38.58%, and the average coherence value rises to 0.64. This indicates that the inclusion of Arctic sea ice variability further strengthens the synergistic effects among ENSO, PDO, AO, and precipitation, demonstrating a more pronounced modulation of precipitation variability across a wider range of time–frequency scales.

Previous studies, such as by Hussain et al. [43], showed that climate models, like ENSO, PDO, and AO, intermittently influence regional precipitation across multiple time scales. Consistent with their findings, our study reveals significant associations between China's precipitation variability and major climate modes at both short (4–16 months) and long (32–64 months) periods. Nalley et al. [44] emphasized the importance of jointly considering multiple climate drivers to better explain hydroclimatic variability. In comparison, our study also takes into account Arctic sea ice variability (SIC), suggesting that the inclusion of polar processes may further enhance the coherence and extent of climate influences on regional precipitation patterns.

5. Conclusions

This study investigated the time-lagged statistical relationships between winter Arctic sea ice anomalies and spring precipitation variability in China, based on precipitation data from the National Tibetan Plateau Data Center and Arctic Ocean reanalysis data from the EU Copernicus Program for 1991–2022. Based on wavelet analysis and Granger causality testing, a strong correlation is found between winter sea ice variations and precipitation in the following spring. SVD identifies the Barents Sea, Kara Sea, East Siberian Sea, and Chukchi Sea as key regions influencing precipitation in China. A decrease in sea ice within these key areas is associated with reduced spring precipitation in coastal South China and parts of Northeast China.

The mechanism analysis indicates that winter sea ice changes in key Arctic regions influence turbulent heat fluxes, inducing atmospheric energy fluctuations that propagate toward East Asia through atmospheric wave trains. Increased sea ice in the East Siberian and Chukchi Seas generates positive geopotential height anomalies over the Arctic, accompanied by negative anomalies over Northeast Asia. This pattern adjusts the upper-level jet stream, influencing frontal activity and precipitation in northern China. Conversely, increased sea ice in the Barents–Kara Seas produces persistent negative geopotential height anomalies extending from the Barents Sea toward South China, strengthening jet stream activity over southern China and enhancing warm–moist air transport at 850 hPa, thereby promoting precipitation in southern China.

Further analysis based on multi-wavelet coherence reveals that the inclusion of Arctic sea ice variability further enhances the synergistic effects between precipitation and ENSO, PDO, and AO, exhibiting a more pronounced modulation of precipitation variability across a broader range of time–frequency scales. Although this study identifies significant statistical associations between winter Arctic sea ice anomalies and spring precipitation variability in China, several limitations should be acknowledged. First, the analyses relied on reanalysis and observational datasets, which may introduce inherent uncertainties. Second, the identified relationships are statistical in nature, and the underlying physical mechanisms remain complex and are not fully established. Therefore, future research combining high-resolution dynamical modeling and multi-source observational validation is needed to further explore and verify potential causal pathways linking Arctic sea ice changes to regional climate variability.

Author Contributions: Conceptualization, W.W. and H.W.; data curation, H.W.; methodology, W.W., H.W. and F.G.; writing—original draft, H.W.; writing—review and editing, W.W., H.W. and F.G. All authors have read and agreed to the published version of the manuscript.

Funding: This research was funded by The National Key Research and Development Program of China (grant number: 2023YFC3209201).

Data Availability Statement: All data used in this study are freely accessible. The Arctic Ocean Physics Reanalysis dataset "ARCTIC_MULTIYEAR_PHY_002_003", provided by the Arctic Monitoring and Forecasting Centre (ARC MFC) through the Copernicus Marine Service, is available at http://marine.copernicus.eu/ (accessed on 21 March 2024). Reanalysis data from ERA5, produced by the European Centre for Medium-Range Weather Forecasts (ECMWF), were used to analyze the mechanisms through which Arctic sea ice influences precipitation in China, and are accessible at https://cds.climate.copernicus.eu/ (accessed on 2 December 2024). Precipitation data were obtained from the CHM_PRE dataset hosted by the National Tibetan Plateau Data Center at https://data.tpdc.ac.cn/ (accessed on 11 December 2024). The Niño 3.4 index, PDO index, and AO index were sourced from the Earth System Research Laboratory of NOAA at http://www.esrl.noaa.gov/ (accessed on 2 December 2024).

Conflicts of Interest: The authors declare no conflicts of interest.

References

1. Screen, J.A.; Simmonds, I. The Central Role of Diminishing Sea Ice in Recent Arctic Temperature Amplification. *Nature* **2010**, *464*, 1334–1337. [CrossRef] [PubMed]
2. Davy, R.; Griewank, P. Arctic Amplification Has Already Peaked. *Environ. Res. Lett.* **2023**, *18*, 084003. [CrossRef]
3. Dai, A.; Luo, D.; Song, M.; Liu, J. Arctic Amplification Is Caused by Sea-Ice Loss under Increasing CO_2. *Nat. Commun.* **2019**, *10*, 121. [CrossRef]
4. Matveeva, T.A.; Semenov, V.A. Regional Features of the Arctic Sea Ice Area Changes in 2000–2019 versus 1979–1999 Periods. *Atmosphere* **2022**, *13*, 1434. [CrossRef]
5. Wang, X.; Li, S.; Zhao, Y.; Wang, Y.; Yang, Z. Comparison of Arctic and Antarctic Sea Ice Spatial–Temporal Changes during 1979–2018. *J. Hydrol.* **2024**, *632*, 130966. [CrossRef]
6. Guo, Y.; Wang, X.; Xu, H.; Hou, X. Spatiotemporal Variation and Freeze-Thaw Asymmetry of Arctic Sea Ice in Multiple Dimensions during 1979 to 2020. *Acta Oceanol. Sin.* **2024**, *43*, 102–114. [CrossRef]
7. Jahn, A. Reduced Probability of Ice-Free Summers for 1.5 °C Compared to 2 °C Warming. *Nat. Clim. Change* **2018**, *8*, 409–413. [CrossRef]
8. Wu, B.; Li, Z. Possible Impacts of Anomalous Arctic Sea Ice Melting on Summer Atmosphere. *Intl J. Climatol.* **2022**, *42*, 1818–1827. [CrossRef]
9. Guo, D.; Gao, Y.; Bethke, I.; Gong, D.; Johannessen, O.M.; Wang, H. Mechanism on How the Spring Arctic Sea Ice Impacts the East Asian Summer Monsoon. *Theor. Appl. Clim.* **2014**, *115*, 107–119. [CrossRef]

10. Zhang, S.; Zeng, G.; Yang, X.; Iyakaremye, V.; Hao, Z. Connection between Interannual Variation of Spring Precipitation in Northeast China and Preceding Winter Sea Ice over the Barents Sea. *Intl J. Climatol.* **2022**, *42*, 1922–1936. [CrossRef]
11. Li, X.; Sun, J.; Zhang, M.; Zhang, Y.; Ma, J. Possible Connection between Declining Barents Sea Ice and Interdecadal Increasing Northeast China Precipitation in May. *Intl J. Climatol.* **2021**, *41*, 6270–6282. [CrossRef]
12. Han, T.; Zhang, M.; Zhu, J.; Zhou, B.; Li, S. Impact of Early Spring Sea Ice in Barents Sea on Midsummer Rainfall Distribution at Northeast China. *Clim. Dyn.* **2021**, *57*, 1023–1037. [CrossRef]
13. Zhao, P.; Zhang, X.; Zhou, X.; Ikeda, M.; Yin, Y. The Sea Ice Extent Anomaly in the North Pacific and Its Impact on the East Asian Summer Monsoon Rainfall. *J. Clim.* **2004**, *17*, 3434–3447. [CrossRef]
14. Wu, B.; Zhang, R.; Wang, B. On the Association between Spring Arctic Sea Ice Concentration and Chinese Summer Rainfall: A Further Study. *Adv. Atmos. Sci.* **2009**, *26*, 666–678. [CrossRef]
15. Wu, Z.; Li, X.; Li, Y.; Li, Y. Potential Influence of Arctic Sea Ice to the Interannual Variations of East Asian Spring Precipitation*. *J. Clim.* **2016**, *29*, 2797–2813. [CrossRef]
16. Zuo, J.; Ren, H.-L.; Wu, B.; Li, W. Predictability of Winter Temperature in China from Previous Autumn Arctic Sea Ice. *Clim. Dyn.* **2016**, *47*, 2331–2343. [CrossRef]
17. He, S.; Gao, Y.; Furevik, T.; Wang, H.; Li, F. Teleconnection between Sea Ice in the Barents Sea in June and the Silk Road, Pacific–Japan and East Asian Rainfall Patterns in August. *Adv. Atmos. Sci.* **2018**, *35*, 52–64. [CrossRef]
18. Ricker, R.; Hendricks, S.; Kaleschke, L.; Tian-Kunze, X.; King, J.; Haas, C. A Weekly Arctic Sea-Ice Thickness Data Record from Merged CryoSat-2 and SMOS Satellite Data. *Cryosphere* **2017**, *11*, 1607–1623. [CrossRef]
19. Han, J.; Miao, C.; Gou, J.; Zheng, H.; Zhang, Q.; Guo, X. A New Daily Gridded Precipitation Dataset for the Chinese Mainland Based on Gauge Observations. *Earth Syst. Sci. Data* **2023**, *15*, 3147–3161. [CrossRef]
20. Hersbach, H.; Bell, B.; Berrisford, P.; Hirahara, S.; Horányi, A.; Muñoz-Sabater, J.; Nicolas, J.; Peubey, C.; Radu, R.; Schepers, D.; et al. The ERA5 Global Reanalysis. *Quart. J. R. Meteoro Soc.* **2020**, *146*, 1999–2049. [CrossRef]
21. Da Costa, E.D. On the Invariance of the First EOF of North Atlantic Sea Level Pressure to Temporal Filtering. *Geophys. Res. Lett.* **2003**, *30*, 2003GL017312. [CrossRef]
22. Bretherton, C.S.; Smith, C.; Wallace, J.M An Intercomparison of Methods for Finding Coupled Patterns in Climate Data. *J. Clim.* **1992**, *5*, 541–560. [CrossRef]
23. Takaya, K.; Nakamura, H. A Formulation of a Phase-Independent Wave-Activity Flux for Stationary and Migratory Quasi-geostrophic Eddies on a Zonally Varying Basic Flow. *J. Atmos. Sci.* **2001**, *58*, 608–627. [CrossRef]
24. Hu, W.; Si, B. Technical Note: Improved Partial Wavelet Coherency for Understanding Scale-Specific and Localized Bivariate Relationships in Geosciences. *Hydrol. Earth Syst. Sci.* **2021**, *25*, 321–331. [CrossRef]
25. Su, L.; Miao, C.; Duan, Q.; Lei, X.; Li, H. Multiple-Wavelet Coherence of World's Large Rivers With Meteorological Factors and Ocean Signals. *JGR Atmos.* **2019**, *124*, 4932–4954. [CrossRef]
26. Grinsted, A.; Moore, J.C.; Jevrejeva, S. Application of the Cross Wavelet Transform and Wavelet Coherence to Geophysical Time Series. *Nonlin. Process. Geophys.* **2004**, *11*, 561–566. [CrossRef]
27. Hu, W.; Si, B.C. Technical Note: Multiple Wavelet Coherence for Untangling Scale-Specific Andlocalized Multivariate Relationships in Geosciences. *Hydrol. Earth Syst. Sci.* **2016**, *20*, 3183–3191. [CrossRef]
28. Hamed, K.H.; Ramachandra Rao, A. A Modified Mann-Kendall Trend Test for Autocorrelated Data. *J. Hydrol.* **1998**, *204*, 182–196. [CrossRef]
29. Parkinson, C.L.; Cavalieri, D.J. Arctic Sea Ice Variability and Trends, 1979–2006. *J. Geophys. Res. Ocean.* **2008**, *113*. [CrossRef]
30. Cavalieri, D.J.; Parkinson, C.L. Arctic Sea Ice Variability and Trends, 1979–2010. *Cryosphere* **2012**, *6*, 881–889. [CrossRef]
31. Meier, W.; Stroeve, J. An Updated Assessment of the Changing Arctic Sea Ice Cover. *Oceanog* **2022**. [CrossRef]
32. Parkinson, C.L.; Comiso, J.C. On the 2012 Record Low Arctic Sea Ice Cover: Combined Impact of Preconditioning and an August Storm. *Geophys. Res. Lett.* **2013**, *40*, 1356–1361. [CrossRef]
33. Mukherjee, A.; Ravichandran, M. Role of Atmospheric Heat Fluxes and Ocean Advection on Decadal (2000–2019) Change of Sea-Ice in the Arctic. *Clim. Dyn.* **2023**, *60*, 3503–3522. [CrossRef]
34. Juszak, I.; Iturrate-Garcia, M.; Gastellu-Etchegorry, J.-P.; Schaepman, M.E.; Maximov, T.C.; Schaepman-Strub, G. Drivers of Shortwave Radiation Fluxes in Arctic Tundra across Scales. *Remote Sens. Environ.* **2017**, *193*, 86–102. [CrossRef]
35. Tietsche, S.; Hawkins, E.; Day, J.J. Atmospheric and Oceanic Contributions to Irreducible Forecast Uncertainty of Arctic Surface Climate. *J. Clim.* **2016**, *29*, 331–346. [CrossRef]
36. Jaiser, R.; Dethloff, K.; Handorf, D.; Cohen, J. Impact of Sea Ice Cover Changes on the Northern Hemisphere Atmospheric Winter Circulation. *Tellus A Dyn. Meteorol. Oceanogr.* **2012**, *64*, 11595. [CrossRef]
37. Zou, C.; Zhang, R. Arctic Sea Ice Loss Modulates the Surface Impact of Autumn Stratospheric Polar Vortex Stretching Events. *Geophys. Res. Lett.* **2024**, *51*, e2023GL107221. [CrossRef]
38. Petrie, R.E.; Shaffrey, L.C.; Sutton, R.T. Atmospheric Response in Summer Linked to Recent Arctic Sea Ice Loss. *Quart. J. R. Meteoro Soc.* **2015**, *141*, 2070–2076. [CrossRef]

39. Outten, S.D.; Esau, I. A Link between Arctic Sea Ice and Recent Cooling Trends over Eurasia. *Clim. Change* **2012**, *110*, 1069–1075. [CrossRef]
40. He, S.; Gao, Y.; Li, F.; Wang, H.; He, Y. Impact of Arctic Oscillation on the East Asian Climate: A Review. *Earth-Sci. Rev.* **2017**, *164*, 48–62. [CrossRef]
41. Xiao, M.; Zhang, Q.; Singh, V.P. Influences of ENSO, NAO, IOD and PDO on Seasonal Precipitation Regimes in the Yangtze River Basin, China: Influences of ENSO Regimes on Precipitation. *Int. J. Clim.* **2015**, *35*, 3556–3567. [CrossRef]
42. Chen, G.; Li, X.; Xu, Z.; Liu, Y.; Zhang, Z.; Shao, S.; Gao, J. PDO Influenced Interdecadal Summer Precipitation Change over East China in Mid-18th Century. *npj Clim. Atmos. Sci.* **2024**, *7*, 114. [CrossRef]
43. Hussain, A.; Cao, J.; Ali, S.; Ullah, W.; Muhammad, S.; Hussain, I.; Abbas, H.; Hamal, K.; Sharma, S.; Akhtar, M.; et al. Wavelet Coherence of Monsoon and Large-scale Climate Variabilities with Precipitation in Pakistan. *Intl J. Climatol.* **2022**, *42*, 9950–9966. [CrossRef]
44. Nalley, D.; Adamowski, J.; Biswas, A.; Gharabaghi, B.; Hu, W. A Multiscale and Multivariate Analysis of Precipitation and Streamflow Variability in Relation to ENSO, NAO and PDO. *J. Hydrol.* **2019**, *574*, 288–307. [CrossRef]

MDPI AG
Grosspeteranlage 5
4052 Basel
Switzerland
Tel.: +41 61 683 77 34

Water Editorial Office
E-mail: water@mdpi.com
www.mdpi.com/journal/water

www.ingramcontent.com/pod-product-compliance
Lightning Source LLC
Chambersburg PA
CBHW040839250926
51850CB00007BA/1360
9783725857395